# Organic Thin-Film Transistor Applications

## *Materials to Circuits*

# Organic Thin-Film Transistor Applications

## *Materials to Circuits*

Brajesh Kumar Kaushik • Brijesh Kumar
Sanjay Prajapati • Poornima Mittal

**CRC Press**
Taylor & Francis Group
Boca Raton  London  New York

CRC Press is an imprint of the
Taylor & Francis Group, an **informa** business

CRC Press
Taylor & Francis Group
6000 Broken Sound Parkway NW, Suite 300
Boca Raton, FL 33487-2742

First issued in paperback 2020

© 2017 by Taylor & Francis Group, LLC
CRC Press is an imprint of Taylor & Francis Group, an Informa business

No claim to original U.S. Government works

ISBN 13: 978-0-367-57461-1 (pbk)
ISBN 13: 978-1-4987-3653-4 (hbk)

**Library of Congress Cataloging-in-Publication Data**

Names: Kaushik, Brajesh Kumar, author.
Title: Organic thin-film transistor applications : materials to circuits / Brajesh Kumar Kaushik, Brijesh Kumar, Sanjay Prajapati, and Poornima Mittal .
Description: Boca Raton : Taylor & Francis, CRC Press, 2017. | Includes bibliographical references and index.
Identifiers: LCCN 2016015291 | ISBN 9781498736534 (hardcover : alk. paper)
Subjects: LCSH: Thin film transistors. | Organic semiconductors. | Organic thin films.
Classification: LCC TK7871.96.T45 K38 2017 | DDC 621.3815/28--dc23
LC record available at https://lccn.loc.gov/2016015291

**Visit the Taylor & Francis Web site at**
**http://www.taylorandfrancis.com**

**and the CRC Press Web site at**
**http://www.crcpress.com**

# Contents

## SECTION I — Organic Device Physics and Modeling

## SECTION II — Organic Device Applications

# Preface

Significant developments have been made in organic thin-film transistors (OTFTs) since the mid-1990s. Over the last decade, aggressive efforts have been made to develop high-performance, lightweight, flexible organic electronic circuits. Undoubtedly, organic electronics foresee a range of vital and high-end applications, such as flexible displays, static random access memory (SRAM), light-emitting diodes (LEDs), radio frequency identification (RFID) tags, sensors, solar cells, photovoltaic cells, and flexible and disposable organic integrated circuits. This became possible due to their cost-effective, low temperature, and easy fabrication process for the production of large-area electronic applications. Since organic materials can be deposited at lower temperature, they provide a strong compatibility with the flexible substrates including plastic, paper, fiber, foil, and even smart cloth in comparison to their inorganic counterparts. OTFT, the most prominent device among the organic devices, is now exhibiting much higher or comparable performance to hydrogenated amorphous silicon TFTs (a-Si:H TFTs) commonly used as the pixel drivers in active matrix flat-panel displays (AMFPDs). These additional degrees of freedom for the OTFTs raise two main concerns: first, how to select the most suitable OTFT platform for a specific application and, second, how to estimate its potential in organic analog and digital circuits. These queries have been dealt thoroughly in this book.

This book provides a comprehensive review of the theory behind organic electronics development, covering most recent aspects from material to device physics. It discusses various organic materials used in different layers of OTFT, charge transport phenomenon, and analytical models of single gate in top- and bottom-contact OTFT structures. Moreover, this book discusses the impact of thickness variation of organic semiconductor and dielectric materials. Additionally, single gate (SG), dual gate (DG), cylindrical gate (CG), and vertical channel OTFT devices are reviewed in terms of their structure, characteristics, and performance parameters. The SG and DG OTFTs are analytically modeled, and their application as organic inverters, logic gates, flip-flops, and multiplexers are presented. This book also discusses organic all-$p$, complementary, and hybrid complementary configurations of inverter circuits as well as SRAM cell designs in terms of their static noise margin (SNM), gain, propagation delay, read stability, and write ability. This book is classified in two main sections: Section I covers organic device physics and modeling conceptions, and Section II discusses their digital circuit applications using SG and DG OTFT structures along with different configurations. The book consists of 10 chapters dealing with different aspects of organic materials, devices, circuits, and applications.

Chapter 1 introduces organic TFTs in terms of their operation, advantages, and applications. Chapter 2 presents recent advancements in organic high-performance materials used for different layers of OTFTs, device structures, charge transport models, fabrication techniques, electrical characteristic influencing factors, and some vital applications. Chapter 3 deals with organic device physics and the development of analytical models for SG OTFT devices with top and bottom contact structures. The analytical model is validated with reported experimental results of fabricated OTFT devices by research laboratories. Chapter 4 analyzes the performance of top and bottom contact OTFTs mainly in terms of drain current ($I_{ds}$), mobility ($\mu$), on/off current ratio ($I_{on}/I_{off}$), threshold voltage ($V_t$), and subthreshold slope ($SS$) with the scaling of semiconductor thickness ($t_{osc}$) and dielectric thickness ($t_{ox}$). In addition, a dependence of contact resistance on the $t_{osc}$ and $t_{ox}$ is demonstrated at different gate voltages for both top and bottom contact OTFTs. Also, Chapter 4 explains the TCAD simulation flow with necessary illustrations. Chapter 5 provides an insight into the organic light-emitting transistor (OLET). Moreover, basic operation of OLETs, usage of different active organic materials, and different optical, electrical, and thermal properties of OLETs are discussed.

At the commencement of Section II, Chapter 6 analyzes inverter circuits using different organic–inorganic semiconductor-based single gate thin-film transistors (SG TFTs) for driver and load. Furthermore, the static and dynamic responses of different inverters using TFTs based on pentacene–$C_{60}$, CuPc–$F_{16}$CuPc, pentacene–a-Si:H, pentacene–ZnO, and pentacene–pentacene are compared. The analytical results of switching threshold voltage and propagation delay are also provided to validate circuit simulation. Chapter 7 compares the performance of single and dual gate OTFT-based all-$p$ inverter circuits in diode load logic (DLL) and zero-$V_{gs}$ load logic (ZVLL) configurations. In addition, the performance of the dual gate inverter circuit is analyzed with different back gate (BG) biasing at driver and load transistors. A bootstrapping technique is also applied to the DG inverters, which boosts the performance of both load logics. Furthermore, static and dynamic behaviors of the universal logic gates NAND and NOR are realized using dual gate OTFTs in DLL and ZVLL configurations. Chapter 8 demonstrates implementation of combinational circuits such as the multiplexer (MUX) and various benchmark sequential circuits such as RS, D, JK, and T latches using SG and DG OTFTs in DLL and ZVLL configurations. Realizations of these combinational and sequential circuits using OTFTs demonstrate the possibilities of producing complex systems on flexible substrates that may become a steppingstone toward producing a low-cost organic embedded system. Chapter 9 analyzes and compares the performance of $p$-type organic and $n$-type organic–inorganic TFT-based SRAM cells in terms of SNM, read stability, and write ability with different cell ratios and pull-up ratios using various TFT combinations, such as all-$p$ (pentacene), organic complementary (pentacene–$C_{60}$), and hybrid (pentacene–ZnO). To achieve an overall improved performance, the organic complementary SRAM cell design is analyzed by replacing $n$-type access TFTs with $p$-type. The proposed SRAM cell demonstrates improved performance with substantially lower width of $p$-type access OTFTs. Finally, Chapter 10 concludes the journey with an outline of the applications of organic devices and a discussion about future perspectives and thrust areas of research in organic electronics.

All the chapters are supported with multiple choice, short and descriptive answer type questions, and numerical problems (given at the end of the chapter). The Appendix includes complete simulation codes for realizing single and dual gate OTFT based inverters, 2-to-1 multiplexer, SR latch, and SRAM cell.

We would like to express our gratitude to Professor Ramgopal Rao, Director, Indian Institute of Technology Delhi; Professor Y. S. Negi, Indian Institute of Technology Roorkee; Deepa Saini, Hemwati Nandan Bahuguna Garhwal University; Anil Kumar Baliga, Shubham Negi, Arun Pratap Singh Rathod, Srishti, Yamini Pandey, Aanchal Verma, Akanksha Uniyal, and Kamlesh Kukreti, Graphic Era University, for their kind support in completing this book.

**Brajesh Kumar Kaushik**
**Brijesh Kumar**
**Sanjay Prajapati**
**Poornima Mittal**

# Authors

**Brajesh Kumar Kaushik** earned a bachelor of engineering in electronics and communication engineering at D.C.R. University of Science and Technology (formerly *C. R. State College of Engineering*) Murthal, Haryana, in 1994. He earned a master of technology in engineering systems at Dayalbagh Educational Institute–Agra in 1997 and a doctor of philosophy (PhD) in 2007 under the AICTE-QIP scheme at the Indian Institute of Technology Roorkee, India. He served Vinytics Peripherals Pvt. Ltd., Delhi, as a research and development engineer in microprocessor, microcontroller, and DSP processor-based system design. He joined the Department of Electronics and Communication Engineering, Govind Ballabh Pant Engineering College, Pauri Garhwal, Uttarakhand, India, as a Lecturer in July 1998, where later he served as an assistant professor from May 2005 to May 2006 and an associate professor from May 2006 to December 2009. He joined the Department of Electronics and Communication Engineering, Indian Institute of Technology, Roorkee, as an assistant professor in December 2009; since April 2014 he has been an associate professor. He has extensively published in several national and international journals and conferences. He is a reviewer of many international journals belonging to various organizations and publishers, including IEEE, IET, Elsevier, Springer, Taylor & Francis, Emerald, ETRI, and PIER. He has also served as general chair, technical chair, and keynote speaker at many reputed international and national conferences. He is a senior member of IEEE and a member of many expert committees constituted by government and nongovernment organizations.

He holds the position of editor and editor-in-chief of various journals in the field of VLSI and microelectronics. He is editor-in-chief of *International Journal of VLSI Design and Communication Systems* (*VLSICS*), AIRCC Publishing Corporation. He also holds the position of editor of *Microelectronics Journal* (*MEJ*), Elsevier Inc.; *Journal of Engineering, Design and Technology* (*JEDT*), Emerald Group Publishing Limited; and *Journal of Electrical and Electronics Engineering Research* (*JEEER*), Academic Journals. He has received many awards and recognitions from the International Biographical Center (IBC), Cambridge. His name has been listed in Marquis Who's Who in Science and Engineering® and Marquis Who's Who in the World®. His research interests are in the areas of high-speed interconnects, low-power VLSI design, carbon nanotube-based designs, organic electronics, FinFET device circuit co-design, electronic design automation (EDA), and spintronics-based devices and circuits.

**Brijesh Kumar** earned a BTech in electronics engineering in 1999 at Bundelkhand Institute Engineering and Technology (BIET), Jhansi, India; an MTech in VLSI design in 2007 at UP Technical University, Lucknow, India; and a PhD at the Indian Institute of Technology Roorkee, India, in 2014. His thesis was titled "Organic Thin Film Transistor Device Modeling and Circuit Co-Design and Performance Analysis." He has more than fifty national and international publications in reputed journals and conferences. His research interests include organic devices and circuits, conducting polymers, organic electronics, microelectronics, and VLSI design and technology.

He joined the Department of Electronics Engineering, BIT, M. Nagar, Uttar Pradesh, India, as a lecturer in 2000, where later he became an assistant professor from 2005 to 2006. Afterward, he joined the Department of Electronics and Communication Engineering, Graphic Era University, Dehradun, Uttarakhand, India, as an assistant professor in September 2006, where later he was promoted to associate professor in 2009. He has more than 15 years of experience in the field of research and academic. He has received various awards and certificates of appreciation for his academic and research activities. He has attended and organized many national-level workshops, seminars, and faculty development programs. His name has been listed in Marquis Who's Who in the World, USA. He is a life member of the Indian Society of Technical Education (ISTE), International Association of Engineers (IAENG), and a member of IEEE.

**Sanjay Prajapati** earned a BE in electronics and communication engineering in 1996 at Government Engineering College, Modasa (GEC), Gujarat, India, and an MTech in VLSI design in 2010 at Nirma University, Ahmedabad, Gujarat, India. His MTech dissertation was titled "Modelling of Devices and Design of Circuits in Emerging Technologies—Fin-FET and OTFT." He is pursuing a PhD at Indian Institute of Technology, Roorkee, India. His research interests include spintronics, organic electronics, organic device modeling and circuit designing, and analog VLSI design.

He joined the Department of Electronics and Communication Engineering, Government Polytechnic, Surat, Gujarat as a lecturer in 1998. He became an assistant professor at Vishwakarma Government Engineering College, Ahmedabad, Gujarat, India, in 2004 and later promoted to associate professor in September 2012 at Government Engineering College, Dahod, affiliated with Gujarat Technological University, Ahmedabad, Gujarat, India. He has attended many national-level workshops, seminars, and faculty development programs. He has more than 16 years of research and academic experience.

**Poornima Mittal** earned a BE in electronics engineering at Kolkata, India, in 2004; an MTech at Uttarakhand Technical University, Dehradun, India, in 2008; and a PhD at Uttarakhand Technical University. Her thesis focused on organic material-based devices and their analog and digital circuit applications.

She has more than 35 national and international journal and conference publications. Her research interests include solid-state devices, modeling and simulation, organic electronics, and VLSI circuit design. She has attended many national-level workshops, seminars, and faculty development programs. She has received various awards and certificates of appreciations for her academic and research activities. She received a research award in 2014 and 2015 for her dedicated research outcome at Graphic Era University Dehradun. She is a life member of the Indian Society for Technical Education (ISTE), International Association of Engineers (IAENG), and a member of IEEE. She has more than 11 years of experience in the field of research and teaching. She has worked for more than 3 years with the Department of Electronics Engineering, BIT, M. Nagar, Uttar Pradesh, India. She is an associate professor in the Department of Electronics and Communication Engineering at Graphic Era University.

# Organic Device Physics and Modeling

# Introduction

## 1.1 INTRODUCTION

Over the last decade, there has been a significant increase in the efforts to implement and develop electronic components on flexible and stretchable substrates. With the advent of soluble organic semiconductors (OSCs) and conductors, printed electronics became feasible. Printed electronics is usually linked with organic electronics where its characteristic feature is the usage of organic materials for realizing devices and circuits. The devices, incorporating organic materials are on the verge of commercialization. Organic devices that have emerged over several generations demonstrate advantages in terms of mechanical flexibility, lightweight, low cost, and straightforward fabrication at low temperature that allow cost-efficient production. Cost-effective fabrication of these devices on flexible substrates can eventually lead to huge benefits on various fronts.

Organic thin-film transistors (OTFTs) are predicted to have a range of imperative and high-end applications, such as flat panel display, radio-frequency identification (RFID) tag [1], sensors [2], static random access memory (SRAM) [3], e-paper [4], solar cell [5], differential amplifier [6], ring oscillator [7], and flexible integrated circuits [8]. Moreover, these transistors have turned out to be a promising backplane driver in organic light-emitting diode (OLED)-based large area flexible displays [9]. Cost-effective fabrication of OTFTs on the flexible substrate can ultimately benefit the RFID domain to replace the bar code technology intended for product and inventory identification. Recently, researchers have found them capable enough

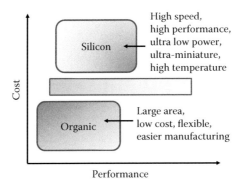

**FIGURE 1.1**   Cost versus performance of organic and inorganic semiconductors.

in realizing the circuits for smart textiles due to their good bending stability. Perhaps in the near future, technology may be developed through combined efforts in the areas of electronics, chemistry, physics, and material science leading to some modern and high-speed applications in graphics, animation, and the video games.

The performance of organic transistors is lower in comparison to the conventional transistors due to inferior mobility ($\mu$); however, the production cost and flexibility of silicon-based devices are obvious constraints. Organic transistors provide an ideal solution as they are inexpensive, flexible, and promising enough to realize large-area electronic circuits. The OTFTs fabricated on flexible substrate have demonstrated comparable or even improved characteristics in comparison to hydrogenated amorphous silicon (a-Si:H)-based TFTs. On steady improvement, the mobility of organic transistors has been augmented by several orders, now in excess of 15 cm$^2$/Vs [10] for single crystal and 3.2 cm$^2$/Vs [11] for thin film.

With optimization of fabrication methodology and synthesis of novel materials, the mobility can be undoubtedly further increased. For a comparative study of organic and inorganic semiconductors applications, Figure 1.1 compares their performance and cost characteristics. Though, the performance of the organic transistor is not comparable to the silicon transistor, it still finds utilization in certain innovative applications that are not possible with conventional semiconductors, or if feasible they are too expensive to be realized commercially.

## 1.2  ORGANIC SEMICONDUCTOR MATERIALS FOR ORGANIC DEVICES

Initially organic polymers with carbon backbones were insulators and used for encapsulation. In 1977, discovery of highly conductive polymers marked the beginning of a new field of electronics called organic electronics. Organic semiconductors can be classified into two categories: polymers and small molecules. The mobility of solution-processed polymer semiconductors is lower than that of small molecules deposited by vacuum evaporation technique [12].

## 1.2.1 POLYMERS

During the 1960s, all organic polymers were considered to be nonconducting. They were used for packaging and insulation. Later, in 1977 Shirakawa *et al.* discovered the highly conductive polymer polyacetylene using oxidative doping with iodine [13]. Improvement in electrical characteristics was observed after the discovery of polyphenylene-vinylene (PPV) and polythiophene (PT) group conducting polymers [14]. The highest mobility of 0.1 cm$^2$/Vs was reported for the solution-processed regioregular material poly(3-hexylthiophene) (P3HT) in comparison to poly(3-alkylthiophene) (P3AT) and poly(3-octylthiophene) (P3OT) conducting polymers [15,16].

Between 1998 and 2001, improvement in the mobility of P3HT was achieved by selecting appropriate solvents, thermal annealing, optimization of chain length, and substrate surface treatment before deposition of polymer films [17–19]. In 2000, Sirringhaus *et al.* demonstrated an air-stable polymer with mobility of 0.01–0.02 cm$^2$/Vs called poly(9,9'-dioctyl-fluorene-co-bithiophene) (F8T2), using interface preparation and high-temperature annealing [20].

In 2004, Ong *et al.* developed poly(3,3'''-dialkyl-quaterthiophene)s (PQTs) with increased oxygen resistance, solution processability, and self assembly [21]. In 2006, McCulloch *et al.* analyzed poly(2,5-bis(3-alkylthiophen-2-yl)) thieno[3,2-b]thiophene) (PBTTT) displaying higher mobility and stability [22]. In 2010, Zschieschang *et al.* fabricated organic thin-film transistor (OTFT) on a flexible substrate using dinaphtho-[2,3-b:2',3'-f]thieno[3,2-b] thiophene (DNTT) exhibiting mobility of 0.6 cm$^2$/Vs and *on/off* current ratio of 10$^6$ [23]. In 2011, Suraru *et al.* discovered a new derivative of diketopyrrolopyrrole exhibiting the hole mobility of 0.7 cm$^2$/Vs and a current *on/off* ratio of 10$^6$ [24].

## 1.2.2 SMALL MOLECULES

A widely studied *p*-type small molecule organic semiconductor is pentacene due to its chemical and thermal stability along with high mobility. Between 1998 and 2001, research was carried out on vacuum-deposited pentacene TFTs showing mobilities in the range of 0.5 to 1.5 cm$^2$/Vs [25,26]. In 2002, Cantatore *et al.* observed solution-processable pentacene based on the precursor route showing initial mobility of 0.02 cm$^2$/Vs [27]. In 2005, Kawasaki *et al.* developed an inkjet-printed pentacene active layer with mobility of 0.15 cm$^2$/Vs and current *on/off* ratio of 10$^5$ [28]. In 2011, Yu *et al.* fabricated pentacene OTFT using 1,1'-bis(di-4-tolylaminophenyl) cyclohexane (TAPC) as the hole transport layer (HTL) exhibiting high mobility of 0.97 cm$^2$/Vs [29].

### 1.2.3 SEMICONDUCTOR BLENDS

Semiconductor blends are prepared using two or more organic semiconductors (OSCs) so as to combine the beneficial properties of each component. A polymer–small molecule blend can provide control of crystallization and improved film uniformity while maintaining the high charge carrier mobility. A polymer-small molecule blend based on oligothiophene and acenes were solution-processed for OTFTs. A blend of 2,8-difloro-5,11-bis (triethylsilylethynl) anthradithiophene and poly(triarylamine) has shown mobility of 2 cm²/Vs [30]. In 2014, Feng *et al.* fabricated OTFT using 6,13-bis(triisopropylsilylethynyl)-pentacene blended with polystyrene as the channel material showing mobility of 0.8 cm²/Vs, a subthreshold swing of 100 mV/decade and an *on/off* ratio of about $10^4$ [31].

## 1.3 ORGANIC THIN-FILM TRANSISTORS (OTFTs)

The concept of forming a transistor by means of thin film was primarily suggested and developed by Weimer [32] in 1962 by producing a semiconductor film of cadmium sulfide (inorganic semiconductor). Later in 1979, Le Comber *et al.* [33] reported a TFT based on a-Si:H that opened the possibilities of forming a thin film at a comparatively lower temperature. Subsequently, in the 1980s, several researchers, including Tsumura *et al.* [34], Kudo *et al.* [35], and Ebisawa *et al.* [36], demonstrated organic transistors on glass and plastic foil. Since then, enormous efforts have been made to enhance the performance of OTFTs.

An OTFT device is generally a layered structural design consisting of a thin film of OSC; dielectric; and three electrodes named source (S), drain (D), and gate (G), as shown in Figure 1.2. The source and drain electrodes inject into and extract the charge carriers, making contact with the active organic semiconducting layer. On the other hand, the gate is separated from the semiconductor

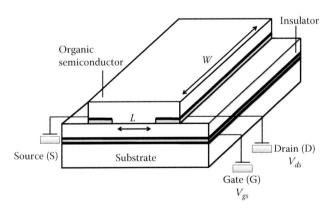

**FIGURE 1.2**    Schematic of the device configuration of an OTFT.

film through dielectric that controls the conductivity of the channel. The TFT structure with gate on the top of a semiconductor is similar to the conventional metal-oxide-semiconductor field-effect transistor (MOSFET); however, the majority of OTFTs are built as the bottom gate structure due to an ease in deposition of active material on the top of the dielectric instead of at the bottom. In these structures, methods pertaining to thermal treatment can be safely employed to produce the insulating layer without creating any impairment in the OSC layer. This bottom gate structure can be further categorized as top contact and bottom contact, based on relative position of S/D electrodes with respect to the OSC layer. The OTFT in bottom gate top contact (BGTC) structure outperforms the bottom gate bottom contact (BGBC) due to a large carrier injection area leading to a low contact resistance. However, BGBC structure is promising for low-end applications, since it can be fabricated through simple cost-effective printing techniques.

### 1.3.1 OPERATING PRINCIPLE

The operating principle of organic transistors is very much similar to the MOSFET; however, the concept of channel formation is quite different. The channel in OTFT is formed by accumulation of the charge as in the bulk semiconductors; whereas in MOSFET, the inversion process takes place to form a layer of charge carriers. Similar to the MOSFET, an organic transistor operates as a voltage-controlled current source; wherein, on applying a bias between the gate and source ($V_{gs}$), a sheet of mobile charge carriers is accumulated near the semiconductor-dielectric interface that allows the flow of current through the active layer on applying a suitable drain to source potential ($V_{ds}$).

Despite the dissimilarity in charge transport physics of organic transistors, the current-voltage characteristics from linear to the saturation regime can be expressed similarly to that of a MOSFET. As long as the drain voltage remains lower than the overdrive voltage (e.g., $|V_{ds}| < |V_{gs} - V_t|$), the drain current builds up linearly due to the presence of carriers all along the channel. Furthermore, the point $V_{ds} = V_{gs} - V_t$ is the onset of the saturation exhibiting the pinch-off phenomena of the channel. Any further increase in $V_{ds}$ does not contribute more toward the enhancement of the magnitude of drain current.

Fundamentally, an OTFT operates like a capacitor that produces an electric field in the dielectric at negative/positive $V_{gs}$ for $p/n$ type OSCs. It results in accumulation of holes/electrons by means of aligning the metal's Fermi level ($E_F$) near highest occupied molecular orbital (HOMO) or lowest unoccupied molecular orbital (LUMO) levels of $p$- or $n$-type semiconductors, respectively.

Figure 1.3 shows a combined energy diagram for a typical combination of gold (Au) S/D contacts and a pentacene semiconductor. Since, the Fermi level is distant from the LUMO edge; therefore, the injection of an electron is insufficient on applying a positive gate bias. On the other hand, holes are injected through the source by reversing the gate voltage owing to the closeness of

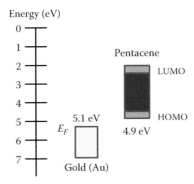

**FIGURE 1.3** Energy band diagram for a typical combination of gold (S/D) and pentacene (OSC).

HOMO level to the Fermi level. Due to formation of positively charged channel, pentacene is said to be a *p*-type organic material that exhibits the highest mobility among the other *p*-type small molecules.

## 1.4 FUTURE PERSPECTIVES

During last two decades, gradual and noteworthy advancements on the organic semiconductor front have motivated designers for OTFT-based designs. OTFT characteristics have undergone remarkable improvements over the passage of time; however, numerous challenges are still to be resolved in order to make organic devices practically viable. These devices undergo various constraints at materials and device physics levels. At the outset, the understanding of charge transport in organic devices is critical and important for continuing the miniaturization of electronic devices. Regardless of the enormous amount research, there is huge scope for development of compact models to analyze the operating conditions and charge transport phenomena so that one can predict and optimize the device performance prior to the fabrication.

The performance of organic transistors depends on the materials used for different layers, device dimensions, and, most important, the device structure. A bigger challenge is to increase the performance of organic devices so as to expand their usage in real-time commercial applications. Necessary prerequisites for a superior organic transistor are high mobility ($\mu$), elevated drain current ($I_{ds}$), low threshold voltage ($V_t$), large *on/off* current ratio ($I_{on}/I_{off}$), high transconductance ($g_m$), and steeper subthreshold slope (*SS*), and low S/D resistance ($R_S/R_D$). These parameters are significantly affected by materials, device geometry, trap states, charge carrier injection, and thickness of different layers. The understanding of the impact of materials and device geometry on the performance of organic devices and circuits still represents a formidable challenge.

Organic circuits are deemed to be ideal for flexible electronics, but they consume a significant amount of energy. Inclusion of novel OTFT structures with an increase in gate controllability over the channel is necessary for high-performance organic electronic circuits. Realizing novel structures through high-performance materials and their circuits may open new dimensions in organic electronics, thereby satisfying the much sought after applications of highly sophisticated integrated circuits.

With the advent of novel high-mobility organic materials, cost-effective process techniques, device structures, circuit design conceptions, and relentless research efforts make organic devices strong contenders for producing flexible large-area electronic applications. Organic-material-based device fabrication, modeling, and circuit applications are rapidly emerging research areas. Recognizing this fact, this book provides in-depth understanding of applied materials for fabrication of OTFT devices, their molecular structures, basics of pi-conjugated semiconducting materials and their properties, OTFT charge transport phenomena and fabrication techniques, basic to advanced OTFT structures, modeling and extraction of performance parameters, performance analysis of benchmark combinational and sequential circuits, and static memory designs.

## 1.5  SCOPE OF THE BOOK

All the work described between the two covers of this book can be highlighted as follows:

- Development of an analytical model to accurately predict the electrical characteristics and performance parameters of top and bottom contact OTFTs and subsequent validation with reported experimental results.
- Improvement in the performance of top and bottom contact OTFTs by optimizing the organic semiconductor and dielectric thicknesses.
- A detailed description of the unipolar and ambipolar organic light-emitting transistors (OLETs) with necessary classifications and various properties of OLETs. A comparative analysis of OLETs and OLEDs is presented, based on the device geometry, emission efficiency, and device lifetime.
- Analysis of inverter circuits using different combinations of organic- and inorganic-material-based single gate (SG) TFTs for driver and load.
- Realization of dual gate (DG) OTFT-based all $p$-type inverter, universal logic gates, combinational and sequential circuits in diode load logic (DLL), and zero-$V_{gs}$ load logic (ZVLL) configurations.
- Analysis and comparison among the performance of all-$p$, hybrid, and organic complementary SRAM cell designs mainly in terms of static noise margin (SNM), read stability, and write ability at different cell ratio ($\beta_c$) and pull-up ratio ($\beta_p$).

## PROBLEMS
## MULTIPLE CHOICE

1. What made printed electronics feasible?
   a. Soluble organic semiconductor
   b. Silicon
   c. Amorphous silicon
   d. ZnO
2. Why is the performance of an OTFT lower than a conventional transistor?
   a. Contact resistance
   b. Channel length
   c. Inferior mobility
   d. Flexibility
3. What made an OTFT preferable?
   a. Flexibility
   b. High throughput, low temperature fabrication
   c. Both a and b
   d. Only b
4. What is the operating principle of an OTFT?
   a. Inversion
   b. Accumulation
   c. Depletion
   d. All the above
5. Why is the BGBC structure promising for low-end applications over BGTC?
   a. Stability
   b. Low contact resistance
   c. High speed
   d. Fabrication through cost-effective printing techniques

## ANSWER KEY

1. a; 2. c; 3. c; 4. b; 5. d

## SHORT ANSWER

1. List the various applications of OTFTs.
2. Enumerate various advantages of an OTFT over conventional transistors.
3. Describe various elemental structures of OTFTs in brief.
4. Explain the operating principle of an OTFT with a diagram.
5. What are the prerequisites for a superior OTFT?
6. Explain the difference and advantages of small molecules over conducting polymers.

# REFERENCES

1. Cantatore, E.; Geuns, T. C. T.; Gelinck, G. H.; Veenendaal, E. V.; Gruijthuijsen, A. F. A.; Schrijnemakers, L.; Drews, S.; De Leeuw, D. M. "A 13.56-MHz RFID system based on organic transponders," *IEEE J. Solid-State Circuits* **2007**, 42(4), 84–92.
2. Brianda, D.; Opreab, A.; Courbata, J.; Barsanb, N. "Making environmental sensors on plastic foils," *Materials Today* **2011**, 14(9), 416–423.
3. Takamiya, M.; Sekitani, T.; Kato, Y.; Kawaguchi, H. "An organic FET SRAM with back gate to increase static noise margin and its application to braille sheet display," *IEEE J. Solid-State Circuits* **2007**, 42(1), 93–100.
4. Tobjork, D.; Osterbacka, R. "Paper electronics," *Adv. Mater.* **2011**, 23(17), 1935–1961.
5. Kumar, P.; Jain, S. C.; Kumar, V.; Chand, S.; Tandon, R. P. "A model for the J-V characteristics of P3HT:PCBM solar cells," *J. Appl. Phy.,* **2009**, 105(10), 104507-1–104507-5.
6. Marien, H.; Steyaert, M. S. J.; Veenendaal, E. V.; Heremans, E. V. P. "A fully integrated $\Sigma$ADC in organic thin film transistor technology on flexible plastic foil," *IEEE J. Solid-State Circuits* **2011**, 46, 276–286.
7. Nausieda, I.; Ryu, K. K.; He, D. D.; Akinwande, A. I.; Bulovi, V.; Sodini, C. G. "Mixed signal organic integrated circuits in a fully photolithographic dual threshold voltage technology," *IEEE Trans. Electron Devices* **2011**, 58(3), 865–873.
8. Guerin, M.; Daami, A.; Jacob, S.; Bergeret, E.; Benevent, E.; Pannier, P.; Coppard, R. "High gain fully printed organic complementary circuits on flexible plastic foils," *IEEE Trans. Electron Devices* **2011**, 58(10), 3587–3593.
9. Ohode, Y.; Negi, Y. S.; Suzuki, Y.; Kawamura, I.; Yamamoto, N.; Yamada, Y. "Polyimide, polyamide-imide, polyamide-liquid crystal orienting film and display using same," Patent No. JP 4-055495 A2, Jpn Kokai Tokkyo Koho, **1992**.
10. Lee, K. S.; Smith, T. J.; Dickey, K. C.; Yoo, J. E.; Stevenson, K. J.; Loo, Y. L. "High-resolution characterization of pentacene/polyaniline interfaces in thin-film transistors," *Adv. Funct. Mater.* **2006**, 16(18), 2409–2414.
11. Schon, J. H.; Batlogg, B. "Trapping in organic field-effect transistors," *J. Appl. Phys.* **2001**, 89(1), 336–341.
12. Horowitz, G. "Organic field effect transistors," *Adv. Mater.* **1998**, 10(5), 365.
13. Shirakawa, H.; Louis, E. J.; MacDiarmid, A. G.; Chiang , C. K.; and Heeger, A. J. "Synthesis of electrically conducting organic polymers: Halogen derivatives of polyacetylene, $(CH)_x$," *J.Chem. Soc., Chem. Commun.* **1977**, 16, 578–580.
14. Higgins, S. J.; Eccelston, W.; Sedgi, N.; Raja, M. http://www.rsc.org/lap/education /eic/2003/higgins_may03.htm (2008).
15. Bao, Z.; Dodabalapur, A.; Lovinger, A. J. "Soluble and processable regioregular poly(3-hexylthiophene) for thin film field-effect transistor application with high mobility," *Appl. Phys. Lett.* **1996**, 69, 4108.
16. Bao, Z.; Feng, Y.; Dodabalapur, A.; Raju, V. R.; Lovinger, A. J. "High performace plastic transistors fabricated by printing techniques," *Chem. Mater.* **1997**, 9, 1299.
17. Sirringhaus, H.; Brown, P. J.; Friend, R. H.; Nielsen, M.M.; Bechgaard, K.; Langeveld-Voss, B. M. W.; Spiering, A. J. H.; Janssen, R. A. J.; Meijer, E. W.; Herwig, P. T.; de Leeuw, D. M. "Two dimensional charge transport in self-organized, high mobility conjugated polymers," *Nature* **1999**, 401, 685.

18. Dimitrakopoulos, C. D.; Mascaro, D. J. "Organic thin film transistor: A review of recent advances," *IBM J. Res. Dev.* **2001**, 45(1), 111.

19. Sirringhaus, H.; Tessler, N.; Friend, R. H. "Integrated optoelectronic devices based on conjugated polymers," *Science* **1998**, 280, 1741.

20. Sirringhaus, H.; Kawase, T.; Friend, R. H.; Shimoda, T.; Indasekaran, M.; Wu, W.; Woo, E. P. "High-resolution inkjet printing of all-polymer transistor circuits," *Science* **2000**, 290, 2123.

21. Ong, B. S.; Wu, Y.; Liu, P.; Gardner, S. "High performance semiconducting polythiophenes for organic thin-film transistors," *J. Am. Chem. Soc.* **2004**, 126(11), 3378.

22. McCulloch, I.; Heeney, M.; Bailey, C.; Genevicius, K.; MacDonald, I.; Shkunov, M.; Sparrowe, D.; Tierney, S.; Wagner, R.; Zhang, W.; Chabinyc, M. L.; Kline, R. J.; McGehee, D.; Toney, M. F. "Liquid crystalline semiconducting polymers with high charge-carrier mobility," *Nat. Mater.* **2006**, 5, 328–333.

23. Zschieschang, U.; Ante, F.; Yamamoto, T.; Takimiya, K.; Kuwabara, H.; Ikeda, M.; Sekitani, T.; Someya, T.; Kern, K.; Klauk, H. "Flexible low-voltage organic transistors and circuits based on a high-mobility organic semiconductor with good air stability," *Adv. Mater.* **2010**, 22, 982–985.

24. Suraru, S.; Zschieschang, U.; Klauk, H.; Wurthner, F. "Diketopyrrolopyrrole as a p-channel organic semiconductor for high performance OTFTs," *Chem. Commun.* **2011**, 47, 1767–1769.

25. Jackson, T. N.; Lin, Y. Y.; Gundlach, D. J.; Klauk, H. "Organic thin-film transistors for organic light emitting flat-pannel display backplanes," *IEEE J. Sel. Top. Quantum Electron.* **1998**, 4(1), 100.

26. Bao, Z.; Dadabalapur, A.; Katz, H. E.; Raju, R. V.; Rogers, J. A. "Organic semiconductors for plastic electronics," *IEEE TAB New Technology Directions Committee (NTDC), 2001 IEEE Workshop on New and Emmerging Technologies* **2001**.

27. Cantatore, E.; Gelinck, G. H.; de Leeuw, D. M. "Polymer electronics: From discrete transistors to integrated circuits and actrive matrix displays," *IEEE Procedings of the Bipolar/BiCMOS Circuits and Technoly Meetings.* **2002**, p. 167.

28. Kawasaki, M.; Ando, M.; Imazeki, S.; Sekiguchi, Y.; Hirota, S.; Sasaki, H.; Uemura, S.; Kamata, T. "Printable organic TFT technologies for FPD applications," *Proc. SPIE* **2005**, 5940, 59400 Q1-5940010.

29. Yu, X.; Yu, J.; Zhou, J.; Wang, H.; Cheng, L.; Jiang, Y. "Performance improvement of pentacene organic thin film transistor by inserting 1,1'-bis(di-4-tolylaminophenyl) cyclohexane hole transport buffer layer," *Jpn. J. Appl. Phys.* **2011**, 50, 104101.

30. Parylene conformal coatings specification and properties, Technical notes, Specialities Coating Systems, http://www.scscookson.com/parylene-knowl edge/specifications.cfm (2010).

31. Feng, L.; Tang, W.; Zhao, J.; Cui, Q.; Jiang, C.; Guo, X. "All-solution-processed low-voltage organic thin-film transistor inverter on plastic substrate," *IEEE Trans. Electron Devices* **2014**, 61, 1175–1180.

32. Weimer, P. K. "The TFT—A new thin-film transistor," *Proc. IRE*, **1962**, 50, 1462–1469.

33. LeComber, P. G.; Spear, W. E.; Ghaith, A. "Amorphous-silicon field-effect device and possible application," *Electron. Lett.* **1979**, 15(6), 179–181.

34. Tsumura, A.; Koezuka, H.; Ando, T. "Macromolecular electronic device: Field effect transistor with a polythiophene thin film," *Appl. Phys. Lett.* **1986**, 49(18), 1210–1212.

35. Kudo, K.; Yamashina, M.; Moriizumi, T. "Field effect measurement of organic dye films," *Jpn. J. Appl. Phys.* **1984**, 23, 130–130.

36. Ebisawa, F.; Kurokawa, T.; Nara, S. "Electrical properties of polyacetylene/polysiloxane interface," *J. Appl. Phys.* **1983**, 54(6), 3255-1–3255-6.

# OTFT Parameters, Structures, Models, Materials, Fabrication, and Applications

*A Review*

## 2.1 INTRODUCTION

Organic electronics has been a field of intense academic and commercial interest for the past three decades. Steady improvement in the electrical performance and stability of organic semiconductors (OSCs) has opened an era of low-cost and large-area electronic applications. Prior to the existence of conjugated polymers in the late 1970s, these materials were mainly known as insulators. Shirakawa *et al.* in 1976 first introduced the conducting organic materials that opened a new research domain, bridging the fields of condensed matter physics and chemistry [1]. Tsumura *et al.* reported the first organic transistor in 1986 that consisted of OSC material for facilitating the flow of electric current [2]. Since then, organic thin-film transistor (OTFT) characteristics have undergone spectacular improvements.

Consistent advancements in fabrication techniques based on organic materials has motivated researchers to exploit various flexible substrates, such as paper, plastic, and fiber for low cost and lightweight flexible electronic applications. This chapter presents the recent advancements in the performance of OTFTs with a motive to present an overview of high-performance organic materials of individual OTFT layers, charge transport models, OTFT structures, performance parameters and their influencing factors, fabrication techniques, and some vital applications. An extensive literature review pertaining to issues of OTFTs and their circuit applications is presented in order to be aware of the timeline of the work being carried out as well as to understand the various technical gaps.

This chapter is organized in nine sections, including the current introductory section. The characteristic parameters of organic transistors are discussed in Section 2.2, and Section 2.3 covers the study of major developments in OTFT structures. Section 2.4 deals with a survey of efforts made to develop analytical models for analyzing device performance. The charge transport models for organic semiconductors are presented in Section 2.5. Furthermore, Section 2.6 enumerates the efforts made for different materials used in individual OTFT layers along with their chemical structures and performance, mainly in terms of field effect mobility and *on/off* current ratio. Major processing steps involved in the fabrication are illustrated in Section 2.7, and performance-influencing factors of organic transistors are discussed in Section 2.8. Finally, concluding remarks based on the elaborated work have been put forth in Section 2.9.

## 2.2 PARAMETERS OF ORGANIC THIN-FILM TRANSISTORS (OTFTs)

The performance of an OTFT is characterized in terms of several conventional parameters, including field effect mobility, threshold voltage, *on/off* current ratio, and subthreshold slope. These performance parameters determine the applicability of an organic transistor, for example, high mobility is often required for high-speed applications, and a large *on/off* current ratio is necessary for realizing memory and display circuits. Furthermore, a low $V_t$ is useful in reducing the device power consumption and is, thereby, beneficial in producing portable devices. However, a steeper subthreshold slope is essentially required for realizing a switching element. These parameters are influenced by several factors that will be discussed in subsequent sections.

### 2.2.1 MOBILITY

The mobility of a device is described as the average charge carrier drift velocity per unit electric field. It is a measure of how efficiently charge carriers can move along the conducting path. It is an important parameter in determining the processing speed of a device. In fact, high mobility is a key factor in obtaining a large *on*-current that is essentially required for memory applications. The mobility of an organic transistor is enhanced with an increase in the gate over-drive voltage and is, thereby, named as field dependent mobility. Horowitz *et al.* [3] and Deen *et al.* [4] demonstrated the variation in mobility by means of a mobility enhancement factor, $\alpha$. Based on the alpha power law function, gate bias dependent mobility is expressed as

$$\mu = \mu_0 \, (V_{gs} - V_t)^\alpha \qquad (2.1)$$

where, $\mu_0$ is the band mobility of an OSC determined at very low $V_{gs}$ (~0.5 V). The parameter $\alpha$ that usually lies in the range of 0.2 to 0.5 is dependent on the conduction mechanism of the device, the doping density, and the dielectric permittivity of active material [5]. Evidence of enhancement in mobility with respect to gate bias was reported by Dimitrakopoulos *et al.* [6] for a pentacene transistor, where it was ranging from 0.02 to 1.26 cm$^2$/Vs for a variation in $V_{gs}$ from −14 to −146 V.

Another factor that strongly affects the mobility is grain size of the active layer, which depends on how perfectly the layer is deposited. Horowitz and Hajlaoui reported a grain size dependent mobility for octithiophene TFT [7]. Similarly, Knipp *et al.* demonstrated the impact of grain size on the mobility of a pentacene-based TFT [8]. Significant improvements in the mobility of organic materials are obtained with the passage of time by synthesizing novel high-performance materials and optimizing the fabrication techniques that would increase the possibility of realizing organic devices for high-speed applications.

## 2.2.2 THRESHOLD VOLTAGE

Threshold voltage is the minimum gate voltage required for accumulating the charge carriers at the OSC–dielectric interface forming a conducting path between the source and the drain. It determines the switching behavior of a device; therefore, needs to be controlled to ensure proper operation of the devices and thereby the circuits. It shows a strong dependence on the doping concentration, dielectric constant of the insulator ($k$), channel length ($L$), and the thicknesses of active and the dielectric layers. Kano *et al.* reported that the devices with smaller length and the larger OSC thickness are liable to have reduced threshold voltages [9]. In addition to this, a decrease in the thickness of the insulating layer results in a significant reduction in the threshold voltage due to high gate capacitance ($C_{ox}$).

Organic material constitutes the trap states that are caused by the noncrystalline structure and defects. Horowitz *et al.* first investigated the presence of shallow traps in the active layer [10]. The filling of these traps is essentially required before accumulating the carriers at the OSC–dielectric interface. Later, Pernstich *et al.* verified an increase in the trap states due to existence of the charge states or dipoles at the surface of SiO$_2$ material [11]. These dipoles can be substantially reduced by employing a surface treatment to the dielectric layer that helps to build a good interface between the semiconductor and dielectric.

### 2.2.3 *ON/OFF* CURRENT RATIO

The performance of a transistor strongly depends on the *on/off* current ratio, which is defined as the ratio of accumulation to depletion region currents. This current ratio is strongly influenced by the channel conductivity, charge density, mobility, and thickness of the semiconductor and dielectric. Additionally, a higher $I_{on}/I_{off}$ can be obtained with a high dielectric constant of the insulator, a high gate capacitance. Moreover, short channel devices are employed for producing a higher current ratio. The ratio ought to be above $10^6$ for display and memory applications. Lowering the thickness of the dielectric and semiconducting layers results in an increase of $I_{on}$ and decrease of $I_{off}$, respectively, that increases the $I_{on}/I_{off}$ ratio.

Resendiz *et al.* reported an increase in the $I_{on}/I_{off}$ from 10 to $5 \times 10^{10}$ for P3HT-based TFT, upon scaling down the active film thickness from 160 to 20 nm [12]. The current ratio was found to be more dependent on $t_{osc}$ due to a dominant impact of the *off*-current over the *on*-current. The *off*-current can be substantially reduced by using a thinner film of the semiconductor. Recently, Islam demonstrated a reduction in *off*-current by 6 orders of magnitude when the film thickness was reduced from 45 to 10 nm [13]. A high current ratio exceeding $10^8$ is an essential requirement for display applications. Usually, short channel devices are fabricated to produce a high *on/off* current ratio.

### 2.2.4 SUBTHRESHOLD SLOPE

Subthreshold slope (*SS*) is a ratio of change in the gate biasing to the change in the drain current in logarithmic scale that can be expressed as

$$SS = \frac{\partial V_{gs}}{\partial \log_{10}(I_{ds})} \tag{2.2}$$

With the change in drain and gate biasing, large variations can be observed in this slope due to a change in the conductivity of the channel. The subthreshold operation of an OTFT is closely related to the mobility enhancement for carrier hopping. Lower trap density results in a steeper slope that shows better switching behavior. Besides this, an increase in the carrier injection density from the source contact can also help to attain a high switching response. It can be achieved by improving the metal-semiconductor interface or selecting an appropriate combination of the semiconductor and contact metal; wherein, the HOMO/LUMO level of *p/n* type OSC aligns well to the Fermi level of metal.

Subthreshold slope is an important parameter that determines an efficient usage of the transistor as a switch. The quality of an active thin-film achieved during the fabrication process substantially affects this slope. Cosseddu and Bonfiglio demonstrated a discontinuity in the semiconductor layer that led to the accumulation of defects and an increase in the trap states, which resulted in a high *SS* [14]. To circumvent this, Cui and Liang reported a reduction of 35% in *SS* by placing an additional gate and dielectric to the single gate device [15]. Furthermore, Schon *et al.* reported a decrease in *SS* by improving the interface between the semiconductor and dielectric layer [16]. In fact, a self-assembled monolayer (SAM) of a dielectric is preferably one of the solutions to achieve a better switching response. Klauk *et al.* illustrated a low subthreshold slope of about 100 mV/dec by producing a SAM of aluminum-oxide ($Al_2O_3$) on the aluminum (Al) gate [17].

## 2.3  OTFT STRUCTURES

OTFTs are differentiated based on the ordering of the layers regardless of the materials and dimensions. Even the structures are classified based on the relative position of source (S), drain (D), and gate (G) contacts with regard to the OSC layer. Single gate TFTs were first proposed and made in 1960s. Since then, enormous efforts have been made toward the development of novel structures, such as dual gate, cylindrical gate, and vertical channel, that led to the improved electrostatic control of gate over the channel. Certain merits and demerits are associated with each of them. The following sections present the performance of different OTFT structures.

### 2.3.1  SINGLE GATE STRUCTURE

Primarily, the structure of a thin-film transistor is distinguished on the basis of the gate position that can be either on the top or at the bottom, accordingly named as top gate and bottom gate structures, respectively. The performance of an OTFT in top gate structures severely degrades if the underlying OSC layer is contaminated during the deposition of a metal gate at high temperature; therefore, bottom gate structures are preferred over top gate. The position of a source and drain contacts with respect to the active layer further classifies them into the top contact and bottom contact structures, while keeping the gate electrode at the same position.

Figure 2.1 shows the OTFT structures in top gate top contact (TGTC) and top gate bottom contact (TGBC) configurations; whereas, bottom gate top contact and bottom gate bottom contact structures are plotted in Figure 2.2. The performance of an OTFT strongly depends on its structures, regardless of the

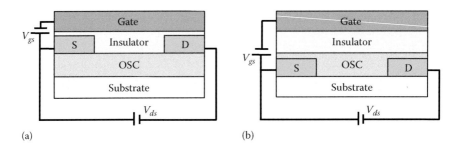

**FIGURE 2.1** Top gate OTFTs (a) TGTC and (b) TGBC structures.

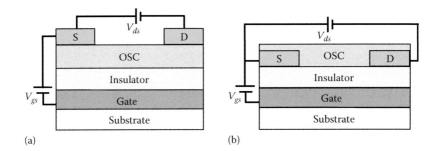

**FIGURE 2.2** Bottom gate OTFTs (a) BGTC and (b) BGBC structures.

materials and dimensions. It is due to a dissimilar path traversed by the charge carriers between the source and drain. A top contact structure exhibits a higher current due to the large injection area for the charge carriers resulting in a lower contact resistance. However, the bottom contact structure demonstrates an inferior performance due to high metal–OSC contact resistance that is an effect of contact barriers as well as a nonuniform deposition of the semiconductor around the prepatterned S/D contacts. In spite of low performance, bottom contact structures are promising for cost-effective flexible electronic applications, since they can be fabricated through simple printing techniques that makes them highly suitable for large-area display applications.

The performance in terms of $I_{ds}$, $\mu$, $I_{on}/I_{off}$, $V_t$, and $SS$ for different single gate structures is summarized in Table 2.1. The top and bottom contact structures are characterized by large performance variation due to existence of an energy barrier at the metal–OSC interface. Shim *et al.* reported a strong influence of the Schottky barrier in bottom contact structure; wherein, an increase of 0.4 eV in the barrier height resulted in a corresponding rise of 1 KΩ in the contact resistance [18]. However, the top contact structure showed a negligible dependence on the barrier height. The effect of an energy barrier can be substantially reduced by improving the surfaces of active layer thin-film and the S/D contacts. Furthermore, Gupta *et al.* reported almost 30 times higher current in the top contact structure at 70% reduced width (*W*) in comparison to the bottom

**TABLE 2.1   Performance of Single Gate OTFTs in Different Configurations**

| Materials of Different Layers | Structure | $I_{ds}$ (µA) | $\mu$ (cm²/Vs) | $I_{on}/I_{off}$ | $V_t$(V) | SS (V/dec) | Biasing (V) $V_{ds}$ | $V_{gs}$ |
|---|---|---|---|---|---|---|---|---|
| OSC: Pentacene, | TGTC | −4.9 | 0.307 | $9 \times 10^6$ | −1.0 | 0.096 | −1.5 | 0 to −3 |
| S/D: Au, I: $Al_2O_3$, | TGBC | −2.5 | 0.246 | $2 \times 10^3$ | −0.5 | 0.155 | | |
| G: Al, Sub: Glass | BGTC | −4.5 | 0.395 | $5 \times 10^7$ | −1.2 | 0.094 | | |
| | BGBC | −3.4 | 0.301 | $4 \times 10^7$ | −1.0 | 0.098 | | |
| OSC: Pentacene, | BGTC | −12 | 0.085 | NR | −3.2 | NR | −25 | 0 to −20 |
| S/D: Au, I: $SiO_2$, | BGBC | −0.4 | 0.0014 | NR | −8.5 | NR | | |
| G: $n^+$ Si, Sub: Si | | | | | | | | |
| OSC: Pentacene, | BGTC | −32 | 0.01 | $10^5$ | −32 | 7.2 | −100 | 0 to −100 |
| S/D/G: PEDOT/ | BGBC | −8 | 0.004 | $10^5$ | −30 | 14.5 | | |
| PSS, I: PET, Sub: | | | | | | | | |
| Plastic | | | | | | | | |
| OSC: Pentacene, | BGTC | −1.2 | 0.45 | $1.6 \times 10^3$ | −0.9 | 0.18 | −5 | 0 to −2 |
| S/D: Au, I: $Al_2O_3$, | BGBC | −0.1 | 0.15 | $8.5 \times 10^4$ | −0.1 | 0.79 | | |
| G: Al, Sub: Glass | | | | | | | | |
| OSC: P3HT, S/D: Ti, | TGBC | −1.4 | 0.015 | $8.4 \times 10^6$ | −1.3 | NR | −10 | 10 to −10 |
| I: PMMA/$TiO_2$, | | | | | | | | |
| G: Au, Sub: Si | | | | | | | | |

*Note:* NR, not reported.

contact [19]. Besides this, the mobility of BGBC structure was found to be lower by 2 orders of magnitude due to the smaller injection area for the carriers.

Doping density is another important factor that significantly affects the device behavior. Ishikawa *et al.* demonstrated a large difference in the performance of top and bottom contact devices based on the doping concentration in the active layer [20]. Compared to bottom contact, the current in the top contact structure was higher by 7 orders at a carrier concentration of $10^{14}$ cm$^{-3}$; whereas, both devices exhibited almost an equal current when concentration was increased to $10^{17}$ cm$^{-3}$. This is due to the availability of sufficient charge carriers in the bottom contact structure even after filling the trap states completely.

## 2.3.2  DUAL GATE STRUCTURE

Organic transistors are realized in a dual gate configuration to achieve better charge carrier modulation in the semiconductor layer. In 1981, the first DG-TFT based on the cadmium selenide (CdSe) was demonstrated by Luo *et al.* [21]. Subsequently, Tuan *et al.* [22] and Kaneko *et al.* [23] reported a-Si:H based DG-TFTs in 1982 and 1992, respectively. Later in 2005, Cui and Liang developed the first pentacene-based dual gate OTFT [15]. An organic dual gate transistor demonstrates numerous advantages such as higher *on*-current, steeper

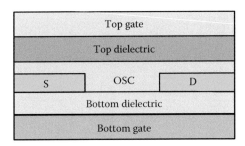

**FIGURE 2.3**   Dual gate OTFT structure.

subthreshold slope, and, most important, better control on the threshold voltage in comparison to single gate.

A dual gate OTFT structure, shown in Figure 2.3, consists of a gate in the front (or bottom) along with its front dielectric (FD), source/drain contacts, organic semiconductor, and a back (or top) gate with a back dielectric (BD). The front gate (FG) accumulates the carriers in the channel, while a bias on the back gate (BG) further increases the conductivity of the channel electrostatically. External control on $V_t$ can bring out a highly controlled operation of the device.

The performance of different OTFTs in single and dual gate configurations is summarized in Table 2.2. Koo *et al.* reported a shift in $V_t$ from 1.95 to −9.8 V with variation in the top gate bias of pentacene TFT from −10 to 10 V [24]. In addition to the significant deviation in $V_t$, an improvement of 70% in threshold voltage was reported. The mobility and *on/off* current ratio were also increased by a factor of 2 in the dual gate mode as compared to the single gate. Similarly, Ha *et al.* reported an improvement in current and mobility by two times for the diketopyrrolopyrrole-naphthalene (PDPP-TNT) based dual gate OTFT compared to the single gate [25]. Similarly, Cui and Liang [15] reported 5 times higher mobility and an improvement of 35% in the subthreshold slope for the dual gate OTFT as compared to the single gate. The DG OTFT shows a significant enhancement in $I_{on}/I_{off}$, especially with the control of *off*-current that makes it more reliable and suitable for display applications.

### 2.3.3 VERTICAL CHANNEL STRUCTURE

The performance of a conventional OTFT is limited by the morphological disorders of thin-film, low mobility of carriers, and the long channel length. In top contact, it is extremely challenging to achieve short channel length by using a low-cost shadow masking technique. However, it is imperative to reduce the driving voltage without compromising the output driving capability that can be achieved by reducing the channel length. To fulfill this gap, Nishizawa *et al.* investigated a vertical structure for the OTFT that has proven its potential for fabricating smaller length devices [26]. A vertical transistor consists of five

**TABLE 2.2    Performance of Different Dual Gate OTFTs**

| Materials of Different Layers | Mode | $I_{ds}$ (µA) | µ (cm²/Vs) | $I_{on}/I_{off}$ | $V_t$(V) | SS (V/dec) | Biasing (V) $V_{ds}$ | $V_{gs}$ |
|---|---|---|---|---|---|---|---|---|
| OSC: PDPP-TNT, S/D: Au BD: D139, FD: SiO$_2$, BG: Al, FG: Si, Sub: Si | BG | −40 | 0.42 | $6.4 \times 10^5$ | −2.3 | 0.83 | −40 | 20 to −40 |
| | FG | −40 | 0.42 | $6.2 \times 10^6$ | −2.5 | 0.75 | −40 | 20 to −40 |
| | DG | −82 | 0.90 | $1.8 \times 10^7$ | −0.3 | 0.42 | −40 | 20 to −40 |
| OSC: Pentacene, S/D: Au, BD/FD: SiO$_2$, BG: Al, FG: n$^+$ Si, Sub: Si | FG | −0.7 | 0.02 | $3.2 \times 10^3$ | −2.0 | 2.0 | −3 | 5 to −15 |
| | DG | −1.5 | 0.1 | $3.8 \times 10^3$ | −2.2 | 1.3 | −3 | 5 to −15 |
| OSC: Pentacene, S/D: Au, BD/FD: Al$_2$O$_3$, BG/FG: Titanium, Sub: Glass | BG | −0.1 | 0.005 | $2.3 \times 10^4$ | 12 | 3.3 | −20 | 20 to −20 |
| | FG | −1 | 0.03 | $5 \times 10^5$ | −2.8 | 0.48 | −20 | 20 to −20 |
| | DG | −10 | 0.06 | $1 \times 10^6$ | −0.8 | 0.47 | −20 | 20 to −20 |
| OSC: PTAA, S/D: Au, BD: Polyisobutylmeth-acrylate, FD: SiO$_2$, BG: Au, FG: n$^+$ Si | DG | −10 | 0.0017 | NR | −2.6 | NR | −20 | 30 to −30 |

*Note:* NR, not reported.

different layers that include metallic layers of the source, drain, and gate in conjunction with two semiconductor layers as shown in Figure 2.4.

Kudo *et al.* proposed a vertical static induction transistor (SIT) that showed high performance due to formation of a Schottky contact at the interface between the active layer and gate [27]. Furthermore, Chen and Shih reported a P3HT-based vertical top and the bottom contact transistors with a channel length of 5 µm; wherein, the mobility was increased by a factor of 3.3 and $I_{on}/I_{off}$ by 11 for the vertical TC structure compared to planar OTFT due to a significant reduction in the contact resistance [28]. Interestingly, the mobility and the *on/off* current ratio of vertical BC structure also increased by 1.5 and 3.6

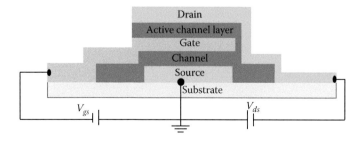

**FIGURE 2.4**    Schematic of the vertical organic thin-film transistor.

times in comparison to the planar BC structure due to less surface contamination in the vertical structure [28]. Furthermore, Naruse *et al.* demonstrated a self-aligned vertical multichannel pentacene-based organic transistor with a channel length of 100 nm [29]. Due to the multichannel, it demonstrated a high saturation current of 22 μA at −20 V of gate and drain bias. The performance of various reported vertical OTFTs is compared in Table 2.3, mainly in terms of $I_{ds}$, $\mu$, $I_{on}/I_{off}$, and $V_t$.

The performance of the vertical transistor can be further improved by inserting a semiconductor layer at the contact–OSC interface. Watanabe and Kudo reported a high performance vertical SIT by inserting an additional layer of copper phthalocyanine (CuPc) OSC between the ITO source and the pentacene active layer to improve the carrier injection from the source [30]. By adding an ultrathin CuPc layer of 1 nm, the device exhibited a larger current of 40 μA magnitude, even at low drain and gate voltages of −3 and −1 V, respectively.

Tanaka *et al.* demonstrated a comparison between the pentacene-based planar BGBC and vertical transistor [31]. A mobility of 0.2 cm²/Vs was achieved for the vertical transistor; whereas, for BGBC it was limited up to 0.0018 cm²/Vs. In the vertical channel devices, it is difficult to determine the behavior of ultrashort channel devices due to the tunneling effect. The performance can be improved substantially by increasing controllability of the channel. It can be achieved by employing a meshed structure for the source electrode. It allows the gate electric field to penetrate into the channel from the array of small pinholes on the source that can substantially reduce the drive voltage.

---

**TABLE 2.3   Performance of Vertical TFTs with Different Combinations of Materials**

| Materials of Different Layers | Structure | $I_{ds}$ (μA) | $\mu$ (cm²/Vs) | $I_{on}/I_{off}$ | $V_t$ (V) | Supply Voltage (V) $V_{ds}$ | $V_{gs}$ |
|---|---|---|---|---|---|---|---|
| OSC: P3HT, S/D: Gold, I: SiO$_2$, G: $n^+$ Si, Sub: Si | TC | −3.2 | 0.0083 | 164 | −1 | −40 | 20 to −20 |
|  | BC | −1.8 | 0.0038 | 55 | +1 | −40 | 20 to −20 |
| OSC: Pentacene, S-ITO, D: Gold, G: Al, Sub: Glass | SIT | −40 | NR | 10³ | −1 | −3 | 0 to −1 |
| OSC: Pentacene, S/D: IZO, I: Tantalum oxide, G-Tantalum, Sub: Glass | Multichannel | −22 | NR | NR | NR | −20 | 0 to −20 |

*Note:* NR, not reported.

### 2.3.4 CYLINDRICAL GATE STRUCTURE

Organic materials are receiving immense attention as they exhibit a unique combination of electronic and mechanical properties that make them applicable for smart textiles. Recently, cylindrical gate (CG) OTFTs have turned out to be promising enough to realize the circuits for wearable electronics due to their hysteresis-free operation and good bending stability. Moreover, cylindrical structures are intended for size reduction, thereby aiming for higher packing density.

Fabrication of CG-TFTs begins with a metal core of yarn that works as the gate electrode, thereafter casing it with a thin insulating layer. Later, the OSC layer is deposited, and finally S/D contacts (metal or conductive polymer) are formed through either thermal evaporation or soft lithography methods. The structure and cross-sectional view of a cylindrical OTFT are shown in Figure 2.5 [32]. Recently, distributed cylindrical transistors have been demonstrated on a stretched fiber-like structure for application in wearable electronics [33]. These transistors can be arranged on a single fiber substrate or at the intersection of two isolated fibers. Performance of a few cylindrical OTFTs is shown in Table 2.4.

Recently, Jang *et al.* reported pentacene-based cylindrical OTFTs with two different polymer gate dielectrics: (1) poly(vinyl cinnamate) (PVCN) and (2) poly(4-vinyl phenol) (PVP) with a high bending stability [33]. They observed an increase in mobility by 2.5 times for a transistor with the PVCN insulator as compared to PVP. Additionally, Maccioni *et al.* developed pentacene-based CG-OTFTs with gold and poly(3,4-ethylenedioxythiophene):styrene sulfonic acid (PEDOT:PSS) S/D electrodes [32]. As compared to gold, the device demonstrated an improvement in $\mu$ and $V_t$ by 50% and 45%, respectively, with PEDOT:PSS contacts. Cylindrical transistors can be realized for some innovative applications, such as a medical shirt for the patient's imperative indications and defense sensors for enemy identification. Regardless of their potential applications, they are limited by their mechanical durability, especially in daily wear.

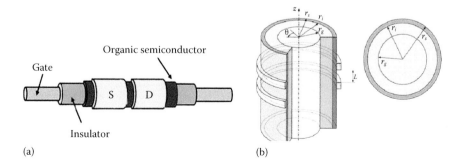

(a)　　　　　　　　　　　　　　　　　　　(b)

**FIGURE 2.5** CG-OTFT (a) basic structure and (b) schematic cross-sectional view.

**TABLE 2.4**    **Performance of Cylindrical OTFTs**

| Materials of Different Layers | $I_{ds}$ (μA) | $\mu$ (cm²/ Vs) | $I_{on}/I_{off}$ | $V_t$ (V) | Supply Voltage (V) | |
|---|---|---|---|---|---|---|
| | | | | | $V_{ds}$ | $V_{gs}$ |
| OSC: Pentacene, S/D: Au, I: PVCN, G: Al wire, Sub: Al | −10 | 0.53 | $4.2 \times 10^3$ | −7.05 | −40 | 0 to −40 |
| OSC: Pentacene, S/D: Au, I: PVP, G: Al wire, Sub: Al wire | −7 | 0.24 | $2.5 \times 10^3$ | −4.78 | −40 | 0 to −40 |
| OSC: Pentacene, S/D: Au, G: Polyimide, I: Polyimide, Sub: metallic fiber | −0.7 | 0.04 | $7 \times 10^3$ | −17.3 | −50 | 0 to −100 |
| OSC: Pentacene, S/D: PEDOT: PSS, G: Polyimide, I: Polyimide, Sub: metallic fiber | −0.34 | 0.06 | $3 \times 10^3$ | −9.6 | −50 | 0 to −100 |

## 2.4 OTFT MODELS

Analytical models are often incorporated in the simulators to predict and optimize the performance of electronic devices and the circuits. These models should be precise enough in the device simulation along with a high degree of convergence in the circuit implementation. Essentially, the model has to take into account the material specifications and the physical bases of a device structure. Moreover, the models should be simple and easily implementable, upgradable, reducible, and modifiable. A few proposed OTFT models are discussed in the following sections.

### 2.4.1 COMPACT DIRECT CURRENT (DC) MODEL

The metal-oxide-semiconductor (MOS) models are adapted and extended to analyze the characteristics of organic transistors due to their similar behavior. Numerous mathematical models were developed for the OTFT primarily based on the classical transistor model by introducing empirical parameters. Based on the MOS model, the drain current in OTFT from linear to saturation regime can be expressed as

$$I_{ds} = \frac{W}{L} \mu C_{ox} (V_{gs} - V_t) V_{ds} \quad \text{for linear regime,} \quad V_{ds} < V_{gs} - V_t \quad (2.3)$$

$$I_{ds} = \frac{W}{2L} \mu C_{ox} (V_{gs} - V_t)^2 \quad \text{for saturation regime,} \quad V_{ds} \geq V_{gs} - V_t \quad (2.4)$$

This model demonstrates the transistor operation above the threshold voltage. Numerous models have been projected to imitate the carrier transport in an organic transistor [5]. Marinov *et al.* [4,34] demonstrated a compact direct current (DC) OTFT model that successfully claimed the transistor operation from ohmic to the saturation regimes. The performance of an OTFT usually deviates from the conventional transistor due to key parameters, such as the bulk leakage current, contact resistance, contact–OSC interface, OSC–dielectric interface, morphological disorders, device configuration, channel length modulation, trap states, gate bias dependent mobility, and many others that raise the difficulty in proposing an unified OTFT model. In using typical MOSFET expressions for these transistors, one needs to consider these parameters up to the maximum extent. A significant variation in the characteristics of two similar devices and their dependence on time make DC modeling somewhat challenging.

## 2.4.2 CHARGE DRIFT MODEL

A few parameters, such as field dependent mobility and the contact resistance, are frequently described to develop analytical models for organic transistors. Among different OTFT models, one important common factor is mobility enhancement at the high gate overdrive voltage, thereby named field dependent mobility [34]. It is an important factor for evaluating OTFT performance in a more realistic way that can be expressed in the form of $\mu \propto (V_{gs} - V_t)^\alpha$. According to a typical charge drift model, the current per unit width is specified as

$$\frac{I_{ds}}{W} = \mu_x q_x |E_x| \tag{2.5}$$

where, $|E_x| = \partial V_x / \partial x$ is the electric field and $q_x$ is the areal charge density that can be expressed as

$$q_x = C_{ox} (V_{gs} - V_t - V_x) \tag{2.6}$$

Furthermore, the field dependent mobility, $\mu_x$, at a point $x$ in the channel $0 \le x \le L$ can be defined

$$\mu_x = \mu_0 (V_{gs} - V_t - V_x)^\alpha \tag{2.7}$$

where, $V_x$ is the voltage at a point $x$ and $\alpha$ is the mobility enhancement factor. Now, the drain current for whole channel length can be simplified by incorporating $q_x$ (Equation 2.6) and $\mu_x$ (Equation 2.7) as

$$I_{ds} = \frac{W \mu_0 C_{ox}}{L(\alpha + 2)} \times \left[ (V_{gs} - V_t - V_s)^{(\alpha+2)} - (V_{gs} - V_t - V_d)^{(\alpha+2)} \right] \tag{2.8}$$

where, $V_s$ and $V_d$ are the voltages at the source and drain, respectively. The characteristic of an OTFT strongly depends on its structure. The bottom contact structure shows a high trap density, thereby a high alpha value in comparison to the top contact.

Cosseddu and Bonfiglio reported heterogeneities that are produced during the deposition of semiconductor around the contacts in the bottom contact structure [14]. It resulted in an increase in the trap states that, in turn, reduced the mobility and drain current substantially. Additionally, Gupta *et al.* [19] demonstrated a large deviation from the ideal behavior of bottom contact OTFT. This was due to the smaller grain size of the semiconductor near the contacts in comparison to the channel that resulted in a lower mobility. These morphological disorders may possibly vary from device to device, despite the same material and similar dimensions, since they strongly depend on the deposition accuracy of the active layer. It raises complexity in development of the models, especially for bottom contact structures where these morphological disorders are prominent.

### 2.4.3 CHARGE DRIFT MODEL FOR SUBTHRESHOLD REGION

A model should be applicable in all the regimes under which a device can operate. To observe the transistor operation in the subthreshold region, the drain current can be expressed as [4]

$$I_{ds} = \frac{W\mu_0 C_{ox}}{L} \times \frac{\left[f(V_{gs} - V_t - V_s)\right]^{(\alpha+2)} - \left[f(V_{gs} - V_t - V_d)\right]^{(\alpha+2)}}{\alpha+2} \qquad (2.9)$$

where, $f(V_{gs}, V)$ is an asymptotical interpolation function. At $V = V_d$ or $V = V_s$, this function is expressed in terms of an overdrive voltage ($V_{overdrive}$) as

$$V_{overdrive}(V) = f(V_{gs}, V) = V_{sub} \ln\left[1 + \exp\left(\frac{V_{gs} - V_t - V_s}{V_{sub}}\right)\right] \qquad (2.10)$$

By incorporating the overdrive voltage function, the drain current can be simplified as

$$I_{ds} = \frac{W\mu_0 C_{ox} V_{sub}(\alpha+2)}{L}$$

$$\times \frac{\left[\ln\left\{1+\exp\left(\frac{V_{gs}-V_t-V_s}{V_{sub}}\right)\right\}\right]^{(\alpha+2)} - \left[\ln\left\{1+\exp\left(\frac{V_{gs}-V_t-V_d}{V_{sub}}\right)\right\}\right]^{(\alpha+2)}}{\alpha+2}$$

$$(2.11)$$

where, $V_{sub}$ is the subthreshold voltage that corresponds to the steepness of the curve. Another important parameter that must be included in the model is channel length modulation. In the saturation regime, as the drain voltage attains its saturation value, $V_{ds} = V_{ds(sat)} = (V_{gs} - V_t)$, the charge at the drain end ($x = L$) becomes nearly zero that is called a "pinch-off" condition. The portion of the channel that is pinched-off ($\Delta L$) reduces the length of an effective channel ($L - \Delta L$). Since the pinch-off segment depends upon the drain voltage, an empirical relation can be expressed between the $\Delta L$ and the $V_{ds}$ as

$$L\left(1 - \frac{\Delta L}{L}\right) = L\left[1 - \beta(V_{ds} - V_{sat})\right] \approx \frac{L}{\left[1 + \beta(V_{ds} - V_s)\right]} \qquad (2.12)$$

where, $\beta$ is the channel length modulation coefficient. The TFT charge drift model (Equation 2.8) can be modified by incorporating $\beta$ as

$$I_{ds} = \frac{\left[W\mu_0 C_{ox}\left\{1 + \beta(V_{ds} - V_s)\right\}\right]}{L(\alpha + 2)}$$

$$\times \left[(V_{gs} - V_t - V_s)^{(\alpha+2)} - (V_{gs} - V_t - V_d)^{(\alpha+2)}\right] \qquad (2.13)$$

In 1992, Xie *et al.* reported a model by incorporating bulk leakage and contact resistance that was among the first few compact models for OTFTs [35]. Furthermore, Necliudov *et al.* included different structural designs to describe the response of different organic transistors [36]. Another model reported by Natali *et al.* was based on the contact resistance due to a significant amount of the voltage drop across the contacts [37]. Due to high contact resistance in the organic transistors, the internal voltage at the channel terminals appears somewhat less than the external applied voltage, especially in the linear region.

Another important point is the assumption of a symmetric structure for the organic TFT that implies interchangeability between the source and the drain contacts. This makes the compact model simpler, since the model developed for the one contact can be extended for another contact. Jung *et al.* developed a model based on the contact resistance at the source end and then this resistance was doubled by considering the same resistance at the drain end [38]. Contradictorily, Burgi *et al.* considered the low charge carrier injection at the source as a major obstruction in lowering the performance than the extraction of carriers at the drain that resulted in an asymmetry between the two contact resistances [39]. All the proposed models reduced to the basic compact model, while incorporating different parameters for analyzing the actual behavior of the OTFT.

## 2.5 CHARGE TRANSPORT IN ORGANIC SEMICONDUCTORS

The charge transport phenomenon in the amorphous and organic semiconductors is different from the metals and conventional semiconductors. In the former, it is phonon assisted; whereas, in the latter, phonon scattering limits the transport efficiency. Organic materials exhibit $sp^2$-hybridized linear carbon chains that hold a $sp^2 - 2p_z$ configuration in double bond. The carrier movement within the molecular chain is called intrachain, whereas between the adjacent molecules it is known as the interchain. Overlapping of $sp^2$ orbitals through intramolecular interaction, results in formation of the σ bond, as shown in Figure 2.6a.

On the other hand, two $p_z$ orbitals result in π and π* (bonding and anti-bonding) orbitals, where the latter bond possesses a higher energy state as depicted in Figure 2.6b. Molecules of different energy levels constitute the energy bands. The energy difference between the HOMO and the LUMO is called an energy gap ($E_g$) that usually lies between $1 < E_g < 4.9$ eV [40]. Most of the OSCs exhibit a disordered molecular arrangement that raises difficulty in conduction between different molecules.

In organic semiconductors, band formation is rare due to weak intermolecular forces. Therefore, the charge carriers move through hopping between the localized states formed by single or several molecules together. The charge carriers of OSCs exhibit a polaronic nature and interact strongly with the defects. The charge transport in these materials is strongly limited by the surface roughness, doping density, dipole formation, chemical impurity, morphological disorders, and presence of defects leading to the complexity of modeling the charge transport phenomena.

The interaction between the semiconductor and dielectric is another important factor for charge transport in an organic TFT. The influence of a dielectric

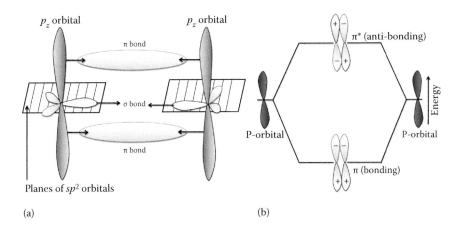

(a)                                              (b)

**FIGURE 2.6**    (a) $sp^2$ Hybridization of two carbon atoms and (b) bonding of $p_z$ orbitals.

is indeed more fundamental than just affecting the morphology of an active layer due to the polarization effect. It strongly affects the charge carrier distribution, surface potential, and the motion of carriers in the conducting channel. Swensen *et al.* reported a direct dependence of the mobility on the dielectric constant of an insulator [41]. Veres *et al.* reported a significant change in the mobility of poly(triarylamine) (PTAA)-based OTFTs by means of different gate dielectrics with *k* ranging from 2.1 to 10.4 [42]. The device showed a mobility of 0.005, 0.0005, and 0.00008 cm²/Vs for CYTOP (*k* = 2.1), polymethyl methacrylate (PMMA) (*k* = 3.5) and poly(vinyl alcohol) (PVA) (*k* = 10.4) organic dielectrics, respectively. The organic materials exhibit the characteristics similar to the amorphous materials. Therefore, the conventional charge transport models can be adapted and extended for them. Charge transport phenomenon in organic materials is generally modeled by the variable range hopping (VRH), multiple trapping and release (MTR), and polaron models, which are discussed in the following sections.

## 2.5.1 VARIABLE RANGE HOPPING (VRH) MODEL

The hopping of charge between the localized states overcomes the energy difference by either emitting or absorbing the phonons. To model the hopping in inorganic semiconductors, Miller and Abrahams described the rate of single phonon jumps [43]. Furthermore, carrier transport dependence on energy distribution was investigated by Vissenberg and Matters for the amorphous transistors that further helped to determine the carrier mobility of the organic semiconductors [44].

The hopping of charge carriers is mainly affected by the energy distribution among the trap states. Moreover, it also depends on the hopping distance. At low bias, very few charge carriers come up to a free state, otherwise, most of them remain entrapped in the localized states. Therefore, the system can be modeled as a setup of resistors along with a conductance $X_{pq} = X_0 \, exp \, (-S_{pq})$ between the hopping sites, say, *p* and *q*. The relation includes a prefactor, $X_0$, for the conductivity and the term $S_{pq}$ is expressed by means of the energy difference as

$$S_{pq} = 2xR_{pq} + \frac{\left|E_p - E_F\right| + \left|E_q - E_F\right| + \left|E_p - E_q\right|}{2k_BT} \tag{2.14}$$

The first right-hand side term illustrates the tunneling process of the carriers between *p* and *q* sites. It includes an effective overlapping parameter, *x*, that depends on the overlap of the sites; whereas, the parameter $R_{pq}$ is the distance between the sites. The second term corresponds to the activation energy required for hopping of the charge carrier from one site to another. The hopping between the two sites strongly depends on the $E_F$ and the respective energies

$E_p$ and $E_q$ of the sites $p$ and $q$. Due to thermal excitation of the charge carriers, the energy difference between the Fermi level and the HOMO level becomes less than the product of Boltzmann constant ($k_B$) and temperature ($T$), that in turn increases the conductance in the vicinity of the semiconductor–dielectric interface.

### 2.5.2 MULTIPLE TRAPPING AND RELEASE (MTR) MODEL

The multiple trapping and release (MTR) model was primarily proposed by Shur and Hack in 1984 for describing the mobility of the a-Si:H semiconductor [45]. It was further extended by Horowitz to understand the mechanism of charge carrier's trapping and releasing phenomena in organic semiconductors [46]. Charge transport in OSCs is completely dominated by a large number of structural and chemical defects. If the energy levels of traps are aligned near the center of $E_g$, then the traps can be deep, otherwise they are shallow if located close to HOMO or the LUMO levels [45], as shown in Figure 2.7.

The model assumes that the extended states are liable to charge transport; however, the majority of injected carriers entrap multiplicatively in the states that are localized in the forbidden gap. Furthermore, releasing the carriers thermally through these states leads to a drift mobility, $\mu_D$, specified as [46]

$$\mu_D = \mu_o a \exp\left( -\frac{E_{tr}}{k_B T} \right) \tag{2.15}$$

where, $E_{tr}$ is the energy level of a single trap. A ratio of effective density of states to trap concentration is expressed by the parameter, $a$. The temperature affects the Fermi function that results in a corresponding change in the number of charge carriers. Therefore, lowering of the temperature reduces drift mobility regardless of an increase in intrinsic mobility.

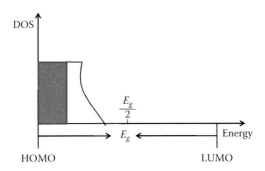

**FIGURE 2.7**  Trap states distribution between HOMO and LUMO.

### 2.5.3 POLARON MODEL

Quasi-particle composition of an electron with its associated field of polarization is called polaron. In organic materials, the charge transport can be modeled by means of these polarons. In 1958, Yamashita and Kurosawa investigated the first model based on the polaron's conduction mechanism in the inorganic semiconductors [47]. Later, in 1959, the same mechanism was demonstrated by Holstein for the single and polycrystalline organic materials [48]. In organic conjugated polymers, the polaron is generated due to deformation of the chain under the action of charge [46]. Organic molecules interact with the polarizing neighbor molecules during condensation, thereby reducing their energy. Furthermore, absorption of a phonon by organic molecule generates an excitation that may dissociate into charge carriers. An attraction between the excited electron and the hole forms a bound electron–hole pair rather than the free charge carriers. This binding energy may be higher by one order than the energy at room temperature. According to the model proposed by Holstein [48], this binding energy is expressed as

$$E_b = \frac{A^2}{2M\omega_0^2} \tag{2.16}$$

where, $\omega_0$ is the frequency, $A$ is the constant, and $M$ is the reduced mass of each molecular site.

Horowitz demonstrated the formation of localized states in the gap between the HOMO and LUMO levels of polythiophene material [46]. These trap states were generated due to self trapping of the charge by means of defects and the chain deformation. The organic molecules are bounded together by means of weak Van der Waal forces that originate from the instantaneous dipole–dipole interaction [6]. A small polaron localized in the single molecule hops from one molecule to another, and thus the charge transport takes place in the organic semiconductors.

## 2.6 MATERIALS

During the last decade, tremendous efforts have been made toward synthesis of new organic materials for improving the performance of transistors. With the advent of high-performance novel organic materials, the applicability of the organic transistors has increased rapidly. These materials urge for a strong push in terms of cost effectiveness, characteristics, properties, and processing. The performance of an OTFT strongly depends on the material constituents for the active layer; however, the selection of materials for other layers, including the dielectric and electrodes, also plays a vital role. Besides this, an appropriate selection of substrate material makes the device suitable for accomplishing flexible circuitry at low cost. This section provides insight into different materials, including semiconductors, dielectrics, electrodes, and substrates.

## 2.6.1 ORGANIC SEMICONDUCTORS

Researchers throughout the world have devoted significant efforts to improve the charge transport properties and characteristics of OSCs for their commercial applications. Organic semiconductors are classified as conducting polymers and small molecules. Mobility of polymers is found to be lower than their counterparts due to their higher molecular weight. For obtaining high mobility, the grains of semiconductors should be larger in size. The possibilities of producing highly ordered thin films result in a significant performance improvement of OTFTs. The mobility of small molecules (>3.2 cm²/Vs) [49] is reported better than the polymers (>0.9 cm²/Vs), but now, this gap is shrinking due to advancements in fabrication methodologies. Numerous conducting polymer and small molecule organic materials have been investigated for possible utilization as the active layer in organic devices that are categorized to $p$- and $n$-type OSCs [50].

### 2.6.1.1 $p$-TYPE

Most of the OSCs investigated so far exhibit the hole as the charge carrier in their inadvertently doped form. Currently, polymers and π-conjugated oligomers are the subject of immense research. Most extensively used $p$-type conducting polymers are poly(3-hexylthiophene) (P3HT), poly(3-octylthiophene) (P3OT), poly(3,3″-dialkylquarterthiophene) (PQT-12), poly(9,9′-dioctylfluorene-co-bithiophene) ($F_8T_2$), and poly(2-methoxy-5-(2′-ethyl-hexyloxy)-1,4-phenylene vinylene) (MEH-PPV) [46]. However, pentacene, copper phthalocyanine (CuPc) and tetracene are the invariably used small molecule organic materials.

Pentacene has proven to be the most widely used $p$-type OSC due to its high hole mobility that is an effect of adequate orbital overlapping among the molecules in the crystal lattice. In addition to high mobility, it shows good chemical stability even in adverse environmental conditions, orderly formation in thin-film structure, and good interface with commonly used electrode metals such as gold and aluminum. Regardless of better performance, it is not a good choice for low-cost printing methods due to its lower solubility.

In comparison to pentacene, P3HT exhibits lower mobility but simultaneously shows outstanding solubility in a range of organic solvents that makes it suitable for fabricating through solution processing techniques. It was initially synthesized by McCullough *et al.* [51] in 1993 that was further utilized by Bao *et al.* [52] to realize the first P3HT-based TFT in 1996.

The performance in terms of $\mu$ and $I_{on}/I_{off}$ of various $p$-type conducting polymers is summarized in Table 2.5. To make the high-performance pentacene material compatible with the solution processing fabrication techniques, Herwig *et al.* [53] synthesized a soluble pentacene precursor in 1999. This was

**TABLE 2.5    Mobility and *On/Off* Current Ratio of *p*-Type Conducting Polymers**

| Material | Structure | $\mu$ (cm²/Vs) | $I_{on}/I_{off}$ | Biasing (V) $V_{ds}$ | Biasing (V) $V_{gs}$ |
|---|---|---|---|---|---|
| P3HT [Poly (3-hexylthiophene)] | BGBC | 0.08 | $7 \times 10^3$ | −40 | 40 to −40 |
| | BGBC | 0.001 | $10^2$ | −30 | 0 to −4 |
| | TGBC | 0.00003 | $5.6 \times 10^6$ | −40 | 0 to −40 |
| | DG | 0.0041 | $6.0 \times 10^6$ | −40 | 0 to −40 |
| $C_{10}$-DNTT{Di-n-decyldinaphtho [2,3-b:2 0, 3 0-f] thieno [3,2-b] thiophene} | BGTC | 2.4 | $10^7$ | −60 | 40 to −60 |
| DHα6T | BGTC | 0.1 | $10^4$ | −50 | 20 to −100 |
| PDBT-co-TT (DPP-Thieno [3,2-b] thiophene) | BGTC | 0.94 | $10^6$ | −75 | 0 to −75 |
| P3OT [Poly (3-octylthiophene)] | BGBC | 0.00016 | $10^4$ | −10 | 15 to −10 |
| MEH-PPV [Poly(2-methoxy, 5 ethyl (2′ hexyloxy) paraphenylenevinylene)] | BGBC | 0.00016 | $10^4$ | −7 | 0 to −10 |
| Poly(4,8-dialkyl-2,6-bis(3-alkylthiophen-2-yl)benzo [1,2-b:4,5-b′] dithiophene) | BGTC | 0.15–0.25 | $10^5–10^6$ | −60 | 20 to −60 |
| $F_8T_2$ [Poly(9,9′-dioctylfluorene-co-bithiophene)] | TGBG | 0.0001 | $10^5$ | −10 | 0 to −10 |
| | BGTC | 0.0001 | $10^6$ | −60 | 0 to −60 |
| PTAA Poly (triarylamine) | BGBC | 0.041 | NR | −60 | 0 to −40 |
| DH4T (Dihexylquaterthiophene) | BGBC | 0.04 | $10^5$ | −20 | 4 to −10 |
| TFB Poly [9,9-dioctyl-fluorene-co-N-(4-butylphenyl)-diphenylamine] | TGTC | 0.02 | NR | −40 | 10 to −40 |
| PBTTT Poly (2,5-bis (3-alkylthiophen-2-yl)thieno [3,2-b] thiophene | BGTC | 0.34 | NR | NR | 0 to −70 |
| PTV Poly (thienylene vinylene) | BGBC | 0.00035 | NR | −1 | 0 to −20 |
| PB16TTT Poly (2,5-bis (3-hexadecylthiophene-2-yl) thieno [3,2-b] thiophene) | BGBC | 0.14 | $10^5$ | −100 | 20 to −100 |

*Note:* NR, not reported.

further explored by Afzali *et al.* [54] and Anthony *et al.* [55] to obtain the thin film of pentacene material with an ease of processing at a low cost. Some of the frequently used pentacene precursors include 6,13-bistriisopropyl-silylethynyl (TIPS) pentacene [55], tetracene, and difluoro-trietetracenethylsilylethynyl anthradithiophene (diF-TESADT). The performance of some *p*-type small molecule OSCs along with their parameters is summarized in Table 2.6. Chemical structures of different *p*-type organic semiconductor materials are shown in Figure 2.8.

**Example 2.1**

Draw chemical structures of the following commonly used *p*-type conducting polymers and small molecules: (a) pentacene, (b) polyacetylene, (c) poly(3-hexylthiophene) (P3HT), (d) 2, 5–dimethylthiophene, (e) polypyrrole, (f) poly(alkylselenophene), (g) polythiophene, (h) copper phthalocyanine (CuPc), and (i) poly)9,9'-dioctylfluorene-co-bithiophene) ($F_8T_2$).
**Solution:** See Figure 2.8.

**TABLE 2.6   Mobility and *On/Off* Current Ratio of *p*-Type Small Molecule OSCs**

| Material | Structure | $\mu$ (cm²/Vs) | $I_{on}/I_{off}$ | Biasing (V) $V_{ds}$ | $V_{gs}$ |
|---|---|---|---|---|---|
| Pentacene | DG | 0.1 | $10^3$ | −2.5 | 0 to −15 |
| | Vertical | NR | $10^3$ | −3 | 0 to −1 |
| | BGBC | 0.5 | NR | −10 | 10 to −10 |
| | BGBC | 0.28 | NR | −30 | 40 to −60 |
| | BGTC | 1.0 | $10^6$ | −100 | 40 to −100 |
| | BGTC | 1.8 | $10^4$ | −6 | 6 to −6 |
| | BGTC | 1.0 | $1.3 \times 10^4$ | −3 | 0 to −3 |
| CuPc (Copper phthalocyanine) | BGBC | 0.02 | $10^4$ | −100 | 0 to −100 |
| | BGTC | 0.02 | NA | −100 | 0 to −100 |
| PDPP-TNT (Diketopyrrolopyrrole- naphthalene copolymer) | TGBC | 0.7 | $10^6$ | −40 | 20 to −40 |
| | BGBC | 0.28 | NR | −40 | 0 to −60 |
| | TGBC | 0.40 | NR | −40 | 0 to −60 |
| TIPS Pentacene (Tris-isopropylsilylethynyl)- pentacene | BGBC | 0.07 | NR | −40 | 20 to −40 |
| | BGBC | 0.02 | $10^4$ | −40 | 40 to −60 |
| DiF-TESADT [2,8-difluoro-5, 11-bis (triethylsilylethynyl) anthra-dithiophene] | BGBC | 5.4 | $10^6$ | −40 | 40 to −40 |
| 2A (2,2-bianthryl) | BGTC | 1.0 | NR | −100 | 0 to −100 |
| | BGBC | 1.0 | NR | −100 | 0 to −100 |
| Sexithiophene ($\alpha$-Sexithienyl) | BGTC | 0.37 | NR | −2 | 0 to 50 |
| Dec- 6T- Dec ($\alpha\,\alpha'$- didecylsexithiophene) | BGTC | 0.1 | $10^4$ | −20 | 20 to −30 |
| | BGBC | 0.5 | $10^5$ | −20 | 20 to −30 |

*Note:* NR, not reported.

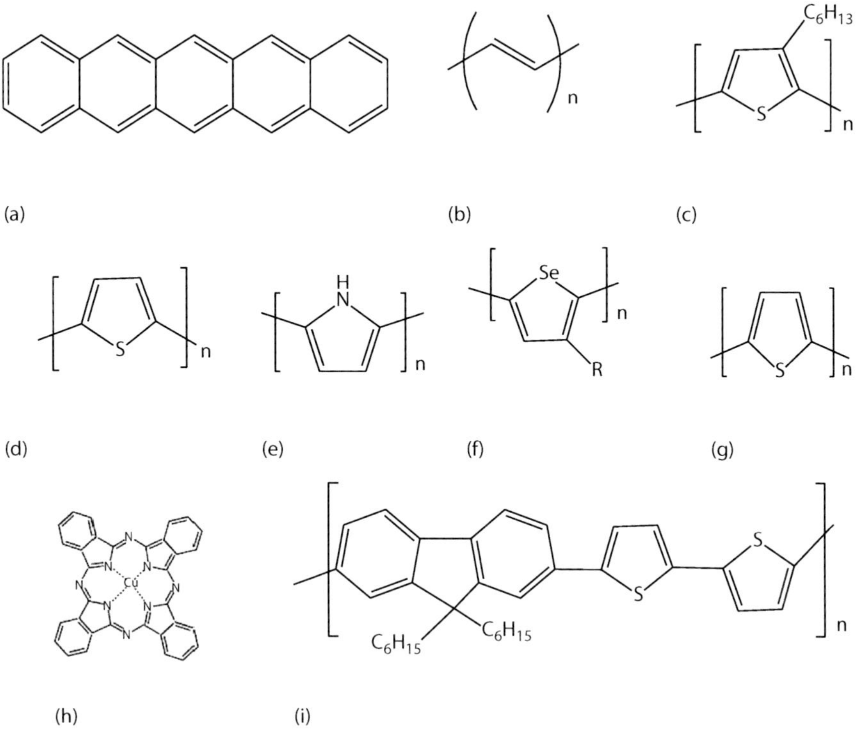

**FIGURE 2.8** Chemical structures of different *p*-type organic semiconductor materials. (a) Pentacene, (b) polyacetylene, (c) P3HT, (d) 2,5-dimethylthiophene, (e) polypyrrole, (f) poly(alkylselenophene), (g) polythiophene, (h) CuPc, and (i) $F_8T_2$.

### 2.6.1.2  *n*-TYPE

The majority of research work has focused on achieving high performance *p*-type OSCs. However, less efforts have been made to synthesize novel *n*-type semiconductors [50]. While aiming to design a complementary inverter, both types of semiconductors are essentially required. Therefore, the development of *n*-type semiconductors is equally important. The chemical structures of some of the commonly used *n*-type OSCs are shown in Figure 2.9.

Nearly, all the *n*-type materials demonstrate instability in the air that strongly depends on the free energy of activation associated with the chemical process/reaction with either water or oxygen. The field effect mobility and the *on/off* current ratio of different *n*-type conducting polymers and small molecule organic materials are outlined in Tables 2.7 and 2.8, respectively.

Performance of *n*-type OSCs is strongly affected by the operational conditions. As an illustration, a very high mobility of 6.2 $cm^2$/Vs was reported for naphthalene diimide in an inert atmosphere. However, under ambient

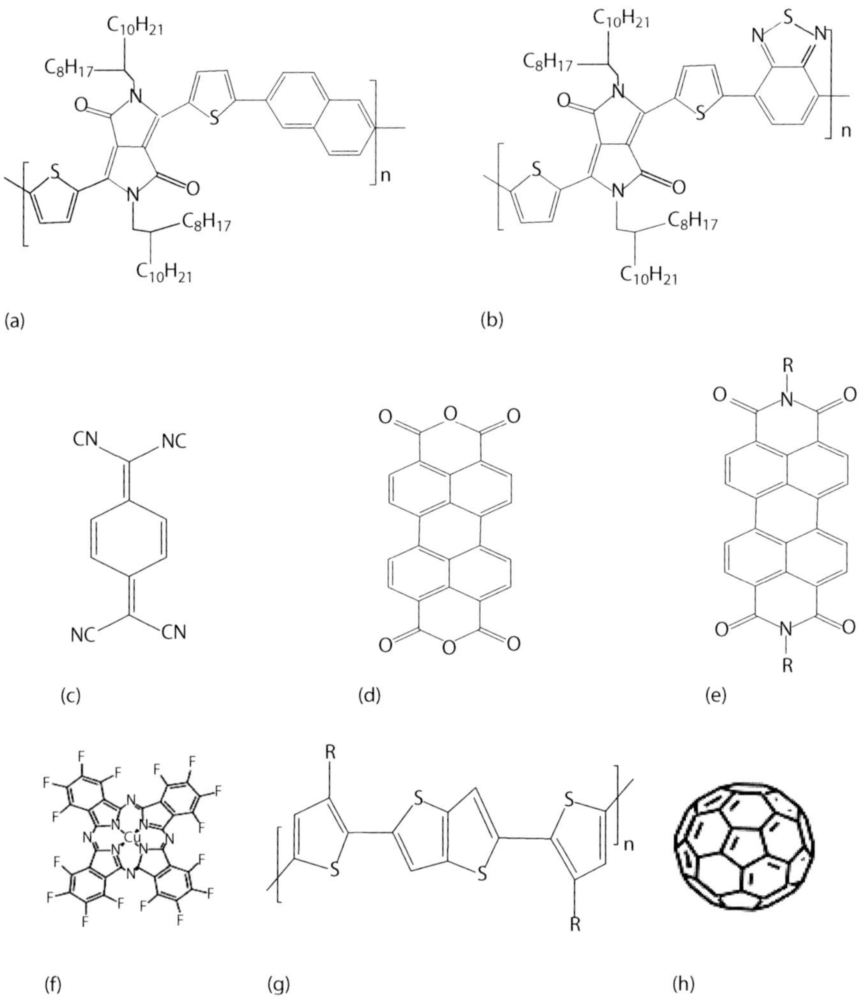

(a)

(b)

(c)

(d)

(e)

(f)

(g)

(h)

**FIGURE 2.9**  Chemical structures of different *n*-type organic semiconductor materials. (a) PDPP-TNT, (b) PDPP-TBT, (c) TCNQ, (d) PTCDA, (e) PTCDI-C8, (f) $F_{16}$CuPc, (g) PBTTT, and (h) $C_{60}$ fullerene.

conditions it was found to be only 0.57 cm$^2$/Vs [56]. Some researchers have shown the synthesis of high-performance *n*-type OSCs by adding –Cl, –CN, and –F groups to the outermost orbital of molecules due to a strong capability of withdrawing the electrons of these groups. Bao *et al.* [57] demonstrated an example of making *n*-type material, copper hexadecafluorophthalocyanine ($F_{16}$CuPc) by adding –F group to the *p*-type material CuPc, that exhibited an electron mobility of 0.03 cm$^2$/Vs. Furthermore, Malenfant *et al.* [58] reported a highly stable *n*-type material; N,N′-dioctyl-3,4,9,10-perylene

**TABLE 2.7    Mobility and *On/Off* Current Ratio of *n*-Type Conducting Polymers**

| Material | Structure | $\mu$ (cm²/Vs) | $I_{on}/I_{off}$ | $V_{ds}$ | $V_{gs}$ |
|---|---|---|---|---|---|
| | | | | **Biasing (V)** | |
| PDI-8CN2 {N,N-bis(n-octyl)- | BGTC | 0.063 | $10^4$ | 100 | −40 to 100 |
| dicyanoperylene-3,4: 9,10-bis | BGBC | 0.33 | $10^5$ | 20 | −30 to 50 |
| (dicarboximide)} | | | | | |
| PTCDI-$C_{13}H_{27}$ (N,N′-ditridecylperylene- | BGTC | 0.6 | $10^7$ | 100 | 0 to 100 |
| 3,4,9,10-tetracarboxylic diimide) | | | | | |
| P(ND12OD-T2) Poly {[n,n9-bis | TGBC | 0.62 | $10^5$ | 8 | 0 to 80 |
| (1-octydodecyl)-naphthalene-1,4,5,8- | TGBC | 0.5 | $10^6$ | 60 | −20 to 60 |
| bis(dicarboximide)-2,6-diyl] | TGBC | 0.1 | NR | 60 | −20 to 60 |
| alt-5,59-(2,29-bithiophene)} | | | | | |
| BBT [Benzobis (thiadiazole) derivative] | BGBC | 0.1 | $10^8$ | 20 | −20 to 40 |
| NDI2-DTP | TGBC | 1.2 | NR | 25 | 0 to 25 |
| PDPP-TBT (Diketopyrrolopyrrole- | TGBC | 0.1 | NR | 40 | −40 to 40 |
| benzothiadiazole copolymer) | | | | | |
| PDI-8CN$_2$ [N, N′ bis-(octyl-)- | BGBC | 0.01 | $10^5$ | 20 | −30 to 50 |
| dicyanoperylene-3,4:9,10- | | | | | |
| bis(dicarboximide)] | | | | | |

*Note:* NR, not reported.

tetracarboxylic diimide (PTCDI C-8) with a high electron mobility of 0.6 cm²/Vs.

**Example 2.2**

Draw the chemical structures of the following commonly used *n*-type conducting polymers and small molecules: (a) diketopyrrolopyrrole-naphthalene copolymer (PDPP-TNT), (b) diketopyrrolopyrrole-benzothiadiazole copolymer (PDPP-TBT), (c) tetracyan-oquinodimethane (TCNQ), (d) perylene tetracarboxylic dianhydride (PTCDA), (e) N,N′-dioctyl-3,4, 9,10-perylene tetracarboxylic diimide (PTCDI C-8), (f) copper hexadeca-fluorophthalocyanine (F16CuPc), (g) poly(2,5-bis (3-alkylthiophene-2-yl) thieno [3,2] thiophene) (PBTTT), and (h) fullerene ($C_{60}$).
**Solution:** See Figure 2.9.

The mobility of *p*- and *n*-type organic materials has improved remarkably with the passage of time, as shown in Figure 2.10. A continuous growth is observed for both types of materials, attributed to the unremitting advancements in the synthesis and fabrication process. Pentacene is the best performing *p*-type material that showed remarkable rise in mobility during the last decade.

**TABLE 2.8**    **Mobility and *On/Off* Current Ratio of *n*-Type Small Molecule OSCs**

| Material | Structure | $\mu$ (cm²/Vs) | $I_{on}/I_{off}$ | Biasing (V) $V_{ds}$ | Biasing (V) $V_{gs}$ |
|---|---|---|---|---|---|
| F$_{16}$CuPc (Copper hexadeca fluorophthalocyanine) | BGTC | 0.012 | NR | 100 | 0 to 100 |
| | BGBC | 0.02 | NR | 100 | 0 to 100 |
| | BGBC | 0.002 | $10^3$ | 8 | −1 to 8 |
| | Vertical | 0.03 | 28 | 10 | 0 to 20 |
| PTCDI C-8 (N,N′-dioctyl-3,4,9,10-perylene tetracarboxylic diimide) | Vertical | 0.6 | 89 | 10 | 0 to 20 |
| NTCDA (2,3,6,7-Naphthalene-tetracarboxylic Dianhydride) | Vertical | 0.003 | 17 | 10 | 0 to 20 |
| PCBM (phenyl-C61-butyric acid methylester) | BGBC | 0.0348 | $10^6$ | 40 | −10 to 40 |
| TEPP {6}-1-(3-(2-thienylethoxycarbonyl)-propyl)-{5}-1-phenyl-[5,6]-C61 | BGBC | 0.0779 | $10^6$ | 40 | −10 to 40 |
| PFP (Perfluoropentacene) | BGTC | 0.22 | $10^5$ | 40 | −50 to 100 |
| F8 [Poly(9,9-dioctylfluorene)] polyfluorene derivative | BGTC | 0.025 | $10^3$ | 50 | 0 to 100 |
| F8BT Poly [(9,9-dioctylfluorenyl-2,7-diyl)-co-(1,4-benzo-{2,1′,3}-thiadiazole)] polyfluorene derivative | BGTC | 0.001 | $10^3$ | 50 | 0 to 100 |
| PDIN1400 (Polyera Activink N1400) | TGBC | 0.037 | NR | 40 | 0 to 60 |
| C$_{60}$ (Fullerene) | BGTC | 5.1 | $10^6$ | 60 | −20 to 60 |
| | BGTC | 1.0 | $10^5$ | 5 | 0 to 5 |
| [60] PCBM (Methanofullerenes [6,6]-phenyl-C61-butyric acid ester) | BGTC | 0.2 | $10^3$ | 80 | −50 to 50 |
| | BGTC | 0.21 | $10^4$ | 80 | 0 to 80 |
| [70] PCBM ([6,6]-phenyl-C71-butyric acid methyl ester) | BGTC | 0.10 | $10^3$ | 80 | 0 to 80 |

*Note:* NR, not reported.

Similarly, fullerene (*n*-type) has also demonstrated a noteworthy improvement. A number of high-performance soluble fullerene derivatives have been investigated for *n*-type OTFTs.

Tiwari *et al.* [59] reported an OTFT with 6,6-phenyl-C$_{61}$-butyric acid methyl ester (PCBM) deposited through solution processing, which exhibited a mobility of 0.13 cm²/Vs. Perfluoro pentacene ($C_{22}F_{14}$) and perfluoro-*p*-sexiphenyl ($C_{36}F_{26}$) are the other novel *n*-type materials employed in OTFTs; whereas, tris-(8-hydroxyquinoline) aluminum (Alq$_3$) is a commonly used material for the electron transport layer in the organic LEDs.

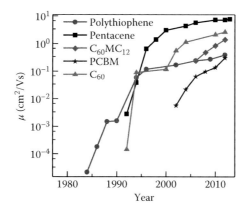

**FIGURE 2.10**    Comparative plot of mobility growth for *p*- and *n*-type OSCs.

## 2.6.2 ELECTRODE MATERIALS

To achieve a high-performance device, the selection of material for making electrodes is of equal importance as the semiconductors. Contact metal for the source and drain electrodes should be selected in such a way that it should not produce high contact resistance. It implies that the contact must possess a low interface barrier with the active layer for enabling a large number of carrier injections. Gold metal is often used in *p*-type OTFTs due to its high work function (WF) (5.1 eV). It is an appropriate metal for usage with *p*-type OSCs as most of them exhibit their HOMO level near 4.9 eV [60]. Thus, a low interface barrier (~0.2 eV) between the gold and *p*-type OSCs makes them suitable for building up contacts with ohmic characteristics. Adding nickel (Ni), titanium (Ti), and chromium (Cr) with gold improves the adhesivity.

The gate electrode material should have good adhesion and patterning capabilities with substrate and gate dielectric, respectively. Moreover, the work function of gate metal should be comparable to the semiconductor in attaining low threshold voltage. Such electrode materials include heavily doped silicon (Si), aluminum, and indium tin oxide (ITO) [17,19,30,50].

The work functions of some of the often used inorganic contact materials are summarized in Table 2.9. Indium tin oxide is commonly used to make the anode terminal in organic display devices because of its high work function (~5 eV) and adequate transparency in the visible spectrum. On the other hand, magnesium, lithium, and calcium with low work function metals are generally used to make the cathode terminal.

A novel series of conducting polymers is also required for fabricating the electrodes, so as to achieve a completely flexible organic device. Some of commonly used organic electrode materials are:

---

**TABLE 2.9**    **Work Functions of Inorganic Electrode Materials**

| Material | WF (eV) |
|---|---|
| Heavily doped silicon | 3.9 |
| Au (Gold) | 5.1 |
| Cu (Copper) | 4.7 |
| Cr (Chromium) | 4.5 |
| Al (Aluminum) | 4.0–4.28 |
| Ni (Nickel) | 4.1–5.0 |
| Ti (Titanium) | 3.84 |
| Pt (Platinum) | 5.65 |
| Ca (Calcium) | 2.87 |
| Co (Cobalt) | 5.0 |
| Fe (Iron) | 5.0 |
| ITO (Indium tin oxide) | 5.3 |

---

- Polyaniline:camphor-sulfonic acid, or PANI:CSA
- Poly(3,4-thylenedioxythiophene)
- Poly(3,4-ethylenedioxythiophene):polystyrene sulfonated, or PEDOT: PSS
- Polypyrrole
- Polythiophene
- Polyanilene

Poly(3,4-ethylenedioxythiophene):styrene sulfonic acid (PEDOT:PSS), PSS (poly(styrene sulfonate)), and PANI:CSA (polyaniline doped with camphor-sulfonic acid) are commonly used conducting polymers for the electrodes [59]. Cosseddu *et al.* [14] reported an OTFT; wherein, source, drain, and gate electrodes were made of PEDOT material. This transistor exhibited a mobility of 0.01 cm²/Vs in BGTC and 0.004 cm²/Vs in BGBC configurations. Additionally, Maccioni *et al.* [32] reported a cylindrical OTFT with S/D contacts of PEDOT:PSS that demonstrated $\mu$ of 0.06 cm²/Vs and $I_{on}/I_{off}$ of $3 \times 10^3$. Chemical structures of commonly used organic electrode materials are shown in Figure 2.11.

**Example 2.3**

Draw the chemical structures of the following commonly used organic electrode materials: (a) poly(3,4-ethylene dioxythiophene) (PEDOT), (b) eumelanin, and (c) 3,4-dihydroxy-l-phenylalanine (l-DOPA).
**Solution:** See Figure 2.11.

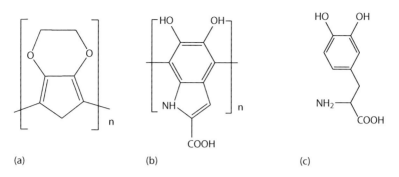

**FIGURE 2.11**   Chemical structures of commonly used organic electrode materials. (a) PEDOT, (b) eumelanin, and (c) 3,4-duhydroxy-L-phenylalanine L-DOPA.

### 2.6.3  GATE DIELECTRIC MATERIALS

Gate dielectric material is used to prevent current leakage between the gate and active layer. This also works as the passivation layer in top gate organic transistors. The accumulation of charge carriers at the semiconductor–dielectric interface strongly depends on the dielectric constant and the thickness of the dielectric layer. The material used for the dielectric layer should properly adhere to the substrate. Moreover, it must produce a good interface with the semiconductor to avoid the generation of dipoles and trap states. Certain desirable characteristics that a dielectric material must exhibit are high dielectric constant and high resistance. High-$k$ materials are beneficial in attaining a steep subthreshold slope and low threshold voltage due to their smaller band gaps as compared to low-$k$ materials. Dielectric constants of some of the frequently used inorganic dielectric materials are summarized in Table 2.10.

Several high-$k$ dielectrics, including $La_2O_3$, $HfSiO_x$, HfLaO, $Pr_6O_{11}$, $HfO_2$ and $Al_2O_3$ were investigated and resulted in a significant decrease in $V_t$ and $SS$ [50]. In addition to the material of high-$k$, the thickness of dielectric layer should be kept low, not only to operate at lower voltages but also to reduce the short channel effects in submicron devices. Moreover, it should possess high breakdown voltage and long-term stability. These requirements are met by a number of dielectric materials, such as $SiO_2$, $Al_2O_3$, PVP, $ZrO_2$, P4VP, PMMA, BZT, PVDF, $HfO_2$, and propylene.

Organic dielectric materials, such as PMMA, PS, PVP, PI, and PVA, demonstrate the ability of processing through low-cost fabrication techniques, good adherence, and fair dielectric properties. Dielectric constants of some of the organic materials are summarized in Table 2.11. In general, a high operating voltage is required for the organic dielectric-based TFTs due to their low

**TABLE 2.10    Dielectric Constants of Inorganic Dielectric Materials**

| Material | Dielectric Constant |
|---|---|
| $SiO_2$ (Silicon dioxide) | 3.5–4.5, 3.9 |
| $Al_2O_3$ (Aluminum oxide) | 8.5–9 |
| $Si_3N_4$ (Silicon nitride) | 6.2, 7.1 |
| $HfO_2$ (Hafnium oxide) | 22, 25 |
| MgO (Magnesium oxide) | 9.8 |
| $LaAlO_3$ (Lanthanum aluminum oxide) | 26 |
| $ZrO_2$ (Zirconium dioxide) | 25, 17.5 |
| $TiO_2$ (Titanium dioxide) | 80–100 |
| $La_2O_3$ (Lanthanum oxide) | 30 |
| $Ta_2O_5$ (Tantalum pentoxide) | 26 |
| $Nd_2O_3$ (Neodymium oxide) | 11.7 |

**TABLE 2.11    Dielectric Constants of Organic Dielectric Materials**

| Material | Dielectric Constant |
|---|---|
| P4VP (Poly (4-vinyl phenol)) | 5.3, 2.56 |
| P3DDT | 3:24 |
| P (VDF-TrFE)/(PVDF)-blend | 10.3 |
| PI (Polyimide) | 2.6 |
| PVA (Polyvinyl alcohol) | 7.8, 8.3 |
| PS (Polystyrene) | 2.6 |
| P4VP-co-PMMA | 4.1 |
| PMMA (Polymethyl methacrylate) | 3.6, 3.3 |
| PVP (Polyvinyl phenol) | 5.3, 3.8, 6.4 |
| Porous PI (Porous polyimide) | 1.67 |
| D139 | 7.4 |

dielectric constant. As a solution, a thin layer can be formed, but it results in a high leakage current due to the presence of defects in a thin organic layer. As an alternative, Klauk *et al.* reported a high-performance OTFT by means of forming bilayers of dielectric films through spin coating of block copolymer on the thermally grown $SiO_2$ surface [61]. Similarly, De Angelis *et al.* [62] and Chou *et al.* [63] demonstrated organic transistors with bi-dielectric layers of $SiO_2$/PMMA and $SiO_2$/polyimide that resulted in the mobility of 1.4 cm$^2$/Vs at −30 V and 2.05 cm$^2$/Vs at −40 V, respectively. Chemical structures of the commonly used organic dielectric materials are shown in Figure 2.12.

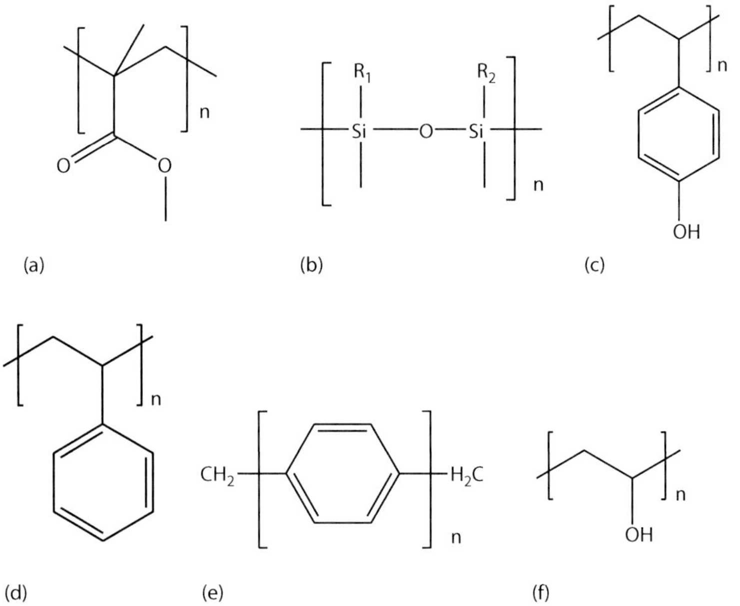

**FIGURE 2.12**    Chemical structures of commonly used organic dielectric materials. (a) Polymethyl methacrylate, (b) silicon network polymer, (c) poly(vinyl phenol), (d) polystyrene, (e) parylene, and (f) poly(vinyl alcohol).

**Example 2.4**

Draw the chemical structures of the following commonly used organic dielectric materials: (a) polymethyl methacrylate (PMMA), (b) silicon network polymer, (c) poly(vinyl phenol) (PVP), (d) polystyrene (PS), (e) parylene, and (f) poly(vinyl alcohol) (PVA).
**Solution:** See Figure 2.12.

## 2.6.4  SUBSTRATE MATERIALS

The selection of a substrate material for any device primarily depends on the kind of application the device is meant for. Silicon finds usage in electronics not only because of its intrinsic properties but also for its ability to produce an oxide layer by a thermal oxidation process. Glass substrate is necessary for fabricating OLED (organic light-emitting diode) displays. On the other hand, organic substrates are essentially required for flexible electronics. In 1990, Peng *et al.* reported the first TFT fabricated on the glass substrate [64]. Thereafter, Garnier *et al.* fabricated a TFT on a flexible polyimide substrate that demonstrated a performance comparable to that of TFTs fabricated on silicon or glass [65].

Later, in 1997 the first fully printed P3HT-based organic transistor was reported by Bao *et al.* [66]; wherein, ITO-coated polyester was used as a substrate. Till then, numerous fully printed organic transistors and circuits were made on flexible substrates such as polyethylene naphthalate (PEN), polyethylene therephthalate (PET), polyimide, polyethylene, plastic, paper, and fiber that opened a new era of flexible low-cost printed electronics.

## 2.7 FABRICATION

The technique chosen to deposit the materials during fabrication plays a crucial role when determining the performance of different OTFTs, even though similar sets of materials are used to fabricate them. Commonly used fabrication techniques include physical vapor deposition, chemical vapor deposition, and solution processing fabrication techniques [17,19,46]. Different fabrication methodologies used in the fabrication process of organic and inorganic devices and circuits are shown in Figure 2.13. The physical vapor deposition (PVD) technique is broadly classified into thermal vacuum evaporation and sputtering fabrication processes; whereas, solution process techniques include spin coating, dip coating, and inkjet screen printing processes. Organic materials allow the fabrication of transistors and circuits up to 120°C, which significantly reduces the manufacturing cost. Small molecule semiconductors, such as pentacene and oligothiophenes, are usually deposited through the vacuum evaporation method, since they exhibit low solubility in the organic solvents. Contradictorily, conducting polymers, such as P3HT and polythiophene, are

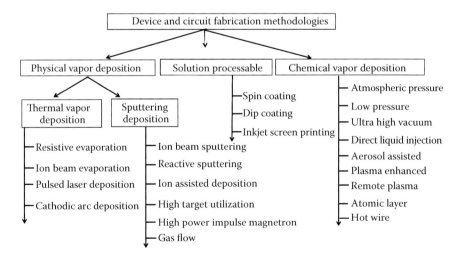

**FIGURE 2.13**   Different fabrication methodologies used in organic and inorganic devices and circuits.

soluble in solvents like chloroform and toluene; therefore, they can be deposited through spin coating or ink-jet printing techniques. Details of these techniques are described in the following sections.

## 2.7.1 PHYSICAL VAPOR DEPOSITION (PVD) TECHNIQUES

The PVD technique is used to deposit thin films of organic semiconducting or dielectric materials. It is generally used in solution phase or vapor phase processes. Unlike the solution phase process, the vapor phase process has to be carried out in vacuum. The PVD technique can be classified broadly as thermal vacuum evaporation and sputtering deposition processes. A detailed discussion of these techniques is presented in the following sections.

### 2.7.1.1 THERMAL VACUUM EVAPORATION TECHNIQUES

The semiconductor film deposited by vacuum evaporation results in superior charge carrier transport properties. This technique is primarily employed for inorganic and small molecule organic materials. Conducting polymers may decompose or even crack at the higher temperature; therefore, this technique is not suitable for them. This technique utilizes natural phenomenon of evaporation and condensation to deposit thin films effectively on any vapor state stable material.

In the thermal vacuum evaporation technique, surroundings are evacuated to a minimum pressure of $10^{-5}$ Torr to pull the evaporated particles directly on the substrate, thereby avoiding any collision with the particles of the background gas. The heat for evaporation is delivered either by Joule heating through a high melting point metal (resistive evaporation) or directly from a focused beam of high energy electrons (electron beam evaporation). The thermal evaporator–Al, thermal evaporator–Cr/Au, 4-target E-beam evaporator, and multipocket electron-beam evaporator (4 target–GaN) are examples of commonly used evaporators.

Butko *et al.* reported a single-crystal-based organic transistor fabricated using the thermal vacuum deposition method exhibiting a hole mobility of 0.3 cm²/Vs and the current *on/off* ratio of $10^5$ [67]. Similarly, Yun *et al.* reported a pentacene OTFT with a mobility of 0.32 cm²/Vs and a high *on/off* current ratio of $2.2 \times 10^6$ [68]. Additionally, Kumaki *et al.* reported a high mobility of 1.0 cm²/Vs for an anthracene oligomer-based transistor; wherein, an active layer of 30 nm was formed through this deposition method [69]. Dimitrakopoulos *et al.* in 1996 demonstrated the deposition of pentacene film through the electron beam evaporation method that yielded a mobility of 0.038 cm²/Vs [70]. Later, in 2008, Cai *et al.* reported a high-performance pentacene-based OTFT deposited through this technique [71]. The transistor exhibited the $I_{ds}$, $\mu$, $V_t$, $I_{on}/I_{off}$, and $SS$ of −80 µA, 1.1 cm²/Vs, −2.71 V, $10^5$ and 0.44 V/dec., respectively.

Vacuum deposition techniques are advantageous for achieving highly ordered films with a precise control on the thickness. However, a highly sophisticated vacuum chamber is required for an adequate flow of the charge carriers. Moreover, deposition of a material at very high temperature makes this technique inappropriate for the flexible devices. Thermal vacuum evaporation fabrication processes are further classified and discussed in the following sections.

2.7.1.1.1 Resistive Evaporation   In this method, the material to be deposited is heated till its fusion due to an electrical current flowing through a filament or metal plate (Joule effect). Thereafter, evaporated material is condensed over the substrate. Additional sources of heating utilized are BN crucible or RF coil surrounding graphite. The assembly of the technique is simple and appropriate for depositing metals and some compounds with low fusion temperature (Al, Ag, Au, $SiO_2$, etc.). Classic metals utilized as resistance heaters are molybdenum (Mo), tantalus (Ta), and wolfram (W), with evaporation-temperature ($T_{evap}$ = 1000°C – 2000°C) and vapor-pressure practically zero [72].

2.7.1.1.2 Ion Beam Evaporation   The ion beam evaporation method utilizes heat generated through bombardment of electron beam of high energy on the material designated for deposition. The electron beam is produced via electron gun using thermionic emission. Electrons thus emitted from incandescent filament (i.e., cathode) are accelerated toward an anode through a high voltage (kilovolts). Either a crucible or an almost perforated disc can be used as an anode. A magnetic field is generally applied to alter the trajectory of electrons, permitting the electron gun to be placed under the line of evaporation. As electrons can be focused, it is likely to get a concentrated heating of the material, with high power of evaporation (many kilowatts). This effectively controls the rate of evaporation and also gives a chance for deposition of high melting point materials like Ta, W, and C. Contamination problems generated by heating and degasification can be removed by cooling the crucible [72].

2.7.1.1.3 Pulsed Laser Deposition (PLD)   In the pulsed laser deposition technique, a pulsed laser beam of very high power is focused inside a vacuum chamber on a target material designated for deposition. The plasma plume from the target material gets deposited as a thin film on the substrate. This process may take place either in the presence of background gas like oxygen for generating oxide layers or in an ultrahigh vacuum environment. In 1987, Dijkkamp *et al.* deposited a better quality thin film of superconductive material $YBa_2Cu_3O_7$ with the help of a laser. For fabricating crystalline films of high quality, the pulsed laser deposition technique is commonly used [73].

2.7.1.1.4 Cathodic Arc Deposition (ARC-PVD)   The cathodic arc deposition technique utilizes an electric arc for vaporizing the material from a target designated as cathode. The material that is vaporized is then allowed to condense on the substrate. Thus, a thin film is formed over the substrate. For the deposition of ceramic, metallic, and composite films, this technique is highly efficient. Initially, the industrial use of cathodic are deposition technique was found in 1960s [74].

### 2.7.1.2  SPUTTERING TECHNIQUES

Sputter involves an ejection of material to be deposited from a source to a target substrate. Ejected atoms have a wide energy distribution, typically up to tens of electron volts (100,000 K). At higher gas pressures, the ions collide with the gas atoms, diffuse to the substrates or vacuum chamber wall, and condense after undergoing a random walk. The dielectric–sputter system and metal–sputter system are commonly used techniques for sputtering [75–78].

*2.7.1.2.1  Ion Beam Sputtering (IBS)*   In the ion beam sputtering technique, an external target is used. A Kaufman source magnetron is used to produce a magnetic field that confines the electrons produced by collision with ions. An electric field is used to accelerate these electrons toward a target. Ions from the source are neutralized by electrons generated by an additional external filament. Ion beam sputtering is advantageous as the flux and energy of ions can be controlled independently. As the flux striking the target is constituted by neutral atoms, targets can be sputtered either using insulating or conducting properties [75,77].

*2.7.1.2.2  Reactive Sputtering*   Thin film can be deposited by means of chemical reaction between a gas and the target material inside the vacuum chamber. The structure of the thin film can be easily controlled with change in the relative pressure of the reactive gases and inert gases. Stoichiometry of the film is one of the major constraints for optimizing functional properties of deposited nitrides and oxides like the stress of SiNx and refractive index in SiOx [77,78].

*2.7.1.2.3  Ion-Assisted Deposition*   In the ion-assisted deposition technique, a secondary ion beam with lower power, compared to the sputter gun, is used to strike the substrate. This technique is used to deposit a diamond-like structure of carbon on a substrate. Misaligned carbon atoms in the crystal lattice have to be removed by the additional beam. Ion-assisted deposition is used for important applications in industries like developing coatings of tetrahedral amorphous carbon surface on platters of hard disk and medical implants with hard transition metal nitride coatings [75,77].

*2.7.1.2.4  High-Target-Utilization Sputtering*   High-target-utilization sputtering can be achieved by isolated generation of very high density plasma inside a sideward chamber with its opening into the main process chamber that contains the target material and the substrate. The target develops the ion current independent of the applied target voltage.

## 2.7.2 SOLUTION PROCESSING TECHNIQUES

Solution processable organic materials are beneficial in realizing large area electronic circuits at considerably low temperature, and therefore, at lower cost. Spin coating, inkjet printing, polymer inking and stamping, transfer printing, and drop casting are regularly used cost-effective techniques. In transfer printing, a thin-film pattern is transferred from a nonadhesive mold to an adhesive substrate. Using this technique, the devices obtained are of longer channel length such as 40 μm as reported by Cho *et al.* [79]. However, Sele *et al.* reported a significant reduction in the channel length (up to 100 nm) by employing the inkjet printing technique [80].

Spin coating is another solution processing technique, often employed for cost effective and large area production of the organic devices and circuits. Assadi *et al.* [81] reported the first solution processed conducting polymer, P3HT, wherein the solubility was improved by adding alkyl chains. Raval *et al.* [82] reported a P3HT based organic transistor fabricated through a spin coating technique that exhibited a mobility of 0.001 cm$^2$/Vs and $V_t$ of −0.9 V. Additionally, Afzali *et al.* [54] reported a mobility of 0.89 cm$^2$/Vs for a spin coated pentacene transistor processed in 1%−2% chloroform solution.

Chen *et al.* [83] reported an OTFT based on polythiophene deposited through the inkjet printing technique. This transistor exhibited a $\mu$, $V_t$, and $I_{on}/I_{off}$ of 0.01 cm$^2$/Vs, −4 V, and 80, respectively. Furthermore, by processing the same materials, they reported an organic *p*-type bootstrap inverter with a gain of 1.8 at 30 V supply. Furthermore, Cho *et al.* reported a BGBC organic transistor fabricated on poly(ether sulfone) (PES) substrate by transfer-printing of gold electrodes and solution processing of the PVP material [79]. The active layer was made by drop casting of the solution of 0.5 wt% TIPS pentacene in toluene. This device demonstrated a saturation mobility of 0.012 cm$^2$/Vs at $V_{ds}$ of −20 V.

Solution processing techniques are the easiest in comparison to all other deposition techniques. They are very cost efficient. Moreover, they can be performed in normal room conditions and a clean room is not compulsory. Solution-based processes are widely used for deposition of polymer organic materials to fabricate OTFTs and organic solar cells. A few solution-processed techniques are discussed in the following sections.

### 2.7.2.1 SPIN COATING

The spin coating process is utilized for depositing uniform and smooth thin films on plane substrates. Generally, a small amount of coating material is on the middle of the substrate. When the substrate is rotated at a high speed, the coating material is uniformly spread over the substrate due to centrifugal force. The device utilized in the spin coating process is known as a spin-coater or spinner. The process of rotation is sustained until the thin film of the required height is obtained. The volatile solvent evaporates simultaneously leaving behind the layer of solute. Therefore, for greater angular velocity of spinning, more thin films are obtained. The concentration and the viscosity of the utilized solution also affects the thickness of the deposited thin films [84,85].

### 2.7.2.2  DIP COATING

Dip coating is an industrial grade thin-film deposition process that is used for manufacturing coated fabrics and cylindrical OTFTs. In the case of flexible planar substrates like fabrics, dip coating is achieved in the form of a nonstop roll-to-roll process. In the case of 3D coating of an object, the object can be simply dipped and withdrawn from the pool of dip-coating solution. For making cylindrical OTFTs, the circular gate is dipped in the solution of coating and removed quickly to deposit a layer of desired material like dielectric material. Thereafter, it is again dipped in the solution of organic material to deposit an organic semiconductor layer. Finally, the electrodes are deposited. Dip-coating processes are generally used to fabricate uniform thin films efficiently and economically over cylindrical and flat substrates [86].

### 2.7.2.3  INKJET PRINTING/SCREEN PRINTING

The inkjet printing method utilizes a mesh to deposit ink over a substrate, other than the areas that are made impervious to ink solution by means of a patterning stencil. A squeegee or blade is progressed over the screen to deposit the solution on the mesh that is not masked by the stencil. Thereafter, a reverse stroke enables the screen to touch the substrate for a moment alongside a contact line. Due to this the ink solution makes the substrate wet and can be taken out of the apertures of mesh when the screen coils back after the passing of the blade [87]. Printed circuit boards (PCBs) and solar cells can be printed using this technique. It can be used in fabricating highly precise metallic tracks on PCBs.

Based on the deposition methods of different organic semiconductors, the performance in terms of drain current, mobility, *on/off* current ratio, and threshold voltage of a few *p*-type OTFTs are compared in Table 2.12.

Over the last two decades, researchers have demonstrated noteworthy improvement in the fabrication methodologies of organic devices [79–81,83]. Figure 2.14 shows a comparative study of growth in mobility for different *p*- and *n*-type organic materials, such as the vacuum processed small molecule (VPSM), solution processed conducting polymer (SPCP), and solution processed small molecule (SPSM). The highest mobility is reported for *p*- and *n*-type small molecule organic materials deposited through the vacuum evaporation method. However, solution processed *p*- and *n*-type conducting polymers have also shown noteworthy progress with the passage of time. In addition to this, the small molecule (*p/n*) organic materials deposited through solution processing have demonstrated continuous growth during the last decade due to synthesis of several precursors like TIPS pentacene [55] and diF-TESADT that showed compatibility with solution processing techniques.

**TABLE 2.12   Performance of Different *p*-Type OTFTs Based on Deposition Method of OSCs**

| Method of OSC Deposition | OSC | $I_{ds}$ (μA) | $\mu$ (cm²/Vs) | $I_{on}/I_{off}$ | $V_t$ (V) | Supply Voltage (V) | |
|---|---|---|---|---|---|---|---|
| | | | | | | $V_{ds}$ | $V_{gs}$ |
| Drop casting | TIPS pentacene | −0.9 | 0.012 | $10^3$ | −1.8 | −20 | 10 to −10 |
| Inking and stamping | Pentacene | −40 | 0.033 | $10^8$ | −1.2 | −5 | 0 to −5 |
| Thermal evaporation | Pentacene | −22 | 0.32 | $10^6$ | −10 | −30 | 20 to −40 |
| | | −3 | 0.19 | $10^7$ | −3 | −20 | 0 to −20 |
| | | −0.8 | 0.038 | NR | −6 | −20 | 20 to −40 |
| Inkjet printing | Polythiophene | −0.1 | 0.01 | 80 | −4 | −40 | 20 to −40 |
| Spin coating | P3HT | −0.4 | 0.001 | NR | −0.9 | −4 | 0 to −4 |
| Electron beam deposition | Pentacene | −80 | 1.1 | $10^5$ | −2.7 | −50 | 0 to −20 |

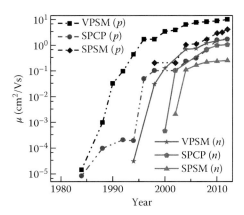

**FIGURE 2.14** Comparative plot for growth in mobility of *p*- and *n*-type transistors.

### 2.7.3 CHEMICAL VAPOR DEPOSITION TECHNIQUE

Chemical vapor deposition (CVD) is a chemical process used to produce high-quality, high-performance solid materials. The process is often used in the semiconductor industry to produce thin films. In typical CVD, the wafer (substrate) is exposed to one or more volatile precursors that react and/or decompose on the substrate surface to produce the desired deposit. Volatile by-products produced are removed by gas flow through the reaction chamber. Microfabrication processes widely use CVD to deposit materials in various forms, including monocrystalline, polycrystalline, amorphous, and epitaxial. These materials include silicon ($SiO_2$, germanium, carbide, nitride, silicon oxynitride), carbon (fiber, carbon nanofibers, nanotubes, diamond, and graphene), fluorocarbons, filaments, tungsten, titanium nitride, and various high-$k$ dielectrics [88]. CVD techniques are costly in comparison to other deposition processes because of high power requirements of the fabrication. Besides this, chemical reactions leave by-products that must be eliminated by a continuous gas flow for longer duration.

### Example 2.5

Draw the flow chart of the basic steps for fabricating bottom gate structure with contacts at the top and bottom of organic thin-film transistors.
**Solution:** The basic fabrication steps for obtaining bottom gate structure with contacts at the top and bottom are shown in Figure 2.15.

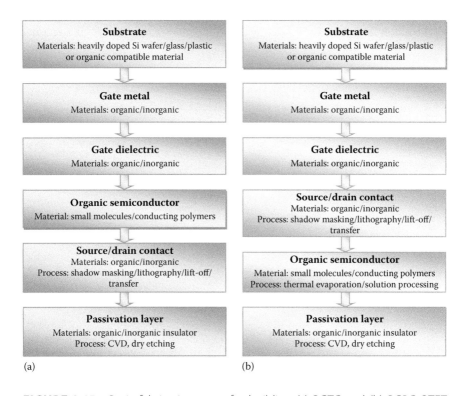

(a)                                                                 (b)

**FIGURE 2.15**   Basic fabrication steps for building (a) BGTC and (b) BGBC OTFT structures.

## 2.8 PERFORMANCE-INFLUENCING FACTORS OF OTFTs

The performance of an OTFT strongly depends on the device dimensions, materials, and fabrication process. In addition to this, the interfaces also play a vital role due to multiple layers. Semiconductor surface interfacing with S/D contacts and gate dielectric affects the performance of an organic transistor to a large extent. This section presents the impact of dimensional parameters and interfaces on the performance of organic transistors.

### 2.8.1 DIMENSIONAL PARAMETERS

Dimensional parameters such as channel length, device width, and the thicknesses of active and insulating layers affect the performance of an organic transistor significantly. The thickness of the dielectric should be small enough to achieve a high capacitance that, in turn, enables high drain current at low switching voltage. In addition to this, the mobility also increases with decreasing thickness of the dielectric. Singh and Baquer reported the rise in mobility by almost 20 times with reduction in thickness of PMMA dielectric from 700 to 210 nm [89].

A thinner film of semiconductor can significantly reduce the *off*-current that, in turn, reduces the leakage power and noise margins. Resendiz *et al.* reported a high *on/off* current ratio of $6 \times 10^9$ at $t_{osc}$ of 20 nm as compared to 10 with 160 nm thickness for P3HT-based OTFTs [12]. Besides this, an increase of 8.5% in mobility and a decrease of 50% in threshold voltage were observed on reducing the thickness from 160 to 20 nm. Similarly, Kano *et al.* demonstrated the effect of dibutylquaterthiophene active layer thickness on device performance [9]. Upon increasing the thickness from 20 to 80 nm, they observed a reduction in mobility from 0.05 to 0.03 cm$^2$/Vs and an increase in subthreshold slope from 3.5 to 5 V/dec due to a proportionate increase in the access resistance. Surprisingly, a large variation from −10 to +18 V in the threshold voltage was observed with an increase in $t_{osc}$ from 20 to 80 nm. Furthermore, Gupta and Hong [90] reported significant improvement in mobility and threshold voltage by scaling down the thickness of the pentacene active layer from 100 nm to 10 nm.

To improve the performance of OTFT, noteworthy efforts have been made to fabricate the devices with shorter channel lengths. Rogers *et al.* in 1999 reported $\alpha$-sexithiophene ($\alpha$-6T)-based *p*-type and $F_{16}$CuPc-based *n*-type OTFTs, each with 100 nm channel length [91]. These devices performed well at low voltage of ±3 V that resulted in mobility of 0.0001 and 0.0008 cm$^2$/Vs and current of −1.5 and 2 μA for *p*- and *n*-type transistors, respectively. Later in 2003, Zhang *et al.* observed a low voltage operation of submicron pentacene TFT of 30 nm channel length that yielded a mobility of 0.02 cm$^2$/Vs and an *on/off* current ratio of 10$^2$ at supply voltage of −2 V [92]. Furthermore, Lee *et al.* in 2005 reported an organic transistor of 10 nm length with reasonably good drain and transfer characteristics at low voltage of −0.3 V [93]. Additionally, they analyzed the performance dependence on the channel length variation; wherein, scaling down the length from 125 to 10 nm resulted in an increase in current by 13 times.

### 2.8.2 CONTACT–SEMICONDUCTOR INTERFACE

The charge carrier injection and extraction to and from the channel is governed by the interface between the metal and semiconductor. Burgi *et al.* reported a large barrier height of 1.2 eV for *n*-type OSCs with commonly used gold, aluminum, and copper metals [39]. The effect of metal–OSC interface is less pronounced in *p*-type materials due to their lower barrier height (0.1–0.2 eV) [94] in comparison to *n*-type OSCs. Wondmagegn *et al.* studied the impact of dipoles at the gold–pentacene interface through numerical simulations [95]. They observed a high contact resistance of 91 kΩ in the presence of dipoles. However, this resistance reduced to half in the absence of dipole at the contact–OSC interface. They observed that the contact resistance extracted in the presence of dipoles was almost equal to the experimental result. This analysis justified the existence of dipoles in the organic devices.

Gupta *et al.* [19] reported the effect of morphological disorders in pentacene-based top and bottom contact organic transistors. The TC structure exhibited uniformity in the grains of the active layer; whereas, diversity in the grain size was observed in the BC, especially at the contact–OSC interface due to the deposition of semiconductor on the prepatterned S/D electrodes. As a result, the mobility reduced by 2 orders of magnitude in the BC structure compared to the TC structure. Several researchers have proposed few solutions to overcome this interface barrier. Tiwari *et al.* reported a considerable improvement in the performance of organic transistor by depositing an additional thin organic active layer between the contacts and the semiconductor [96]. Furthermore, Ishikawa *et al.* demonstrated an improvement in the performance of *p*-type bottom contact TFT by inserting an additional $p^+$ region near the contacts [20]. Moreover, Watanabe and Kudo showed an improvement in $I_{on}/I_{off}$ (540 times) by adding a layer of CuPc organic material between the source (ITO) and the active layer (pentacene) to improve the carrier injection [30].

### 2.8.3 SEMICONDUCTOR–DIELECTRIC INTERFACE

The interface between the semiconductor and gate dielectric should be optimized to achieve an improved ordering of the active film and better charge carrier accumulation at the OSC–dielectric interface. The performance of OTFT strongly depends on the resistivity of a dielectric material. High resistive material reduces the interface trap density that, in turn, increases the mobility and current. Veres *et al.* [42] reported the performance of a PTAA-based *p*-type OTFT with PMMA and PVA materials. The mobility of the transistor with PMMA was observed to be 6 times higher in comparison to the transistor with PVA material. This is due to high resistivity of PMMA dielectric in comparison to its counterpart. Similarly, Jang *et al.* compared the performance of a $C_{60}$-based *n*-type OTFT with $SiO_2$ (with encapsulation) and CYTOP (without encapsulation) gate dielectrics [97]. The transistor with CYTOP material yielded the mobility and *on/off* current ratio of 0.05 cm²/Vs and 6 × 10³, respectively, comparable to the transistor with $SiO_2$ ($\mu$ = 0.08 cm²/Vs and $I_{on}/I_{off}$ = 10⁴) provided with encapsulation. This transistor with CYTOP material improved the stability of the transistor in the air even without encapsulation due to a significant decrease in the interface traps.

Surface treatment is considered to be a viable method to improve the performance of organic transistors. By applying treatment to the surface of dielectric, the trap states can be reduced substantially. This results in an increase in the accumulation of charge at the OSC–dielectric interface that, in turn, improves the performance of device. Lin *et al.* reported a high-performance pentacene transistor by applying surface treatment using octyltrichlorosilane (OTS) material that yielded drain current, mobility, *on/off* current ratio, subthreshold slope and threshold voltage of 70 μA, 1.5 cm²/Vs, 10⁸, 1.6 V/dec, and −8 V, respectively, at the supply voltage of −20 V [98]. Furthermore, Tiwari *et al.* demonstrated a pentacene organic TFT by modifying the surface of $SiO_2$ dielectric with OTS-8 treatment that resulted in a reasonably high *on/off* current ratio of 10⁹ and a mobility of 0.8 cm²/Vs [99]. Additionally, Kobayashi *et al.* observed a significant reduction in the threshold voltage by applying $NH_2$ groups for surface treatment [100]. The shift in $V_t$ can be attributed to the reduction in dipoles or an extra charge generated by the treatment. Recently, Jiang *et al.* reported a P3HT-based transistor with surface modification of $SiO_2$ by octadecyltrimethoxysilane (OTMS) [101]. This transistor exhibited a mobility of 0.24 cm²/Vs and *on/off* current ratio of 2.8 × 10⁵.

The performance of OTFTs has been improved by several researchers by introducing a self-assembled monolayer on the substrate. Klauk *et al.* [17] reported a high-performance pentacene TFT consisting of very thin (5.7 nm) SAM of *n*-octadecylphosphonic material with drain current and mobility of 5 μA and 0.4 cm²/Vs, respectively, at low operating voltage of −3 V. In addition to this, a low leakage current and subthreshold slope of 0.5 pA and 100 mV/dec, respectively, was observed. Additionally, Ito *et al.* reported a SAM of OTMS on the $SiO_2$ surface [102]. By producing this monolayer, a hole mobility of 3.0 cm²/Vs for pentacene and an electron mobility of 5.3 cm²/Vs for $C_{60}$-based transistors were reported. The excellent characteristics of interface between the layers not only reduced the defects but also increased the number of injected carriers to the channel.

## 2.9  CONCLUDING REMARKS

The research in organic electronics has undergone remarkable advancements in recent years. The organic transistor boasts of a bright future with a wide spectrum of applications. Nevertheless, it still faces a number of challenges that need to be resolved to make them a commercially sustainable and viable technology.

Based on the extensive study in this chapter, the following points are identified that can become useful for researchers, academicians, and industry people to improve upon the shortcomings of OTFT technology.

- In recent years, numerous analytical models have been developed for OTFTs, mainly based on the classical MOS transistor model by introducing field-dependent mobility, grain size, contact resistance, and channel length modulation. The majority of models are based on the charge transport within the conducting channel only. However, charge localization and conduction also occurs in the bulk of the organic semiconductor, thereby, bulk sheet resistance starts playing a dominant role. This resistance significantly affects the current conduction and can be minimized using thinner OSC layers but cannot be eliminated completely. Therefore, it needs to be accounted for while developing the models to adequately understand the OTFT's behavior.

- Aggressive efforts are being made to analyze the performance of an OTFT in terms of semiconducting layer thickness. However, the dielectric thickness is also a crucial influencing factor that significantly affects the threshold voltage and mobility. Besides this, the OTFT structures also play a significant role when analyzing the device performance in terms of layers' thicknesses. This is due to a dissimilar path traversed by the charge carriers between the source and drain. Therefore, it is necessary to investigate the impact of the thickness of the active and dielectric layers on the performance of both top and bottom contact structures individually.

- Tremendous efforts have been made for implementation and development of organic devices on flexible substrates. However, only a few attempts have been directed toward organic digital circuits. Inverter and universal logic gates are essentially required in several OTFT-based applications; therefore, significant efforts are required toward designing and enhancing the performance of these circuits. Most of the organic circuits are based on the $p$-type designs only due to higher field effect mobility and better stability of $p$-type organic materials in comparison to $n$-type. An all-$p$-type organic circuit does not provide a full swing at the output due to the large threshold voltage of the organic transistors. This in turn reduces the noise margin and gain, thereby making it less robust. Therefore, it is strongly required to develop techniques that can substantially improve the performance of organic digital circuits.

- The performance of an OTFT strongly depends on the materials at different layers. Noteworthy progress has been made in recent years toward synthesis of several high-performance novel $p$- and $n$-type organic materials. However, little research has actually dealt with the effect of different materials on the performance of inverter circuits. Therefore, to address the effect of different materials, it is necessary to analyze the static and dynamic response of inverter circuits using different materials for driver and load TFTs.

■ Static random access memory (SRAM) is extensively used as on-chip memory in high-performance integrated circuits. The organic TFT-based SRAM cell is essentially required for realization of low-cost memory systems on flexible substrates. Although flexible memory is a key component in many organic electronic applications such as RFID tags, smart cards, and video games, very few organic SRAM cell designs have been reported to date. To the best of our knowledge, attempts to analyze the performance of the SRAM cell using transistors with different materials and configurations have not been reported so far. The major design issue in the SRAM cell is the conflict between the read and write stability. Moreover, it is a big challenge to maintain the performance of the cell with down-scaling of feature size. Therefore, it is essentially required to analyze different configurations to improve the overall performance of the SRAM cell.

## PROBLEMS
## MULTIPLE CHOICE

1. Mobility of an organic transistor enhances with
   a. Increase in overdrive voltage
   b. Supply voltage
   c. Gate to source voltage
   d. Drain to source voltage
2. Reduction in threshold voltage of OTFTs is due to
   a. Contact resistance
   b. Junction capacitance
   c. Gate oxide capacitance
   d. All of the above
3. In an OTFT operating in a linear region, the voltage across the channel is less than the external applied voltage due to
   a. Junction capacitance
   b. Contact resistance
   c. Gate capacitance
   d. None of the above
4. Which structure is used to achieve better charge carrier modulation?
   a. Single gate OTFT
   b. Dual gate OTFT
   c. Vertical OTFT
   d. Cylindrical OTFT

5. Cylindrical gate structure aims for
   a.  High package density
   b.  Low cost
   c.  Better performance
   d.  Simple fabrication
6. Bottom gate structure has a higher mobility enhancement factor than top gate due to
   a.  Low trap density
   b.  Charge carrier injection
   c.  High trap density
   d.  All of the above
7. Why are most organic devices and circuits fabricated using $p$-type OSC materials?
   a.  Higher mobility of $p$-type
   b.  Higher mobility and compatibility with contact electrode and dielectric material
   c.  Low HOMO and LUMO based level of OSC
   d.  None of the above
8. Which organic material group shows high performance of an organic device?
   a.  Small molecule
   b.  Conducting polymer
   c.  Both of the above
   d.  None of the above
9. Reduction in thickness of the organic semiconductor layer can signify
   a.  Reduction in *off*-current
   b.  Minimization of leakage power
   c.  High noise margin
   d.  All of the above
10. Which electrode material is commonly used to make anode terminal in organic displays?
    a.  Aluminum
    b.  Iron
    c.  Indium tin oxide (ITO)
    d.  Copper
11. Smaller channel length devices can be fabricated using
    a.  Single gate OTFT
    b.  Dual gate OTFT
    c.  Vertical OTFT
    d.  Cylindrical OTFT
12. Some advantages organic materials offer over silicon to produce TFTs are
    a.  Large-area electronics at the lower cost
    b.  Ease of processing at considerably lower temperature
    c.  Mechanical flexibility and lightweight
    d.  All of the above

13. Organic thin-film transistors (OTFTs) are operated in
    a. Inversion mode only
    b. Accumulation mode only
    c. Both accumulation and inversion mode
    d. All of the above
14. What is the key limitation of organic material-based devices and circuits?
    a. Mobility not large enough to suit high switching speed applications
    b. Mobility large enough to suit high switching speed applications
    c. High mobility and large drive current
    d. None of the above

**ANSWER KEY**

1. a; 2. c; 3. b; 4. b; 5. a; 6. c; 7. b; 8. a; 9. d; 10. c; 11. c; 12. d; 13. b; 14. a

**SHORT ANSWER**

1. Describe the important characteristic parameters that illustrate the performance of organic thin-film transistors.
2. Draw and explain different structures of organic thin-film transistors.
3. Why is the performance of dual gate organic thin-film transistors better than single gate organic thin-film transistors?
4. Describe different organic thin-film transistor models with suitable equations.
5. Explain charge transport mechanism in organic semiconductors.

**DESCRIPTIVE ANSWER**

1. Describe the following charge transport models: (a) variable range hopping (VRH), (b) multiple-trapping and release (MTR) model, and (c) polaron model.
2. Classify the different organic semiconductor materials. Explain in detail $p$-type and $n$-type organic semiconductor materials along with examples.
3. Write short notes with examples on the following materials used in the fabrication of organic TFTs:
    a. Semiconductor materials
    b. Electrode materials
    c. Gate dielectric materials
    d. Substrate materials
4. Describe the vacuum evaporation and solution processing techniques used in fabrication of different organic thin-film transistors.
5. What are the factors that influence the performance of organic thin-film transistors?
6. Explain advantages, disadvantages, limitations, stability issues, and future scope of organic-material-based devices and circuits.

7. Write short notes on flexible electronics: opportunities and challenges in organic VLSI circuits.
8. Write various recent applications of organic thin-film transistors. Give details about size of market.
9. What are the dopants for polymers?
10. Write various molecular structures along with method of doping and conductivity of conducting polymers useful in organic electronics research.
11. Explain the difference between single gate organic thin-film transistors (SG OTFTs) and dual gate organic thin-film transistors (DG OTFTs). Why are DG OTFTs preferred over SG OTFTs in organic circuit designs?
12. Explain the merits and demerits of organic electronics over inorganic with examples.
13. Describe different structures of organic thin-film transistors along with their cross-sectional view.
14. Describe the drain current equations of organic thin-film transistor in the linear and saturation regime indicating important performance parameters.
15. Explain the operating principle of an organic thin-film transistor. Describe why OTFTs are better candidates in comparison to conventional MOSFETs for some specific applications.
16. Describe the comparative analysis of BGTC, BGBC, TGBC, and TGTC organic field-effect transistor structures. Include a cross-section of transistors in your explanations.
17. Explain the various steps involved in the fabrication process of bottom gate organic thin-film transistors with contacts at the bottom and top.

## EXERCISES

1. Examine the relationship between the current and the terminal voltages for a $p$-type top gate top contact organic thin-film transistor with $\mu = 0.14$ cm$^2$/Vs, $\varepsilon_r = 3.9$, $t_{ox} = 200$ nm, $W = 1$ mm, $L = 30$ μm, and $V_t = -3.2$ V.
2. Using the field-dependent mobility concept in organic thin-film transistor, find zero bias mobility ($\mu_0$). The device mobility is $\mu = 0.02$ cm$^2$/Vs, the enhancement factor is 0.2, with a source-gate voltage ($V_{gs}$) of $-14$ V and threshold voltage ($V_t$) of $-3.2$ V.
3. Extract the current *on-off* ratio ($I_{on}/I_{off}$) of an organic thin-film transistor with $\mu = 0.015$ cm$^2$/Vs, $V_{ds} = -10$ V, $V_{gs} = -10$ V, $V_t = -1.3$ V, $C_{ox} = 800$ nF/cm$^2$, 20 nm of insulator thickness ($t_{osc}$), and $\sigma = 1$ S/cm.
4. Find the operating mode and estimate the drive current of an organic thin-film transistor for the given parameters: $\mu = 1.64$ cm$^2$/Vs, $W = 120$ μm, $L = 10$ μm, $C_{ox} = 800$ nF/cm$^2$, $V_{gs} = 1.6$ V, $V_{ds} = 2$ V, and $V_t = 1.2$ V.

# REFERENCES

1. Shirakawa, H.; Louis, E. J.; MacDiarmid, A. G.; Chiang, C. K.; Heeger, A. J. "Synthesis of electrically conducting organic polymers: Halogen derivatives of polyacetylene, $(CH)_x$," *J. Chem. Soc., Chem. Commun.* **1977**, 16, 578–580.
2. Tsumura, A.; Koezuka, H.; Ando, T. "Macromolecular electronic device: Field effect transistor with a polythiophene thin film," *Appl. Phys. Lett.* **1986**, 49(18), 1210–1212.
3. Horowitz, G.; Lang, P.; Mottaghi, M.; Aubin, H. "Extracting parameters from the current-voltage characteristics of organic field-effect transistors," *Adv. Funct. Mater.* **2004**, 14(11), 1069–1074.
4. Deen, M. J.; Marinov, O.; Zschieschang, U.; Klauk, H. "Organic thin-film transistors: Part II–Parameter extraction," *IEEE Trans. Electron Devices* **2009**, 56(12), 2962–2968.
5. Estrada, M.; Cerdeira, A.; Puigdollers, J.; Resendiz, L.; Pallares, J.; Marsal L. F.; Voz, C.; Iniguez, B. "Accurate modeling and parameter extraction method for organic TFTs," *Solid-State Electron.* **2005**, 49(6), 1009–1016.
6. Dimitrakopoulos, C. D.; Malenfant, P. R. L. "Organic thin film transistors for large area electronics," *Adv. Mater.* **2002**, 14(2), 99–117.
7. Horowitz, G.; Hajlaoui, M. E. "Grain size dependent mobility in polycrystalline organic field-effect transistors," *Synth. Metal* **2001**, 122(1), 185–189.
8. Knipp, D.; Street, R. A.; Volkel, A.; Ho, J. "Pentacene thin film transistors on inorganic dielectrics: Morphology, structural properties, and electronic transport," *J. Appl. Phys.* **2003**, 93(1), 347–355.
9. Kano, M.; Minari, T.; Tsukagoshi, K.; Maeda, H. "Control of device parameters by active layer thickness in organic thin film transistors," *App. Phy. Lett.* **2011**, 98(7), 073307-1–073307-3.
10. Horowitz, G.; Hajlaoui, R.; Bouchriha, H.; Bourguiga, R.; Hajlaoui, M. "The concept of threshold voltage in organic field-effect transistors," *Adv. Mater.* **1998**, 10(12), 923–927.
11. Pernstich, K. P.; Haas, S.; Oberhoff, D.; Goldmann, C.; Gundlach, D. J.; Batlogg, B.; Rashid, A. N.; Schitter, G. "Threshold voltage shift in organic field effect transistors by dipole monolayers on the gate insulator," *J. Appl. Phys.* **2004**, 96(11), 6431-1–6431-8.
12. Resendiz, L.; Estrada, M.; Cerdeira, A.; Iniguez, B.; Deen, M. J. "Effect of active layer thickness on the electrical characteristics of polymer thin film transistors," *Org. Electron.* **2010**, 11(9), 1920–1927.
13. Islam, M. N. "Impact of film thickness of organic thickness on off-state current of organic thin film transistors," *J. Appl. Phys.* **2011**, 110(11), 114906-1–114906-10.
14. Cosseddu, P.; Bonfiglio, A. "A comparison between bottom contact and top contact all organic field effect transistors assembled by soft lithography," *Thin Solid Films* **2007**, 515(19), 7551–7555.
15. Cui, T.; Liang, G. "Dual gate pentacene organic field-effect transistors based on a nanoassembled $SiO_2$ nanoparticle thin film as the gate dielectric layer," *Appl. Phys. Lett.* **2005**, 86(6), 064102-1–064102-3.
16. Schon, J. H.; Kloc, C.; Batlogg, B. "On the intrinsic limits of pentacene field effect transistors," *Org. Electron.* **2000**, 1(1), 57–64.
17. Klauk, H.; Zschieschang, U.; Halik, M. "Low voltage organic thin film transistors with large transconductance," *J. Appl. Phys.* **2007**, 102(7), 074514-1–074514-7.

18. Shim, C. H.; Maruoka, F.; Hattori, R. "Structural analysis on organic thin film transistor with device simulation," *IEEE Trans. Electron Devices* **2010**, 57(1), 195–200.

19. Gupta, D.; Katiyar, M.; Gupta, D. "An analysis of the difference in behavior of top and bottom contact organic thin film transistors using device simulation," *Org. Electron.* **2009**, 10(5), 775–784.

20. Ishikawa, Y.; Wada, Y.; Toyabe, T. "Origin of characteristics differences between top and bottom contact organic thin film transistors," *J. Appl. Phys.* **2010**, 107(5), 053709–053715.

21. Luo, M. F. C.; Chen, I.; Genovese, F. C. "A thin film transistor for flat panel displays," *IEEE Trans. Electron Devices* **1981**, 28(6), 740–743.

22. Tuan, H. C.; Thompson, M. J.; Johnson, N. M.; Lujan, R. A. "Dual gate a-Si:H thin film transistors," *IEEE Electronic Device Lett.* **1982**, 3(12), 357–359.

23. Kaneko, Y.; Tsutsui, K.; Tsukada, T. "Back bias effect on the current voltage characteristics of amorphous silicon thin film transistors," *J. Non-Crystalline Solids* **1992**, 149(3), 264–268.

24. Koo, J. B.; Ku, C. H.; Lim, J. W.; Kim, S. H. "Novel organic inverters with dual gate pentacene thin film transistors," *Org. Electron.* **2007**, 8(4), 552–558.

25. Ha, T. J.; Sonar, P.; Dodabalapur, A. "High mobility top gate and dual-gate polymer thin film transistors based on diketopyrrolopyrrole-naphthalene copolymer," *Appl. Phys. Lett.* **2011**, 98(25), 253305-1–253305-3.

26. Nishizawa, J.; Terasaki, T.; Shibata, J. "Field effect transistor versus analog transistor (static induction transistor)," *IEEE Trans. Electron Devices* **1975**, 22(4), 185–197.

27. Kudo, K.; Wang, D. X.; Iizuka, M.; Kuniyoshi, S.; Tanaka, K. "Organic static induction transistor for display devices," *Thin Solid Film*, **2000**, 111–112, 11–14.

28. Chen, Y.; Shih, I. "Fabrication of vertical channel top contact organic thin film transistors," *Org. Electron.* **2007**, 8(5), 655–661.

29. Naruse, H.; Naka, S.; Okada, H. "Dual self-aligned vertical multichannel organic transistors," *Appl. Phys. Express*, **2008**, 1(1), 011801-1–011801-3.

30. Watanabe, Y.; Kudo, K. "Vertical type organic transistor for flexible sheet display," *Proc. SPIE* **2009**, 7415, 741515-1–741515-10.

31. Tanaka, S.; Yanagisawa, H.; Iizuka, M.; Nakamura, M.; Kudo, K. "Vertical- and lateral-type organic FET using pentacene evaporated films," *Electr. Eng. Jap.* **2004**, 149(2), 43–48.

32. Maccioni, M.; Orgiu, E.; Cosseddu, P.; Locci, S.; Bonfiglio, A. "Towards the textile transistor: Assembly and characterization of an organic field effect transistor with a cylindrical geometry," *Appl. Phys. Lett.* **2006**, 89(14), 143515-1–143515-3.

33. Jang, J.; Nam, S.; Park, J. J.; Im, J.; Park, C. E.; Kim, J. M. "Photocurable polymer gate dielectrics for cylindrical organic field-effect transistors with high bending stability," *J. Mat. Chem.* **2012**, 22, 1054–1060.

34. Marinov, O.; Deen, M. J.; Zschieschang, U.; Klauk, H. "Organic thin film transistors: Part I. Compact DC modeling," *IEEE Trans. Electron Devices* **2009**, 56(12), 2952–2961.

35. Xie, Z.; Abdou, M.; Lu, A.; Deen, M. J.; Holdcroft, S. "Electrical characteristics of poly (3-hexylthiophene) thin film MISFETs," *Can. J. Phys.* **1992**, 70(10), 1171–1177.

36. Necliudov, P.; Shur, M.; Gundlach, D.; Jackson, T. "Modeling of organic thin film transistors of different designs," *J. Appl. Phys.* **2000**, 88(11), 6594–6597.

37. Natali, D.; Fumagalli, L.; Sampietro, M. "Modeling of organic thin film transistors: Effect of contact resistances," *J. Appl. Phys.* **2007**, 101(1), 014501-1–014501-12.

38. Jung, K. D.; Kim, Y. C.; Park, B. G.; Shin, H.; Lee, J. D. "Modeling and parameter extraction for the series resistance in thin-film transistors," *IEEE Trans. Electron Devices* **2009**, 56(3), 431–440.

39. Burgi, L.; Richards, T. J.; Friend, R. H.; Sirringhaus, H. "Close look at charge carrier injection in polymer field effect transistors," *J. Appl. Phys.* **2003**, 94(9), 6129–6137.

40. Locci, S.; Morana, M.; Orgiu, E.; Bonfiglio, A.; Lugli, P. "Modeling of short channel effects in organic thin-film transistors," *IEEE Trans. Electron Devices* **2008**, 55(10), 2561–2567.

41. Swensen, J.; Kanicki, J.; Wang, G.; Heeger, A. J. "Influence of gate dielectrics on electrical properties of $F_8T_2$ polyfluorene thin film transistors," *Proc. SPIE* **2003**, 5217, 159–166.

42. Veres, J.; Ogier, S.; Lloyd, G. "Gate insulators in organic field effect transistors," *Chem. Mater.* **2004**, 16(23), 4543–4555.

43. Miller, A.; Abrahams, E. "Impurity conduction at low concentrations," *Phys. Rev.* **1960**, 120(3), 745–755.

44. Vissenberg, M. C. J. M.; Matters, M. "Theory of the field effect mobility in amorphous organic transistors," *Phys. Rev. B* **1998**, 57(20), 964–967.

45. Shur, M.; Hack, M. "Physics of amorphous silicon based alloy field effect transistors," *J. Appl. Phys.* **1984**, 55(10), 3831–3842.

46. Horowitz, G. "Organic field-effect transistors," *Adv. Mater.* **1998**, 10(5), 365–377.

47. Yamashita, J.; Kurosawa, T. "On electronic current in NiO," *J. Phys. Chem. Solids* **1958**, 5(1–2), 34–43.

48. Holstein, T. "Studies of polaron motion: Part I. The molecular crystal model," *Ann. Phys.* **1959**, 8(3) 325–342.

49. Schon, J. H.; Batlogg, B. "Trapping in organic field-effect transistors," *J. Appl. Phys.* **2001**, 89(1), 336–341.

50. Kumar, B.; Kaushik, B. K.; Negi, Y. S. "Organic thin film transistors: Structures, models, materials, fabrication and applications—A review," *Polymer Rev.*, **2014**, 54(1), 33–111.

51. McCullough, R. D.; Lowe, R. D.; Jayaraman M.; Anderson, D. L. "Design, synthesis, and control of conducting polymer architectures: Structurally homogeneous poly (3-alkylthiophenes)," *J. Org. Chem.* **1993**, 58(4), 904–912.

52. Bao, Z.; Dodabalapur, A.; Lovinger, A. J. "Soluble and processable regioregular poly (3-hexylethiophene) for thin film field effect transistor applications with high mobility," *Appl. Phys. Lett.* **1996**, 69(26), 4108-1–4108-3.

53. Herwig, P. T.; Mullen, K. "A soluble pentacene precursor: Synthesis, solid state conversion into pentacene and application in a field-effect transistor," *Adv. Mater.* **1999**, 11(6), 480–483.

54. Afzali, A.; Dimitrakopoulos, C. D.; Breen, T. L. "High performance, solution processed organic thin film transistors from a novel pentacene precursor," *J. Am. Chem. Soc.* **2002**, 124(30), 8812–8813.

55. Anthony J. E.; Brooks, J. S.; Eaton D. L.; Parkin, S. R. "Functionalized pentacene: Improved electronic properties from control of solid-state order," *J. Am. Chem. Soc.* **2001**, 123(38), 9482–9483.

56. Shukla, D.; Nelson, S. F.; Freeman, D. C.; Rajeswaran, M.; Ahearn, W. G.; Meyer, D. M.; Carey, J. T. "Thin film morphology control in naphthalene-diimide-based semiconductors: High mobility *n*-type semiconductor for organic thin-film transistors," *Chem. Mater.* **2008**, 20(24), 7486–7491.

57. Bao, Z.; Lovinger, A. J.; Brown, J. "New air stable *n*-channel organic thin film transistors," *J. Amer. Chem. Soc.* **1998**, 120(1), 207–208.

58. Malenfant, P. R. L.; Dimitrakopoulos,C. D.; Gelorme, J. D.; Kosbar, L. L.; Graham, T. O.; Curioni, A.; Andreoni, W. "*n*-Type organic thin film transistor with high field effect mobility based on a N,N'-dialkyl-3,4,9,10-perylene tetracarboxylic diimide derivative," *Appl. Phys. Lett.* **2002**, 80(14), 2517–2519.

59. Tiwari, S. P.; Namdas, B.; Rao, V. R.; Fichou, D.; Mhaisalkar, S. G. "Solution processed *n*-type organic field effect transistors with high ON/OFF current ratios based on fullerene derivatives," *IEEE Electron Device Lett.* **2007**, 28(10), 880–883.

60. Li, C.; Pan, F.; Wang, X.; Wang, L.; Wang, H.; Wang, H.; Yan, D. "Effect of the work function of gate electrode on hysteresis characteristics of organic thin-film transistors with $Ta_2O_5$/polymer as gate insulator," *Org. Electron.* **2009**, 10(5), 948–953.

61. Klauk, H.; Halik, M.; Zschieschang, U.; Schmid, G.; Radik, W. "High-mobility polymer gate dielectric pentacene thin film transistors," *J. Appl. Phys.* **2002**, 92(9), 5259–5263.

62. De Angelis, F.; Cipolloni, S.; Mariucci L.; Fortunato, G. "High-field-effect-mobility pentacene thin-film transistors with polymethylmetacrylate buffer layer," *Appl. Phys. Lett.* **2005**, 86(20), 203505-1–203505-3.

63. Chou, W. Y.; Kuo, C. W.; Cheng, H. L.; Chen; Y. R.; Tang, F. C.; Yang, F. Y.; Shu, D. Y.; Liao, C. C. "Effect of surface free energy in gate dielectric in pentacene thin-film transistors," *Appl. Phys. Lett.* **2006**, 89(11), 112126-1–112126-3.

64. Peng, X. Z.; Horowitz, G.; Fichou, D.; Garnier, F. "All-organic thin-film transistors made of alpha-sexithienyl semiconducting and various polymeric insulating layers," *Appl. Phys. Lett.* **1990**, 57(19), 2013-1–2013-3.

65. Garnier, F.; Horowitz, G.; Peng, X. Z.; Fichou, D. "An all-organic soft thin film transistor with very high carrier mobility," *Adv. Mater.* **1990**, 2(12), 592–594.

66. Bao, Z.; Feng, Y.; Dodabalapur, A.; Raju, V. R.; Lovinger, J. "High-performance plastic transistors fabricated by printing techniques," *Chem. Mater.* **1997**, 9(6), 1299–1301.

67. Butko, V. Y.; Chi, X.; Lang, D. V.; Ramirez, A. P. "Field-effect transistor on pentacene single crystal," *Appl. Phys. Lett.* **2003**, 83(23), 4773–4775.

68. Yun Y.; Pearson C.; Petty, M. C. "Bootstrapped inverter using a pentacene thin-film transistor with a poly(methyl methacrylate) gate dielectric," *IET Circuits Devices Syst.* **2009**, 3(4), 182–186.

69. Kumaki, D.; Umeda, T.; Suzuki, T.; Tokito, S. "High mobility bottom contact thin film transistors based on anthracene oligomer," *Org. Electron.* **2008**, 9(5), 921–924.

70. Dimitrakopoulos, C. D.; Brown, A. R.; Pomp, A. "Molecular beam deposited thin films of pentacene for organic field effect transistor applications," *J. Appl. Phys.* **1996**, 80(4), 2501–2508.

71. Cai, Y. G.; Zheng, X.; Ling, Z. S.; Jun, Z. F.; Wei, J. W.; Dan, S. D.; Na, Z. H. *et al.* "Characteristics of pentacene organic thin film transistor with top gate and bottom contact," *Chin. Phys. B* **2008**, 17(5), 1887–1892.

72. Scheffel, B.; Modes, T.; Metzner, C. "Reactive high-rate deposition of titanium oxide coatings using electron beam evaporation, spotless arc and dual crucible," *Surf. Coat. Tech.* **2016**, 287, 138–144.

73. Sansone, M.; De Bonis A. *et al.* "Pulsed laser ablation and deposition of niobium carbide," *Appl. Surf. Sci.* **2015**, APSUS-31537, pp. 1–5.

74. Valleti, K.; Murali, D.; Reddy P. M.; Joshi, S. V. "High temperature stable solar selective coatings by cathodic arc PVD for heat collecting elements," *Sol. Energ. Mat. Sol. Cells* **2016**, 145(3), 447–453.

75. Li, Y.; Fan, P.; Zheng, Z.; Luo, J.; Liang, G.; Guo, S. "The influence of heat treatments on the thermoelectric properties of copper selenide thin films prepared by ion beam sputtering deposition," *J. Alloy. Compd.* **2016**, 658, 880–884.

76. Dong, Y.; He, J.; Li, X.; Chen, Y.; Sun, L.; Yang, P.; Chu, J. "Study on the preheating duration of $Cu_2SnS_3$ thin films using RF magnetron sputtering technique for photovoltaics," *J. Alloy. Compd.* **2016**, 665, 69–75.

77. Gruber, W.; Baehtz, C.; Horisberger, M.; Ratschinski, I.; Schmidt, H. "Micro structure and strain relaxation in thin nanocrystalline platinum films produced via different sputtering techniques," *Appl. Surf. Sci.* 2016, (In press).

78. Jayaraman, V. K.; Kuwabara, Y. M.; Álvarez, A. M.; Olvera-Amador, M. de la L. "Importance of substrate rotation speed on the growth of homogeneous ZnO thin films by reactive sputtering," *Mater. Lett.* **2016**, 169, 1–4.

79. Cho, H.; Yoon, H.; Char, K.; Hong, Y.; Lee, C. "Organic thin film transistors with transfer printed Au electrodes on flexible substrates," *Jpn. J. Appl. Phys.* **2010**, 49(5), 05EB08-1–05EB08-4.

80. Sele, C. W.; Von Werne, T.; Friend, R. H.; Sirringhaus, H. "Lithography-free, self-aligned inkjet printing with sub-hundred-nanometer resolution," *Adv. Mater.* **2005**, 17(8), 997–1001.

81. Assadi, A.; Svensson, C. M.; Willander, O. I. "Field effect mobility of poly(3-hexylthiophene)," *Appl. Phys. Lett.* **1988**, 53(3), 195-1–195-3.

82. Raval, H. N.; Tiwari, S. P.; Navan, R. R.; Mhaisalkar, S. G.; Rao, V. R. "Solution processed bootstrapped organic inverters based on P3HT with a high-k gate dielectric material," *IEEE Electron Device Lett.* **2009**, 30(5), 484–486.

83. Chen, H. C.; Kung, C. P.; Houng, W. G.; Peng, Y. R.; Hsien, Y. M.; Chou, C. C.; Kao, C. J.; Yang, T. H.; Hou, J. "Polymer inverter fabricated by inkjet printing and realized by transistors arrays on flexible substrates," *IEEE/OSA J. Disp. Technol.* **2009**, 5(6), 216–223.

84. Jong, B.; Han, S. H.; Park, J. S. "Properties of CNTs coated by PEDOT: PSS films via spin-coating and electrophoretic deposition methods for flexible transparent electrodes," *Surf. Coat. Tech.* **2014**, 271, 22–26.

85. Zhang, Z.; Wei, D.; Xie, B.; Yue, X.; Li, M.; Song, D.; Li, Y. "High reproducibility of perovskite solar cells via a complete spin-coating sequential solution deposition process," *Solar Energy* **2015**, 122, 97–103.

86. Ghanbarian, M.; Nassaj, E. T.; Kariminejad, A. "Synthesis of nanostructural turbostratic and hexagonal boron nitride coatings on carbon fiber cloths by dip-coating," *Surf. Coat. Tech.* **2016**, 288, 185–195.

87. Kang, B. J.; Oh, J. H. "Influence of substrate temperature and overlap condition on the evaporation behavior of inkjet-printed semiconductor layers in organic thin film transistors," *Thin Solid Films* **2016**, 598, 219–225.

88. Wang, B. B.; Zhu, K.; Feng, J.; Wu, J. Y.; Shao, R. W.; Zheng, K.; Cheng, Q. J. "Low-pressure thermal chemical vapour deposition of molybdenum oxide nanorods," *J. Alloy. Compd.* **2016**, 661, 66–71.

89. Singh, V. K.; Baquer, M. "Impact of scaling of dielectric thickness on mobility in top contact pentacene organic thin film transistors," *J. Appl. Phys.* **2012**, 111(3), 034905-1–034905-6.

90. Gupta, D.; Hong, Y. "Understanding the effect of semiconductors thickness on device characteristics in organic thin film transistors by way of two-dimensional simulations," *Org. Electron.* **2010**, 11(1), 127–136.

91. Rogers, J. A.; Dodabalpur, A.; Bao, Z.; Katz, H. E. "Low-voltage 0.1 µm organic transistors and complementary inverter circuits fabricated with a low-cost form of near-field photolithography," *Appl. Phy. Lett.* **1999**, 75(7), 1010-1–1010-3.

92. Zhang, Y.; Petta, J. R.; Ambily, S.; Shen, Y.; Ralph, D. C.; Malliaras, G. G. "30 nm channel length pentacene transistors," *Adv. Mater.* **2003**, 15(19), 1632–1635.

93. Lee, J. B.; Chang, P. C.; Liddle, J. A.; Subramanian, V. "10-nm channel length pentacene transistors," *IEEE Trans Electron Devices* **2005**, 52(8), 1874–1879.

94. Rhee, S. W.; Yun, D. J. "Metal–semiconductor contact in organic thin film transistors," *J. Mater. Chem.* **2008**, 18, 5437–5444.

95. Wondmagegn, W. T.; Satyala, N. T.; Pieper, R. J.; Quevedo-Lopez, M. A.; Gowrisanker, S.; Alshareef, H. N.; Stiegler, H. J.; Gnade, B. E. "Impact of semiconductor/metal interfaces on contact resistance and operating speed of organic thin film transistors," *J. Comput. Electron.* **2011**, 10(1–2), 144–153.

96. Tiwari, S. P.; Potscavage Jr., W. J.; Sajoto, T.; Barlow, S.; Marder, S. R.; Kippelen, B. "Pentacene organic field effect transistors with doped electrode semiconductor contacts," *Org. Electron.* **2010**, 11(5), 860–863.

97. Jang, J;. Kim, J. W.; Park N.; Kim, J. J. "Air stable $C_{60}$ based *n*-type organic field effect transistor using a perfluoropolymer insulator," *Org. Electron.* **2008**, 9(4), 481–486.

98. Lin, Y. Y.; Gundlach, D. J.; Nelson, S. F.; Jackson, T. N. "Stacked pentacene layer organic thin film transistors with improved characteristics," *IEEE Electron Device Lett.* **1997**, 18(12), 606–608.

99. Tiwari, S. P.; Knauer, K. A.; Dindar, A.; Kippelen, B. "Performance comparison of pentacene organic field-effect transistors with $SiO_2$ modified with octyltrichlorosilane or octadecyltrichlorosilane," *Org. Electron.* **2012**, 13(9), 18–22.

100. Kobayashi, S.; Nishikawa, T.; Takenobu, T.; Mori, S.; Shimoda, T.; Mitani, T.; Shimotani, H.; Yoshimoto, N.; Ogawa, S.; Iwasa, Y. "Control of carrier density by self-assembled monolayers in organic field-effect transistors," *Nat. Mater.* **2004**, 3(5), 317-1–317-22.

101. Jiang, Y. D.; Jen, T. H.; Chen, S. A. "Excellent carrier mobility of 0.24 $cm^2$/Vs in regioregular poly(3-hexylthiophene) based field-effect transistor by employing octadecyltrimethoxysilane treated gate insulator," *Appl. Phys. Lett.* **2012**, 100(2), 023304-1–023304-4.

102. Ito, Y.; Virkar, A. A.; Mannsfeld, S.; Oh, J. H.; Toney, M.; Locklin, J.; Bao, Z. "Crystalline ultrasmooth self-assembled monolayers of alkylsilanes for organic field-effect transistors," *J. Am. Chem. Soc.* **2009**, 131(26), 9396–9404.

# Analytical Modeling and Parameter Extraction of Top and Bottom Contact Structures of Organic Thin-Film Transistors

## 3.1 INTRODUCTION

The performance of organic transistors has experienced an impressive improvement in recent years. Noteworthy progress in the fabrication methodology has made them appropriate for commercialization; however, the contact effect still appears as a limiting factor. Parasitic series contact resistance is the major issue that dominates overall device performance especially at the short channel. In using the typical metal-oxide-semiconductor field-effect transistor (MOSFET) expressions for the current-voltage characteristics of organic thin-film transistors (OTFTs), one needs to consider the non-ohmic behavior of the contacts.

In recent years, numerous mathematical models have been developed [1], based mainly on the classical MOS transistor model introduced by the empirical parameters. The contact resistance is generally estimated by employing the transmission line method, Kelvin probe microscopy, and the gated four-probe system that shows the series and the contact resistances as a function of the gate voltage [2]. Charge localization and conduction occur at the semiconductor–dielectric interface, as well as in the bulk of the organic semiconductor [3]. Therefore, the gate–contact overlap region needs to be analyzed separately for an adequate understanding of device operation.

This chapter deals with organic device physics and development of an analytical model with inclusion of contact effects and access resistance for the single gate (SG) OTFT device. The models are developed for the top and bottom contact structures, wherein total device resistance includes the vertical resistance per unit area of the contact region and the sheet resistance of the channel. Besides this, a differential method is employed to extract the parameters of both structures. The analytical model is validated with reported experimental results of the fabricated OTFT devices by eminent research laboratories.

This chapter is arranged in six sections including the current introductory Section 3.1. The device structures and contact effects are discussed in Section 3.2. The analytical model for top and bottom contact OTFT structures is developed in Section 3.3, and the method to extract the performance parameters is illustrated in Section 3.4. The analytical results are validated with respect to the experimental results in Section 3.5. Finally, the presented work of this chapter is concluded in Section 3.6.

## 3.2  DEVICE STRUCTURE AND CONTACT EFFECTS

An OTFT consists of a thin film of organic semiconductor (OSC), usually fabricated as an inverted structure with a gate at the bottom. This can be classified as top and bottom contact, on the basis of relative location of the contacts pertaining to the semiconductor. In a top contact structure, the source and drain contacts remain isolated from the conducting channel, since they lie on the opposite side of the channel from the dielectric. Traversing from the source (S) to drain (D) electrode, the charge carriers first travel downward followed by horizontal movement along the conducting channel and then proceed upward for extraction from the drain, as shown in Figure 3.1a. Injection of charge carriers through a large area, results in a minimal contact resistance for such a top contact device. In a bottom contact structure, contacts and dielectric layer are aligned on the same side of the induced channel. Therefore, the carriers travel in a single plane due to the establishment of S/D connection directly with the channel, as shown in Figure 3.1b.

The performance of a bottom contact OTFT is usually observed to be inferior in terms of high contact resistance and low mobility. It is due to a lower area

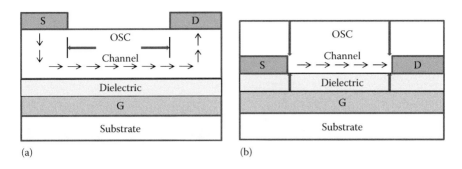

**FIGURE 3.1**    Current flow path in (a) BGTC and (b) BGBC structures.

for carrier injection and morphological disorders of semiconductor film around the prepatterned S/D contacts [4]. Nevertheless, the OTFT in a bottom contact structure is preferred for low-end applications due to fabrication through low-cost solution processing techniques.

The main factor that results in high resistance of the top contact structure is the access resistance. This can be reduced with large peak-to-valley roughness of the OSC film [5], which provides an easy flow of charge carriers, as illustrated in Figure 3.2. Access resistance can also be reduced using thinner OSC layers, but cannot be completely eliminated. Ideally, the contacts should be ohmic to enable the whole applied voltage for current conduction. However, contacts in OTFTs drop a significant amount of the applied voltage, as shown in Figure 3.3. In order to model this non-ohmic characteristic, the charge carrier's movement from S to D contact can be divided among different segments and each segment can be modeled with its respective resistance.

Source and drain contacts' behavior is quite similar in most OTFT devices [2,3,6]. However, Burgi *et al.* demonstrated asymmetry between the two contacts at a large Schottky barrier [7]. The main obstacle is generally observed in charge carrier injection at the source, as compared to the extraction at the drain end. Both contact resistances were found equal at a lower barrier height comparatively. For simplicity, the obstruction to current flow is considered similar at both contacts. Therefore, the segments source-to-channel

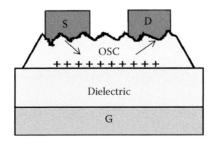

**FIGURE 3.2**    Schematic of top contact OTFT representing the access resistance and metal–OSC interface near the accumulation of holes.

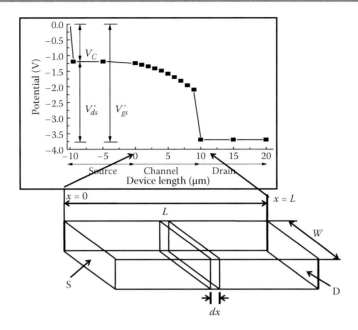

**FIGURE 3.3**   Potential drop across S/D contacts and the conducting channel.

and channel-to-drain are mapped to their equivalent resistances, $R_C/2$ (each). Besides this, the channel resistance is modeled as $R_{CH}$. Hence, the device total resistance, $R_{Total}$, can be expressed as

$$R_{Total} = R_{CH} + R_C \qquad (3.1)$$

The resistances linked to carrier injection and collection regions are collectively called contact resistance, $R_C$. To analyze the combined effect of all three segments, the equivalent model can be represented considering all three resistances connected in series, as shown in Figure 3.4.

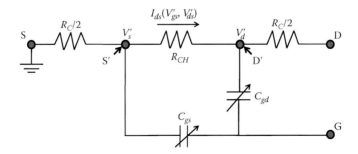

**FIGURE 3.4**   Resistance equivalent circuit of an OTFT along with internal source $(V_s')$ and drain $(V_d')$ voltages at the channel ends.

## 3.3 ANALYTICAL MODELS OF ORGANIC THIN-FILM TRANSISTORS (OTFTs)

Modeling and parameter extraction of an OTFT mainly relies on the trends of output current, $I_{ds}$, in a single crystalline MOSFET. Top and bottom contact OTFTs exhibit the same working principle, but are different in terms of their fabrication and contact/series resistance that results in different electrical characteristics and mobility. Prior to the development of a model, a few points are taken into consideration, including a uniform sheet resistance, $R_{SH}$, across the channel and zero net bulk current in the semiconductor [8]. Accumulation of carriers is assumed at the bottom of the overlapping region as well as in the channel. Usually, the contacts drop a significant amount of voltage, due to which the external applied voltages do not contribute entirely to the current conduction. Accordingly, actual channel ends are represented by terminals S′ and D′ rather than S and D, respectively, as shown in Figure 3.4.

Effective voltages are taken into account after excluding voltage drops across the contacts. Subsequently, the drain current is considered as a function of internal drain-source voltage $V'_{ds}$ (between the terminals D′ and S′) and the gate-source voltage $V'_{gs}$ (between the terminals G and S′), which are somewhat lesser than the external voltages $V_{ds}$ and $V_{gs}$, respectively.

To derive the current-voltage model under these assumptions, the channel and overlap regions are analyzed separately. The channel region is considered first and, thereafter, resistance of the overlapping region is amended. The drain current, analogous to internal voltages, $V'_{ds}$ and $V'_{gs}$, both in the linear $\left[ V'_{ds} < \left( V'_{gs} - V_t \right) \right]$ and saturation $\left[ V'_{ds} \geq \left( V'_{gs} - V_t \right) \right]$ regions, can be expressed as

$$I_{ds}^{Lin} = \frac{W}{L} \mu C_{ox} \left[ \left( V'_{gs} - V_t \right) - \frac{V'_{ds}}{2} \right] V'_{ds} \tag{3.2}$$

$$I_{ds}^{Sat} = \frac{W}{2L} \mu C_{ox} \left( V'_{gs} - V_t \right)^2 \tag{3.3}$$

where, $W$, $L$, $\mu$, $C_{ox}$, and $V_t$ represent the device width, channel length, mobility, gate dielectric capacitance per unit area, and threshold voltage, respectively.

### 3.3.1 MODEL FOR GATE–CONTACT OVERLAP REGION

The gate–source overlap region, shown in Figure 3.5, consists of a variable potential $V(x)$ and current $I_H(x)$ at point $(x)$ along the channel. Current density $J_V(x)$ represents the current per unit area in the vertical direction. Vertical resistance, $R_V$, is expressed in $\Omega$-cm$^2$, which includes both contact and bulk resistances [8].

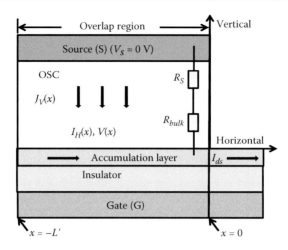

**FIGURE 3.5**   Schematic of top contact OTFT representing the gate–source overlap region and conducting channel.

The gate–source overlap region is analyzed to obtain the voltage drop across the vertical resistance. The gate–drain overlap region is analyzed in the same way. Thereafter, expressions are derived for the voltages $V'_{ds}$ and $V'_{gs}$ in order to find the drain current analytically. Since a uniform accumulation layer is considered in the channel, the sheet resistance of the accumulation layer remains the same as $R_{CH}$, and it is the function of the gate voltage. Total device resistance, $R_{Total}$, can be expressed as [5]

$$R_{Total} = R_{CH} + R_C = R_{SH}\frac{L}{W} + \frac{R'_C}{W} \tag{3.4}$$

where, $R'_C$ ($\Omega$-cm) is the specific contact resistance dependent on the gate voltage. Equation 3.4 facilitates analyzing the impact of channel dimensions, $L$ and $W$, on the relative magnitudes of the contact and the channel resistances. The contact resistance scales as $1/W$ only, therefore, it does not depend on $L$, whereas the channel resistance is a function of the channel length.

A small change in $V(x)$ and $I_H(x)$ by means of change in length, $dx$, along the channel can be expressed in terms of $R_{SH}$ as

$$V(x+dx) = V(x) - I_H(x)R_{SH} \times \frac{dx}{W} \tag{3.5}$$

Traversing from the source to the accumulation layer, the change in potential along the vertical direction resistance, $R_V$, is determined as

$$V(x) + (J_V(x)Wdx)\frac{R_V}{Wdx} = 0 \tag{3.6}$$

The current $I_H(x)$ can be obtained by integrating the current density $J_V(x)$ in a vertical direction; it contributes to the current across the accumulation layer

$$I_H(x) = W \int_{-L'}^{x} J_V(x)\,dx \tag{3.7}$$

Furthermore, a differential equation can be obtained with the help of Equations 3.5 and 3.6 by assuming $L'$ to $\infty$ [6]. Thereafter, it can be solved for $J_V(x)$, $I_H(x)$, and $V(x)$ as

$$\frac{d^2 J_V(x)}{dx^2} = \frac{R_{SH}}{R_V} J_V(x) \tag{3.8}$$

$$J_V(x) = J_{VC} \exp\left(\frac{x}{L_C}\right) \tag{3.9}$$

Equation 3.9 represents the solution for the aforesaid differential equation along with the integration constant $J_{VC}$. The maximum current density is $J_{VC}$ at the boundary ($x = 0$). $L_C$ is the characteristic length and depends on the gate voltage. It is a function of the vertical direction resistance of the overlapping region and the sheet resistance, which can be expressed as [3]

$$L_C = (R_V/R_{SH})^{0.5} \tag{3.10}$$

The solution of the differential equation can be further used to simplify expressions for the current $I_H(x)$ along the channel and the potential drop $V(x)$ as

$$I_H(x) = W \int_{-L_C}^{x} J_{VC} \exp\left(\frac{x}{L_C}\right) dx \tag{3.11}$$

$$V(x) = -R_V J_{VC} \exp\left(\frac{x}{L_C}\right) \tag{3.12}$$

Total current and voltage drop across the overlapping region can be obtained at the boundary ($x = 0$) as

$$I_{ds} = I_H(x = 0) = WL_C J_{VC} \tag{3.13}$$

$$V(x = 0) = -R_V J_{VC} \tag{3.14}$$

Similar analysis can be accomplished for the drain overlapping region by assuming symmetry between the S and the D electrodes. Thus, total contact resistance $R_C$ can be defined as

$$R_C = -2 \frac{V(x=0)}{I_H(x=0)} = \frac{2R_V}{WL_C} \tag{3.15}$$

The contact resistance shows an inverse relationship with $L_C$. The characteristic length is a function of $R_{SH}$, the sheet resistance of the accumulation layer, which depends on the gate voltage [8]. Therefore, $L_C$ also becomes a function of $V_{gs}$ and increases with an increase in the gate voltage. It results in a corresponding decrease in the contact resistance.

### 3.3.2 MODEL FOR CURRENT IN LINEAR AND SATURATION REGION

The internal voltages $V_s'$ and $V_d'$ are different from the externally applied voltages $V_s$ and $V_d$, respectively. These can be expressed according to the circuit, presented in Figure 3.4, and by applying the Kirchhoff's voltage law as

$$V_s' = V(x=0) = -R_V J_{VC} \tag{3.16}$$

$$V_d' = V_d - V(x=0) = V_d + R_V J_{VC} \tag{3.17}$$

It can be observed from Equation 3.16 that voltage at terminal $S'$ is not zero, therefore, the internal channel voltages determined with respect to the voltage $V_s'$ are expressed as

$$V_{ds}' = \Delta V = V_{ds} - 2V(x=0) \tag{3.18}$$

$$V_{gs}' = V_{gs} - V(x=0) \tag{3.19}$$

where, $\Delta V$ is a potential drop across the channel.

Finally, the expression for drain current that flows through the channel in the linear region can be solved by amending the expressions of $V_{ds}'$ and $V_{gs}'$ in Equation 3.2 as

$$I_{ds}^{Lin} = \frac{\mu C_{ox} \dfrac{W}{L} \left[ (V_{gs} - V_t)V_{ds} - \dfrac{1}{2}V_{ds}^2 \right]}{1 + \mu C_{ox} \dfrac{R_V}{LL_C} \left[ 2(V_{gs} - V_t) - V_{ds} \right]} \tag{3.20}$$

It is known that the mobility of organic semiconductors enhances with an increase in gate overdrive voltage ($V_{gs} - V_t$) and is, therefore, named the field dependent mobility, $\mu$ [9]. It varies in accordance with the mobility enhancement factor, $\alpha$ [10,11], as

$$\mu = \mu_0(V_{gs} - V_t)^\alpha \tag{3.21}$$

where, $\mu_0 = \mu_{00}/(V_a)^\alpha$ is the mobility at the gate overdrive voltage ($V_{gs} - V_t$) = $V_a$. Its unit deviates from cm$^2$/Vs, but at the voltage $V_a$ normalized to 1 V, its value will correspond to $\mu_{00}$, which is the mobility at low field [1]. The parameter $\alpha$ is associated with the conduction mechanism of the device and it depends on doping density and dielectric permittivity of the OSC material. Natali *et al.* reported the analogous effect of contact resistance for field dependent ($\alpha > 0$) or constant ($\alpha = 0$) mobility [12]. Therefore, with the same consideration, the expression obtained for the drain current can be further expressed in terms of field dependent mobility as

$$I_{ds}^{Lin} = \frac{K\left[(V_{gs} - V_t)^{\alpha+1}V_{ds} - \dfrac{1}{2}V_{ds}^2(V_{gs} - V_t)^\alpha\right]}{1 + K\dfrac{R_V}{WL_C}\left[2(V_{gs} - V_t)^{\alpha+1} - V_{ds}(V_{gs} - V_t)^\alpha\right]} \tag{3.22}$$

where,

$$K = \mu_0 C_{ox}\frac{W}{L} \tag{3.23}$$

Similarly, the drain current expression in the saturation region can be simplified by inserting the value of $V'_{gs}\left(V'_{gs}\right)$ from Equation 3.19 into Equation 3.3 and can be further modified in terms of $\mu$ as

$$I_{ds}^{Sat} = \frac{W^2 L_C^2}{K R_V^2 (V_{gs} - V_t)^\alpha}\left[1 + \frac{K(V_{gs} - V_t)^{\alpha+1}R_V}{WL_C} - \sqrt{1 + \frac{2K(V_{gs} - V_t)^{\alpha+1}R_V}{WL_C}}\right] \tag{3.24}$$

The top and the bottom contact structures of OTFT can be differentiated on the basis of series resistance due to their respective architectures. In the bottom contact structure, a discontinuity arises in the pentacene layer between the channel and near the contacts owing to deposition of pentacene onto the prepatterned metal contacts. It leads to an accumulation of defects and increase in the trap states, which affects the contact resistance of the device. The bottom contact structure exhibits higher contact resistance and lower mobility than its counterpart, whereas the conduction mechanism is similar for both structures. Therefore, analytical expressions derived from the top contact are also applied to the bottom contact structure.

## 3.4  DIFFERENTIAL METHOD FOR PARAMETER EXTRACTION

The drain current as per Equations 3.22 and 3.24 can be obtained once the parameters, such as, $\alpha$, $V_t$, $\mu_0$, $L_C$, and $R_V$, are known. To extract these parameters in a linear regime, it has been assumed that $V_{ds} \ll V_{gs}$. Since, the voltage drop across the contact resistance remains lower than $V_{ds}$, this contact's voltage drop can be neglected with respect to $V_{gs}$. This, in turn, simplifies the drain current expression for $V'_{gs} = V_{gs}$ [12]. The drain current for lower values of $V_{ds}$ can be expressed as

$$I_{ds} = \frac{W}{L}\mu C_{ox}(V_{gs} - V_t)V_{ds} \tag{3.25}$$

The expression can be solved for the internal voltage $V'_{ds}$ in place of externally applied voltage $V_{ds}$, and it can be expressed in terms of field dependent mobility as

$$I_{ds}^{Lin} = \frac{\dfrac{W}{L}\mu_0 C_{ox}(V_{gs} - V_t)^{\alpha+1}V_{ds}}{1 + \dfrac{2\mu_0 C_{ox}R_V}{LL_C}(V_{gs} - V_t)^{\alpha+1}} \tag{3.26}$$

Equation 3.26 can be understood in the form of a function $f(x)$ as follows

$$f(x) = \frac{pg(x)}{1 + qg(x)} \tag{3.27}$$

where, $p = \dfrac{W}{L}\mu_0 C_{ox}V_{ds}$; $g(x) = (V_{gs} - V_t)^{\alpha+1}$; $q = \dfrac{2\mu_0 C_{ox}R_V}{LL_C}$ and $f(x) = I_{ds}^{Lin}$.

The first derivative of the function $f(x)$ can be expressed as

$$f'(x) = \frac{pg'(x)}{[1 + qg(x)]^2} \tag{3.28}$$

Furthermore, if the first derivative $f'(x)$ is divided by $f(x)^2$, the simplified ratio $r_1$ shows its dependency on factor $p$ but remains independent of the term $q$. Thus, a ratio $r_1$ can be expressed as

$$r_1 = \frac{f(x)^2}{f'(x)} = \frac{pg(x)^2}{g'(x)} \tag{3.29}$$

If this property is applied to the expression of $I_{ds} - V_{gs}$ characteristic curve in the linear regime, Equation 3.26, the first derivative of $g(x)$ and the ratio $r_1$ can be expressed as

$$g'(x) = (\alpha + 1)(V_{gs} - V_t)^\alpha \tag{3.30}$$

$$r_1 = \frac{pg(x)^2}{g'(x)} = \frac{\left(I_{ds}^{Lin}\right)^2}{I_{ds}'^{Lin}} = \frac{W\mu_0 C_{ox}}{L(\alpha+1)}(V_{gs} - V_t)^{\alpha+2}V_{ds} \tag{3.31}$$

The above expression does not contain the resistance term and thus the ratio $r_1$ is independent from the term $R_V/L_C$. This is true for constant mobility or any dependence of mobility on the gate voltage [12]. Now, the relation holds three unknown variables ($\alpha$, $V_t$, and $\mu_0$). Figure 3.6 shows the plot of ratio $r_1 = \left(\left(I_{ds}^{Lin}\right)^2 / I_{ds}'^{Lin}\right)$ versus $V_{gs}$, where $I_{ds}'^{Lin}$ is the numerical derivative of the experimental current $I_{ds}^{Lin}$, which has been adopted from Gupta *et al.* [4].

In order to further reduce the unknown terms, the ratio $r_1$ can be related to the function, which is expressed as

$$f(x) = \delta(x - x_0)^\varepsilon \tag{3.32}$$

This function holds the relation only for $x > x_0$ and it becomes zero when $x < x_0$. Subsequently, the terms can be expressed by relating the ratio $r_1$ to the function $f(x)$ as

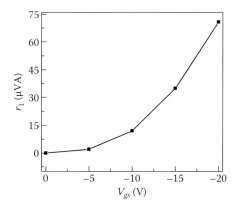

**FIGURE 3.6**  Characteristic plot of the ratio $r_1 = \left(\left(I_{ds}^{Lin}\right)^2 / I_{ds}'^{Lin}\right)$ versus $V_{gs}$ for top contact, where $I_{ds}'^{Lin}$ is the numerical derivative of $I_{ds}^{Lin}$.

$$f(x) = r_1; \; \delta = \frac{W \mu_0 C_{ox} V_{ds}}{L(\alpha + 1)}; \; x = V_{gs}; \; x_0 = V_t \text{ and } \varepsilon = \alpha + 2$$

The integral function of $f(x)$ can be obtained, within the limits $x_0$ and $x$, as

$$\int_{x_0}^{x} f(x') dx' = \frac{\delta (x - x_0)^{\varepsilon + 1}}{\varepsilon + 1} \tag{3.33}$$

Furthermore, another ratio $r_2$ can be obtained by dividing Equation 3.33 by Equation 3.32 as

$$r_2 = \frac{(x - x_0)}{\varepsilon + 1} \tag{3.34}$$

It can be observed from Equation 3.34 that the ratio $r_2$ is a linear function of $(x - x_0)$ and shows its independency from the factor $\delta$. Similarly, the integral function of $r_1$ can be obtained by applying the same strategy but with a lower limit to zero. It is set at zero with an assumption that the conducting channel does not appear for $V_{gs} < V_t$ and, thus, the device remains in a normally *off* mode. The integral function of $r_1$ can be obtained as

$$\int_{0}^{V_{gs}} r_1 \, dV'_{gs} = \frac{W \mu_0 C_{ox}}{L(\alpha + 1)(\alpha + 3)} (V_{gs} - V_t)^{\alpha + 3} \tag{3.35}$$

Subsequently, the ratio $r_2$ can be realized as

$$r_2 = \frac{\displaystyle\int_{0}^{V_{gs}} r_1 \, dV'_{gs}}{r1} = \frac{1}{(\alpha + 3)} (V_{gs} - V_t) \tag{3.36}$$

It can be understood from Equation 3.36 that $r_2$ does not contain the mobility term and, thus, the unknown variables are reduced to $\alpha$ and $V_t$ only. The ratio $r_2$ holds a linear relationship with $V_{gs}$, therefore, the plot of $r_2$ versus $V_{gs}$ is obtained as a straight line.

The parameter, $\alpha$, can be extracted from its slope, while $V_t$ is the $x$-intersection of its abscissa [11], as shown in Figure 3.7. Additionally, the values of $\alpha$ and $V_t$ can be used to extract the mobility, $\mu_0$, as per Equation 3.31, and the contact resistance can be calculated by using the following expression

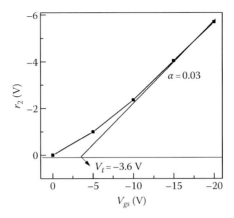

**FIGURE 3.7**   Characteristic plot of ratio $r_2$ versus $V_{gs}$ for top contact, where $r_2$ is obtained by the numerical integration of $r_1$.

$$\frac{2R_V}{WL_C} = R_C = \frac{V_{ds}}{I_{ds}} - \frac{L}{\mu_0 C_{ox} W (V_{gs} - V_t)^{\alpha+1}} \qquad (3.37)$$

Finally, with the help of Equations 3.4, 3.10, and 3.37, the characteristic length can be determined as [3]

$$L_C = \frac{R_C L}{2R_{CH}} \qquad (3.38)$$

where,

$$R_{CH} = \frac{L}{\mu_0 C_{ox} W (V_{gs} - V_t)^{\alpha+1}} \qquad (3.39)$$

## 3.5  RESULTS AND DISCUSSION

This section illustrates the analytical characteristics and the performance parameters of top and bottom contact structures. In order to implement the model, the device width is 1 and 3.6 mm for the top and bottom contact OTFTs, respectively, whereas the channel length is 30 μm for both devices [4]. A layer of dielectric material, SiO$_2$ (200 nm), results in a capacitance value of $1.65 \times 10^{-8}$ F/cm$^2$. Furthermore, the thicknesses of semiconducting layer (pentacene), S/D contacts (Au), and gate electrode ($n^+$ Si) are 50, 45, and 100 nm, respectively.

Figure 3.8 shows a relative energy diagram for different materials at their respective layers.

In order to determine the results analytically, the operating voltages are kept in the same range to facilitate an appropriate comparison with experimental results. Analytical models have been validated by means of their electrical characteristics. Resulting output and transfer characteristics demonstrating analytical and experimental results are shown in Figure 3.9 for the top contact structure. However, the similar plots are presented in Figure 3.10 for the bottom contact structure.

The mobility enhancement factor is extracted as 0.03 and 0.08 for the top and the bottom contact devices, respectively. Additional extracted device performance parameters, such as $\mu$, $V_t$, $R_V$, $R_C$, $L_C$, and $I_{ds}$, are summarized in Table 3.1. The characteristic plots and the extracted parameters demonstrate fairly good match with the experimental results for both top and bottom contact structures.

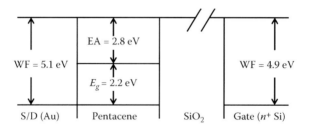

**FIGURE 3.8**    Relative energy diagram demonstrating work functions (WF) of electrodes, energy band gap ($E_g$), and electron affinity (EA) of pentacene semiconductor.

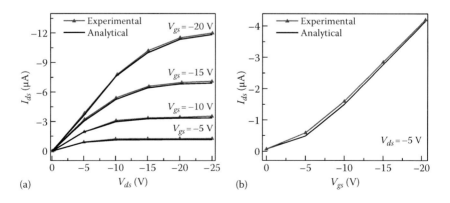

**FIGURE 3.9**    Comparison between the analytical and the experimental results in terms of (a) output and (b) transfer characteristics of the top contact OTFT structure.

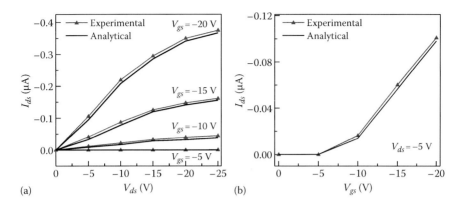

**FIGURE 3.10** Comparison between the analytical and experimental results in terms of (a) output and (b) transfer characteristics of the bottom contact OTFT structure.

**TABLE 3.1    Comparison of Extracted Parameters, Derived from the Analytical Model and through Experiment for Top and Bottom Contact OTFTs**

| Performance Parameters ($V_{ds}$ = −5 V and $V_{gs}$ = −20 V) | Top Contact OTFT | | Bottom Contact OTFT | |
|---|---|---|---|---|
| | **Analytical** | **Experimental[a]** | **Analytical** | **Experimental[a]** |
| $\mu$ (cm²/Vs) | 0.089 | 0.085 | 0.0010 | 0.0014 |
| $V_t$ (V) | −3.6 | −3.2 | −8.9 | −8.5 |
| $R_V$ (Ω-μm²) | 142 × 10⁶ | 20 × 10⁶ | 141 × 10⁹ | 195 × 10⁹ |
| $R_C$ (Ω) | 150K | 57K | 14M | 15.5M |
| $L_C$ (μm) | 1.9 | 0.7 | 5.6 | 7 |
| $I_{ds}$ (μA) | −4.0 | −4.2 | −0.09 | −0.1 |

[a]  Data from Gupta, D.; Katiyar, M.; Gupta, D., "An analysis of the difference in behavior of top and bottom contact organic thin film transistors using device simulation," *Org. Electron.* 2009, 10(5), 775–784.

## 3.6 CONCLUDING REMARKS

This chapter presented an analytical model and device physics to predict the behavior of OTFTs. The proposed model shows the output and the transfer characteristics of pentacene OTFT in the top and the bottom contact configurations. At low $V_{ds}$, the nonlinearity is observed due to the contact resistance effect. To model the overall behavior, the overlapping and the channel regions are considered separately and modeled in terms of the respective resistances.

Besides this, gate bias dependent mobility is incorporated, since it is more realistic and relevant to OTFTs. In addition, expressions for the drain current are simplified in terms of internal voltages that are slightly smaller than the externally applied voltages. The extracted parameters illustrate better performance of the top contact in terms of $\mu_B$, $I_{ds}$, $V_t$, and $R_C$ in comparison to the bottom contact structure. The reason can be attributed to structural difference, intrinsic behavior, and the manner in which the two devices are fabricated. An agreement between the analytical and the experimental results in terms of electrical characteristics and parameters validates the derived current-voltage model for both structures of the OTFT.

## PROBLEMS
## MULTIPLE CHOICE

1. Which limiting factor dominates the overall performance of the short channel device?
   a. Mobility
   b. Power dissipation
   c. Parasitic series contact resistance
   d. Voltage
2. Charge localization and conduction occur at
   a. Gate–dielectric interface
   b. Contact–OSC interface
   c. OSC–dielectric interface
   d. None of the above
3. Mobility of organic transistor enhances with the
   a. Increase in gate overdrive
   b. Decrease in gate overdrive
   c. Increase in gate voltage
   d. Decrease in gate voltage
4. The bottom contact structure exhibits
   a. Higher contact resistance, higher mobility
   b. Higher contact resistance, lower mobility
   c. Lower contact resistance, lower mobility
   d. Lower contact resistance, higher mobility
5. Contact resistance does not depend on
   a. Channel length ($L$)
   b. Channel width ($W$)
   c. $L/W$
   d. $1/W$
6. The bottom contact OTFT is preferred for low-end applications due to
   a. High contact resistance
   b. Fabrication through high-cost solution processing techniques
   c. Fabrication through low-cost solution processing techniques
   d. None of the above

7. The main factor that results in high device resistance of the top contact structure is
   a. Roughness of the OSC layer
   b. Access resistance
   c. Roughness of gate dielectric
   d. All of the above

8. Deposition of pentacene layer over the prepatterned metal contacts in a bottom contact structure results in
   a. Increase in defects
   b. Increase in carrier injection
   c. Increase in mobility
   d. Decrease in carrier injection

9. The access resistance can be reduced using a
   a. Thick layer of OSC
   b. Thick layer of dielectric
   c. Thin layer of OSC
   d. Thin layer of dielectric

10. The mobility enhancement factor ($\alpha$) is associated with the
    a. Device conduction mechanism
    b. Doping density
    c. Dielectric permittivity
    d. All of the above

## ANSWER KEY

1. c; 2. c; 3. a; 4. b; 5. a; 6. c; 7. b; 8. a; 9. c; 10. a

## SHORT ANSWER

1. What are the key differences between bottom contact and top contact OTFT structures?
2. Discuss the OTFT model for current in the linear and saturation regions.
3. Describe the differential method for parameter extraction of OTFTs and its advantages over the OTFT conventional method.
4. Discuss the model for the gate–contact overlap region in the organic thin film transistors.
5. What are the different analytical models for OTFT?
6. Describe the origin of contact resistance in brief.
7. Describe the access resistance concept in an organic thin-film transistor.
8. Show the difference in the current flow path of bottom gate bottom contact (BGBC) and bottom gate top contact (BGTC) with the help of appropriate diagrams.
9. Explain the steps and general strategy of validation of analytical results with experimental results.

10. Describe various contact effects in an organic thin-film transistor.
11. What are the different interfaces that exist in an organic transistor? Also explain their impact on the performance of the OTFT.

## EXERCISES

1. Consider the following top contact OTFT analysis problem, given the following: characteristics length ($L_C$) = 0.7 μm, vertical resistance ($R_V$) = 20 × 10$^6$ Ω-μm$^2$, channel length ($L$) = 30 μm, channel width ($W$) = 1000 μm, and specific contact resistance ($R_C'$) = 150 Ω. Find the channel sheet resistance ($R_{SH}$) and contact resistance ($R_C$).

2. Consider a bottom contact OTFT with the following parameters: characteristics length ($L_C$) = 5.6 μm, vertical resistance ($R_V$) = 141 × 10$^9$ Ω-μm$^2$, channel length ($L$) = 30 μm, channel width ($W$) = 3600 μm, and specific contact resistance ($R_C'$) = 504 Ω. Find the channel sheet resistance ($R_{SH}$) and contact resistance ($R_C$).

3. Calculate the gate-oxide capacitance ($C_{ox}$) and $k$ and also find the region of operation and drain current ($I_{ds}$) for top contact OTFT given drain-source voltage ($V_{ds}$) = −5 V, gate-source voltage ($V_{gs}$) = −20 V, threshold voltage ($V_t$) = −3.2 V, zero field mobility ($\mu_0$) = 0.129 cm$^2$/Vs, mobility enhancement factor ($\gamma$) = 0.03, characteristics length ($L_C$) = 1.9 μm, vertical resistance ($R_V$) = 142 × 10$^6$ Ω-μm$^2$, channel length ($L$) = 30 μm, channel width (W) = 1000 μm, oxide thickness ($t_{ox}$) = 200 nm, and dielectric constant ($\varepsilon_{SiO2}$) = 3.9.

4. Consider a bottom contact OTFT with the following parameters: drain-source voltage ($V_{ds}$) = −5 V, gate-source voltage ($V_{gs}$) = −20 V, threshold voltage ($V_t$) = −8.5 V, zero field mobility ($\mu_0$) = 0.02 cm$^2$/Vs, mobility enhancement factor ($\gamma$) = 0.08, characteristics length ($L_C$) = 5.6 μm, vertical resistance ($R_V$) = 141 × 10$^9$ Ω-μm$^2$, channel length ($L$) = 30 μm, channel width ($W$) = 3600 μm, oxide thickness ($t_{ox}$) = 200 nm, and dielectric constant ($\varepsilon_{SiO2}$) = 3.9. Estimate the gate-oxide capacitance ($C_{ox}$) and $k'$. Also find the region of operation and drain current ($I_{ds}$).

5. Calculate the gate-oxide capacitance ($C_{ox}$), channel resistance ($R_{CH}$), and characteristics length ($L_C$) for a bottom contact OTFT with the following parameters: drain-source voltage ($V_{ds}$) = −5 V, gate-source voltage ($V_{gs}$) = −20 V, threshold voltage ($V_t$) = −8.5 V, zero field mobility ($\mu_0$) = 0.02 cm$^2$/Vs, mobility enhancement factor ($\gamma$) = 0.08, contact resistance ($R_C$) = 14 MΩ, channel length ($L$) = 30 μm, channel width ($W$) = 3600 μm, oxide thickness ($t_{ox}$) = 200 μm, and dielectric constant ($\varepsilon_{SiO2}$) = 3.9. Also estimate the device resistance ($R_{Device}$).

6. Estimate the gate-oxide capacitance ($C_{ox}$), channel resistance ($R_{CH}$), and contact resistance ($R_C$) for a top contact OTFT with the following parameters: drain-source voltage ($V_{ds}$) = −5 V, gate-source voltage ($V_{gs}$) = −20 V, threshold voltage ($V_t$) = −8.5 V, zero field mobility

$(\mu_0)$ = 0.129 cm$^2$/Vs, mobility enhancement factor $(\gamma)$ = 0.03, vertical resistance $(R_V)$ = 3138.75 MΩ-μm$^2$, channel length $(L)$ = 30 μm, channel width $(W)$ = 1000 μm, oxide thickness $(t_{ox})$ = 200 nm, dielectric constant $(\varepsilon_{SiO2})$ = 3.9, and characteristics length $(L_C)$ = 2.79 μm. Also calculate the device resistance $(R_{Device})$.

7. Estimate the gate-oxide capacitance $(C_{ox})$, channel resistance $(R_{CH})$, and contact resistance $(R_C)$ for bottom contact OTFT with the following parameters: drain-source voltage $(V_{ds})$ = −5 V, gate-source voltage $(V_{gs})$ = −20 V, threshold voltage $(V_t)$ = −8.5 V, zero field mobility $(\mu_0)$ = 0.0019 cm$^2$/Vs, mobility enhancement factor $(\gamma)$ = 0.03, vertical resistance $(R_V)$ = 45,000 MΩ-μm$^2$, channel length $(L)$ = 30 μm, channel width $(W)$ = 1000 μm, oxide thickness $(t_{ox})$ = 200 nm, dielectric constant $(\varepsilon_{SiO2})$ = 3.9, and characteristics length $(L_C)$ = 0.20 μm. Also find the device resistance $(R_{Device})$.

8. Consider a top contact OTFT analysis problem with the following parameters: drain-source voltage $(V_{ds})$ = −25 V, gate-source voltage $(V_{gs})$ = −20 V, threshold voltage $(V_t)$ = −3.2 V, zero field mobility $(\mu_0)$ = 0.129 cm$^2$/Vs, mobility enhancement factor $(\gamma)$ = 0.03, characteristics length $(L_C)$ = 2.79 μm, vertical resistance $(R_V)$ = 3138.75 MΩ-μm$^2$, channel length modulation coefficient $(\lambda)$ = 0.035 V$^{-1}$, channel length $(L)$ = 30 μm, channel width $(W)$ = 1000 μm, oxide thickness $(t_{ox})$ = 200 nm, and dielectric constant $(\varepsilon_{SiO2})$ = 3.9. Estimate the gate-oxide capacitance $(C_{ox})$ and $k'$. Also find the region of operation and the drain current $(I_{ds})$.

9. Consider a bottom contact OTFT analysis problem with the given parameters: drain-source voltage $(V_{ds})$ = −25 V, gate-source voltage $(V_{gs})$ = −20 V, threshold voltage $(V_t)$ = −8.5 V, zero field mobility $(\mu 0)$ = 0.0019 cm$^2$/Vs, mobility enhancement factor $(\gamma)$ = 0.03, characteristics length $(L_C)$ = 0.20 μm, vertical resistance $(R_V)$ = 45,000 MΩ-μm$^2$, channel length modulation coefficient $(\lambda)$ = 0.05 V$^{-1}$, channel length $(L)$ = 30 μm, channel width $(W)$ = 1000 μm, oxide thickness $(t_{ox})$ = 200 nm, and dielectric constant $(\varepsilon_{SiO2})$ = 3.9. Estimate the gate- oxide capacitance $(C_{ox})$ and $k'$. Also find the region of operation and the drain current $(I_{ds})$.

10. Estimate total resistance $(R_{tot})$ for bottom contact OTFT given the characteristics length $(L_C)$ = 0.20 μm, channel length $(L)$ = 30 μm, channel width $(W)$ = 1000 μm, and the sheet resistance $(R_{SH})$ = 150 MΩ.

## REFERENCES

1. Marinov, O.; Deen, M. J.; Datars, R. "Compact modeling of charge mobility in organic thin film transistors," *J. Appl. Phys.* **2009**, 106(6), 064501-1–064501-13.
2. Necliudov, P. V.; Shur, M. S.; Gundlach, D. J.; Jackson, T. N. "Contact resistance extraction in pentacene thin film transistors," *Solid-State Electron* **2003**, 47(2), 259–262.

3. Chiang, C. S.; Martin, S.; Kanicki, J.; Ugai, Y.; Yukawa, T.; Takeuchi, S. "Top-gate staggered amorphous silicon thin-film transistors: Series resistance and nitride thickness effects," *Jpn. J. Appl. Phys.* **1998**, 37, 5914–5920.

4. Gupta, D.; Katiyar, M.; Gupta, D. "An analysis of the difference in behavior of top and bottom contact organic thin film transistors using device simulation," *Org. Electron.* **2009**, 10(5), 775–784.

5. Panzer, M. J.; Frisbie, C. D. "Contact effect in organic field effect transistors," in Z. Bao, J. J. Lockli (eds.), *Organic Field Effect Transistors*, CRC Press, Boca Raton, **2007**, 139–156.

6. Jung, K. D.; Kim, Y. C.; Park, B. G.; Shin, H.; Lee, J. D. "Modeling and parameter extraction for the series resistance in thin-film transistors," *IEEE Trans. Electron Devices* **2009**, 56(3), 431–440.

7. Burgi, L.; Richards, T. J.; Friend, R. H.; Sirringhaus, H. "Close look at charge carrier injection in polymer field effect transistors," *J. Appl. Phys.* **2003**, 94(9), 6129–6137.

8. Jung, K. D.; Kim, Y. C.; Kim, B. J.; Park, B. G.; Shin, H.; Lee, J. D. "An analytic current-voltage equation for top contact organic thin film transistors including the effects of variable series resistance," *Jpn. J. Appl. Phys.* **2008**, 47(4), 3174–3178.

9. Marinov, O.; Deen, M. J.; Zschieschang, U.; Klauk, H. "Organic thin film transistors: Part I. Compact DC modeling," *IEEE Trans. Electron Devices* **2009**, 56(12), 2952–2961.

10. Carranza, A. C.; Nolasco, J.; Estrada, M.; Gwoziecki, R.; Benwadih, M.; Xu, Y.; Cerdeira, A. *et al.* "Effect of density of states on mobility in small molecule *n*-type organic thin- film transistors based on a Perylene Diimide," *IEEE Electron Device Lett.* **2012**, 33(8), 1201–1203.

11. Deen, M. J.; Marinov, O.; Zschieschang, U.; Klauk, H. "Organic thin-film transistors: Part II–Parameter extraction," *IEEE Trans. Electron Devices* **2009**, 56(12), 2962–2968.

12. Natali, D.; Fumagalli, L.; Sampietro, M. "Modeling of organic thin film transistors: Effect of contact resistances," *J. Appl. Phys.* **2007**, 101(1), 014501-1–014501-12.

# Impact of Semiconductor and Dielectric Thicknesses on the Performance of Top and Bottom Contact Organic Thin-Film Transistors

4

## 4.1 INTRODUCTION

The performance of an organic transistor is defined by several significant parameters, including drain current, mobility, threshold voltage, *on-off* current ratio, subthreshold slope, and contact resistance. These parameters are often influenced by carrier injection, gate dielectric capacitance, trap states, charge per unit area, interface barrier, location of channel with respect to the organic semiconductor (OSC)–dielectric interface, and bulk access resistance. Understanding of morphological ordering of the semiconducting layer and the interfaces between the layers are extremely important factors in improving the device performance with respect to the charge carrier transport mechanism [1]. One of the main issues related to these effects is the dependence of device performance on the semiconductor and dielectric thicknesses that could significantly alter the device characteristics. In principle, the charge carriers accumulate at the OSC–dielectric interface within a few nanometers in the semiconductor, eliminating the role of rest of the bulk semiconductor in the current conduction. Ideally the thickness of the semiconductor should not affect the device parameters, however, several researchers have reported its effect on the performance of organic thin-film transistors (OTFTs) with a variety of semiconductors, including pentacene, P3HT, poly(9,9-dioctylfluorene-co-N-(4-butylphenyl)-diphenylamine) (TFB), and dibutylquaterthiophene (4QT4). With a thicker active layer (in hundreds of nanometers), the bulk traps play a major role in the behavior of devices. Charge localization occurs at the semiconductor–dielectric interface and within the bulk of the OSC, thereby significantly influencing device performance.

Thickness of the active layer is not the only factor that affects the OTFT's electrical parameters; dielectric thickness is also a crucial influencing factor. A decrease in the thickness of the dielectric layer results in a significant reduction in the threshold voltage due to a high gate capacitance. Lower $V_t$ is useful in reducing the device power consumption and is, therefore, beneficial in producing portable devices. In addition to this, the mobility also increases with reduction of the dielectric thickness. To date, most of the research work has been focused on exploring the impact of active layer thickness on the performance of the OTFT. Moreover, the majority of the studies are based on top contact structure only. However, the impact of the layers' thicknesses is also based on the device structure due to charge carriers traversing on distinct path between the source and the drain. As a result, the thickness of layers affects the OTFT behavior differently while it is realized in top contact (TC) and bottom contact (BC) structures.

This chapter emphasizes the impact of the semiconductor and dielectric thin film layer thicknesses on the performance of both the top and bottom contact structures in terms of various electrical parameters, including $I_{ds}$, $\mu$, $V_t$, $I_{on}/I_{off}$, SS, and $R_C$. Moreover, analysis of additional parameters such as electric field, current density, and charge carrier distribution is carried out from the device physics point of view. The entire analysis, extraction of parameters, and validation are performed at the same structural and operating conditions for both the top and bottom contact OTFT devices. The thickness of the semiconductor primarily affects the *on*-current. However, *off*-current also rises with an increase in $t_{osc}$ due to an increment in the bulk current [2]. A larger magnitude of the *off*-current increases power consumption and adds to the charge leakage through the device that limits switching behavior [3,4]. As discussed earlier, the performance of organic TFTs typically suffers from large contact resistance that is expressed in terms of source/drain resistance and bulk access resistance as

$$\frac{R_C}{2} = R_S + R_{bulk} = R_D + R_{bulk} \qquad (4.1)$$

The $R_S/R_D$ (or barrier) resistance comes into play due to a difference between the Fermi level of metal contact and highest occupied molecular orbital (HOMO) level of the organic semiconductor. However, the access resistance is due to the bulk semiconductor between the metal–OSC interface and the charge accumulation region. The $t_{osc}$ affects the access resistance, thereby resulting in a corresponding variation in $R_C$ that further affects the performance of the OTFT [5]. The top and bottom contact structures are differentiated on the basis of contact resistance due to their dissimilar architecture and different charge carrier injection area. This chapter also analyzes the impact of $t_{osc}$ and $t_{ox}$ on the contact resistance of top and bottom contact structures.

This chapter is organized in six sections, including the present introductory Section 4.1. Independent designs of top and bottom contact structures along with the process/device simulation tool and simulation conditions are discussed in Section 4.2, while Section 4.3 illustrates the electrical characteristics and parameters of these two structures. Section 4.4 investigates the impact of $t_{osc}$ and $t_{ox}$ on the performance of both TC and BC OTFT structures. Section 4.5 deals with the extraction of contact resistance on different $t_{osc}$ and $t_{ox}$. Finally, Section 4.6 summarizes the important outcomes of the presented work.

## 4.2 PROCESS/DEVICE SIMULATION TOOL AND SIMULATION CONDITIONS

OTFT structures and circuits are realized and implemented using the Silvaco Atlas 2-D numerical device simulator with small molecules or conducting polymers as the semiconductor layer. The details of a device simulation tool and simulation conditions are discussed in the following sections.

### 4.2.1 SIMULATION SETUP

This section outlines the Atlas 2-dimensional device simulation tool that was used in this book to model the electric characteristics of different organic devices and circuits. Silvaco Inc. is a provider of electronic design automation (EDA) software and technology computer-aided design (TCAD) process and device simulation software. EDA is a category of software tools that involves various algorithms and models for designing complex electronic devices, circuits, and systems. TCAD is a branch of EDA that models the semiconductor fabrication process and semiconductor device operation. The modeling of the fabrication is termed process TCAD, while the modeling of the device operation is termed device TCAD. Semiconductor process and device simulation results incorporate a diverse set of numerical calculations based on standard physical models such as exponential and double Gaussian built-in density of states distribution models, Poole-Frenkel (PF) electric field dependent mobility model, steady state and transient recombination models, and bimolecular Langevin recombination model. Figure 4.1 shows the integrated structure of Silvaco TCAD software incorporating several tools within one platform. Primarily, Atlas is used for silicon devices but it can be employed for organic devices also by considering Poole-Frenkel and hopping mobility models and organic defect density of states models.

The characteristics of several single gate and dual gate OTFT devices and performance of the circuit designs based on these devices have been simulated in this book. The Poole-Frenkel and Gaussian distributions for the traps are considered for the simulations of the organic devices. Although these models do not take care of morphology of the organic materials directly, the zero-field mobility ($\mu_0$) term in the expressions reflects the dependence of the conduction mechanism on the grain size of the organic materials and operating temperature [6]. Major heads of an Atlas code are shown pictorially in Figure 4.2.

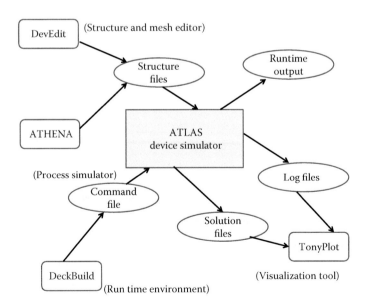

**FIGURE 4.1**   Pictorial view of functioning among various tools of two-dimensional device organic module simulator.

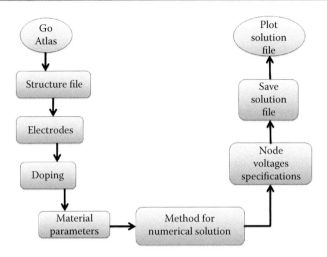

**FIGURE 4.2**    Pictorial view of functioning among various steps of two-dimensional device organic module simulators.

## 4.2.2 DEVICE SIMULATION STEPS

In general, to investigate the behavior of an organic/inorganic device, the structure is converted into a grid of finite elements to define the simulation domain. A set of discrete fundamental equations based on the device physics are applied to each point of the simulation domain and the solution is carried out by the simulator that determines the behavior of the device under investigation. Atlas simulator provides a library of widely used organic materials such as Pentacene, Tetracene, and CuPc. Besides this, it allows the users to incorporate their own defined materials. The complete Atlas simulation code can be divided into the following three segments: (1) defining structural dimensions and mesh specifications, (2) defining material parameters and application of the appropriate physical models, and (3) operational bias conditions and run the simulation [6].

### 4.2.2.1 DEFINING STRUCTURAL DIMENSIONS AND MESH SPECIFICATIONS

The first section defines the dimensions of the individual layers along with the position of electrodes. The layers of insulator and active organic materials are defined. The source/drain (S/D) contacts in the TC structure are positioned at the top, whereas in the BC structure the contacts are placed prior to the OSC layer. After defining all the TFT layers with the appropriate dimensions and materials, individual simulation domains are divided into a finite element grid form called "mesh" that comprises a complex grid of triangles. At these discrete grid points, the model entails the calculations based on elemental device physics equations related to the electric field, electric potentials, and charge carrier densities. The accuracy of simulation results depends on the density of the mesh and at critical points like junctions or interfaces of different materials, the placement of mesh with higher density is required.

## Example 4.1

```
go atlas
#
title Organic Thin Film Transistor simulation
#
mesh smooth=1 space.mult=1.0 width=100
#
x.m          l=0      spac=0.5
x.m          l=10     spac=0.5
x.m          l=20     spac=0.5
x.m          l=30     spac=0.5
#
y.mesh l=-0.030 spacing=0.005
y.mesh l=0.0 spacing=0.005
y.mesh l=0.03 spacing=0.005
y.mesh l=0.0357 spacing=0.01
y.mesh l=0.0557 spacing=0.01
#
region num=1 material=Pentacene y.min=0.0 y.max=0.030
name=Pentacene
region num=2 material=Al2O3 y.min=0.030 y.max=0.0357
region num=3 material=air y.min=-0.030 y.max=0.0
x.min=10 x.max=20
#
elec num=1 material=Gold name=source x.max=10.0 y.min=
-0.030 max=0.0
elec num=2 material=alum name=gate y.min=0.0357
y.max=0.0557
elec num=3 material=Gold name=drain x.min=20.0 y.min=
-0.030 y.max=0.0
```

### 4.2.2.2  DEFINING MATERIAL PARAMETERS AND APPLICATION OF THE APPROPRIATE PHYSICAL MODELS

The second section of the simulator describes the physical models to predict the results with proper boundary conditions. It incorporates the Poole-Frenkel mobility model that effectively evaluates the static and dynamic behavior of the devices and the circuits. This mobility model is expressed [6,7] as

$$\mu = \mu_0 \exp\left[ -\frac{\Delta_h}{k_B T} + \left( \frac{\beta_h}{k_B T} - \gamma \right) \sqrt{E} \right]$$

(4.2)

where, $E$ represents the electric field. The hole activation energy and hole Poole-Frenkel factor are specified by $\Delta_h$ and $\beta_h$, respectively, and $\gamma$ is used as the fitting parameter. The model demonstrates the conduction due to the field enhanced thermal excitation of trapped charge carriers. The mobility increases with an increase in the gate voltage validating the hopping transport phenomenon in OTFTs [8]. Localization of charge carriers around the traps results in drain current reduction in the low field region [9]. The traps that may exist in the semiconductor due to the presence of structural defects or potential barriers at the interfaces need to be modeled for evaluating the OTFT performance in a more realistic way. With the aim of incorporating such traps, a uniform density of trap states $g_A(E)$ is considered that can be modeled by an exponential distribution of acceptor like traps as [10]

$$g_A(E) = N_{TA} \exp\left[(E - E_C)/W_{TA}\right] \tag{4.3}$$

where, $N_{TA}$ and $W_{TA}$ are the effective density of states (DOS) and the characteristic energy width of the exponential trap distribution, respectively, and $E_c$ refers to the energy level of the bottom of the conduction band.

The second segment of the simulation code defines various electrical parameters such as band gap, electron affinity, permittivity, and effective density of states in both conduction ($N_c$) and valence ($N_v$) bands. Effective densities of states in the conduction band ($N_c$) and valence band ($N_v$) are assumed to be of the order of $\sim 10^{21}$ cm$^{-3}$. Zero field mobility and relative permittivity of pentacene are 0.5 cm$^2$/Vs and 4, respectively. Poole-Frenkel model parameters $\Delta E$ and $\beta$ are $1.792 \times 10^{-2}$ and $7.758 \times 10^{-5}$ eV, respectively [6]. $\Delta E$ is the activation energy of holes at zero electric field and $\beta$ is the hole Poole-Frenkel factor.

The activation energy for pentacene material is reported by many researchers in the range of 0.005 to 0.050 eV [7,11]; therefore, it is considered as $1.792 \times 10^{-2}$ eV. Furthermore, the values for $\beta_h$, $N_{TA}$, and $W_{TA}$ are $7.758 \times 10^{-5}$ eV (cm/V)$^{0.5}$, $2.5 \times 10^{18}$ cm$^{-3}$ eV$^{-1}$, and 0.55 eV, respectively [10,12]. However, the energy gap and hole affinity are 2.2 eV and 2.8 eV, respectively. Effective density of states in both conduction and valence band is considered as $10^{21}$ cm$^{-3}$, whereas the relative permittivity of pentacene is 4 [13].

## Example 4.2

```
doping reg=1 uniform conc=3e17 p.type name=pentacene
#
material name=Pentacene \eg300=2.8 nc300=1.0e21
nv300=1.0e21 permittivity=4.0 \ mun=5e-5 mup=.85
#
material name=Al2O3 permittivity=4.5
#
contact name=source workfunc=5.1
contact name=drain workfunc=5.1
contact name=gate workfunc=4.28
```

```
# Application of physical models for mobility and
defects
defects cont dfile=15don.dat afile=15acc.dat \
nta=2.5e18 ntd=1.0e18 wta=0.129 wtd=0.5 \
nga=0.0 ngd=0.0 ega=0.62 egd=0.78 wga=0.15 wgd=0.15 \
sigtae=1.e-17 sigtah=1.e-15 sigtde=1.e-15 sigtdh=1.e-17
\
siggae=2.e-16 siggah=2.e-15 siggde=2.e-15 siggdh=2.e-16
#
mobility deltaep.pfmob=1.792e-2 betap.pfmob=7.758e-5
model pfmob.p print
#
output e.field j.electron j.hole j.conduc j.total
e.velocity h.velocity
ey.field flowlines e.mobility h.mobility qss e.temp
h.temp charge
recomb val.band con.band qfn qfp j.disp photogen impact
tot.doping \u.srh u.rad
#
# Save the structure in a file
outf=hagen.str
tonyplot hagen.str.
```

### 4.2.2.3 OPERATIONAL BIAS CONDITIONS AND RUN THE SIMULATION

The third section of the Atlas code determines the operational bias conditions for the simulation such as $V_{gs}$ and $V_{ds}$ to obtain the electrical characteristics of the device under analysis.

**Example 4.3**

```
# Transfer Characteristics (Ids vs Vgs) Biasing
solve init
#
method carriers=1 hole maxtrap=100
solve vdrain=-1.5
log outf=idvg.log
solve vgate=0.0 vstep=-0.2 vfinal=-3.0 name=gate
log off
tonyplot idvg.log
# Output Characteristics (Ids vs Vds) Biasing
Solve init
#
solve prev
solve vgate=-0.0 name=gate outf=vg-0.0.bin onefile
```

```
solve vgate=-1.5 name=gate outf=vg-1.5.bin onefile
solve vgate=-1.8 name=gate outf=vg-1.8.bin onefile
#
load infile=vg-0.0.bin
solve prev
log outf=vg-0.0mpfhm.log
solve vdrain=0.0 vstep=-0.05 vfinal=-0.5 name=drain
solve vdrain=-0.5 vstep=-0.5 vfinal=-3.0 name=drain
log off
#
load infile=vg-1.5.bin
solve prev
log outf=vg-1.5mpfhm.log
solve vdrain=0.0 vstep=-0.05 vfinal=-0.5 name=drain
solve vdrain=-0.5 vstep=-0.5 vfinal=-3.0 name=drain
log off
#
load infile=vg-1.8.bin
solve prev
log outf=vg-1.8mpfhm.log
solve vdrain=0.0 vstep=-0.05 vfinal=-0.5 name=drain
solve vdrain=-0.5 vstep=-0.5 vfinal=-3.0 name=drain
log off
# Plot the simulation results
tonyplot vg-0.0mpfhm.log -overlay vg-1.5mpfhm.log
vg-1.8mpfhm.log
quit.
```

### 4.2.3 PERFORMANCE PARAMETER EXTRACTION

Electrical parameters are measured to compare the performance of transistors. The SG-OTFT parameters can be extracted from the accumulation and depletion regions of the transfer characteristics. The extracted parameters are mobility ($\mu$), threshold voltage ($V_{th}$), subthreshold slope ($SS$), on-off current ratio ($I_{on}/I_{off}$), and transconductance ($g_m$). Performance parameters are extracted from the slope of the transfer characteristics using standard expressions. OTFT parameters are obtained in the analogous manner as the MOSFET parameters noting the similarity between their operations and characteristics. $I_{on}$ is the maximum saturation current at given $V_{gs}$, whereas $I_{off}$ is the leakage current at $V_{gs} = 0$ V and maximum value of $V_{ds}$ voltage. Threshold voltage, $V_{th}$, can be extrapolated from the transfer characteristics at the maximum value of the slope of current. Subthreshold voltage swing $SS$ is deduced from the weak inversion regime of the transfer characteristics. The in-depth impact of $t_{osc}$ and $t_{ox}$ thickness variations on individual performance parameters of TC and BC OTFTs are discussed in Section 4.4.

**Example 4.4**

Realize the single gate in top contact and bottom contact organic thin-film transistors through the organic module of a two-dimensional numerical device simulator.

1. Draw current flow lines from source (S) to drain (D) for BGTC and BGBC OTFT structures at the biasing of voltages of $V_{ds}$ = −1.5 V and $V_{gs}$ = −3 V.
2. Draw the output characteristics of the OTFT for device dimensions and materials given in Table 4.1. Carry out the simulations for the biasing voltages: $V_{gs}$ = −1.5 to −3 V with step size of 0.3 V at $V_{ds}$ of 0 to −3 V at step size of 1 V.
3. Draw a set of transfer characteristics ($I_{ds} - V_{gs}$) of the TC and BC OTFT in logarithmic scale and TC/BC current ratio for device dimensions and materials given in Table 4.1. Carry out the simulations for the drain voltages: $V_{ds}$ = −1.5 V and $V_{gs}$ = 0 to −3 V with step size of 1 V.
4. Explain the impact of semiconductor and dielectric thicknesses on the top and bottom contact OTFTs with the help of simulations.

**Solutions:**

1. Typical physical dimensions and materials of both structures are summarized in Table 4.1. Primarily, electrical characteristics and parameters of TC structure are validated through reported experimental results [4]. Afterward, BC structure is analyzed with identical dimensions, materials, and operating conditions for an appropriate comparison on a common platform. The conducting channels is formed through pentacene (*p*-type) OSC material with length and width of 10 and 100 μm, respectively,

**TABLE 4.1    Device Dimensions and Materials of TC and BC OTFTs**

| Dimensional Parameter | Value | Material |
|---|---|---|
| Thickness of OSC ($t_{osc}$) | 30 nm | Pentacene |
| Thickness of dielectric ($t_{ox}$) | 5.7 nm | $Al_2O_3$ |
| Thickness of gate electrode ($t_g$) | 20 nm | Aluminum |
| Thickness of S/D contact ($t_s/t_d$) | 20 nm | Gold |
| Channel width ($W$) | 100 μm | – |
| Channel length ($L$) | 10 μm | – |

whereas $Al_2O_3$ of 5.7 nm is used as the dielectric layer. The other parameter of the device, as shown in Figure 4.3, possesses a 30 nm thick film of pentacene as a semiconducting active layer. Pentacene devices show a higher field-effect mobility of up to 1 cm$^2$/Vs. Aluminum was chosen as the gate electrode material of thickness 20 nm. A 20 nm thick layer of gold is used as source and

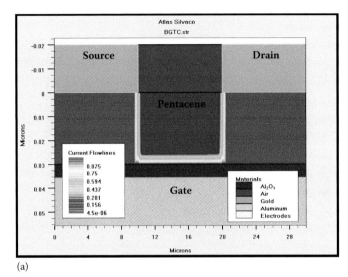

(a)

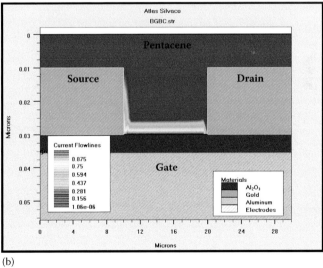

(b)

**FIGURE 4.3** Schematics of (a) BGTC and (b) BGBC OTFTs representing current flow lines from S to D contacts.

drain contact electrodes. Figure 4.3 demonstrates the current flow lines from S to D in top and bottom contact structures, obtained at $V_{ds} = -1.5$ V and $V_{gs} = -3$ V.

2. The solution is discussed in Section 4.3 with the help of the Figure 4.4.

3. The transfer characteristics ($I_{ds}$–$V_{gs}$) in logarithmic scale for TC and BC structures is plotted in Figure 4.5a and the TC/BC current ratio with respect to $V_{gs}$ ($V_{ds} = -1.5$ V) is drawn in Figure 4.5b.

4. The impact of variation in thickness of OSC ($t_{osc}$) and insulator ($t_{ox}$) on behavior of TC and BC structures at $V_{gs} = -3$ V are discussed in Figures 4.6 through 4.9.

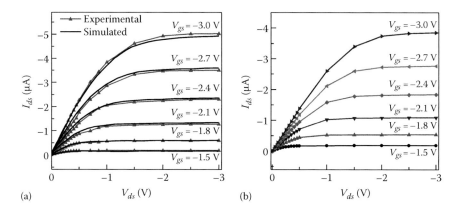

**FIGURE 4.4**    Output characteristics of (a) TC and (b) BC OTFTs.

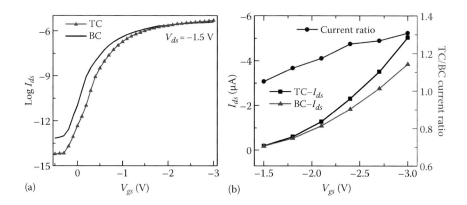

**FIGURE 4.5**    (a) $I_{ds}$–$V_{gs}$ characteristics in logarithmic scale for TC and BC structures and (b) TC/BC current ratio with respect to $V_{gs}$ ($V_{ds} = -1.5$ V).

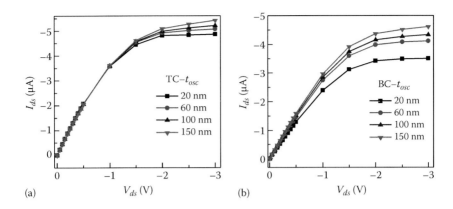

**FIGURE 4.6** $I_{ds}$–$V_{ds}$ characteristics as a function of $t_{osc}$ for (a) TC and (b) BC structures at $V_{gs} = -3$ V.

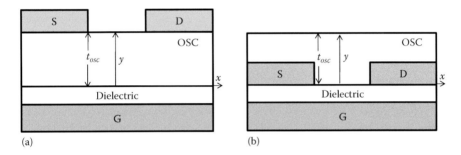

**FIGURE 4.7** Schematics of (a) TC and (b) BC OTFTs showing $x$- and $y$-distance along the OSC.

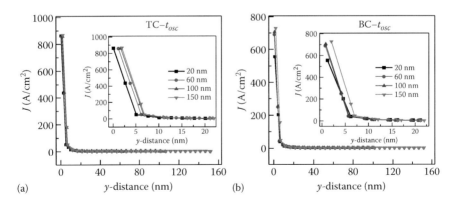

**FIGURE 4.8** Characteristic curves of current density ($J$) along $y$-distance from the OSC–dielectric interface for (a) TC and (b) BC structures. Insets show a close view of current density as a function of $t_{osc}$.

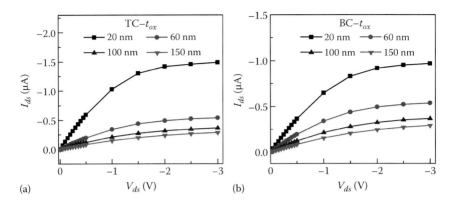

**FIGURE 4.9**   $I_{ds}$–$V_{ds}$ plots as a function of $t_{ox}$ for (a) TC and (b) BC OTFTs at $V_{gs}$ = –3 V.

## 4.3 ELECTRICAL CHARACTERISTICS AND PARAMETERS OF TOP CONTACT AND BOTTOM CONTACT ORGANIC THIN-FILM TRANSISTORS

Drain characteristics of top contact OTFTs as a function of gate voltage is illustrated in Figure 4.4a. The output characteristics and performance parameters of the TC structure obtained by simulations are validated through experimental results [4] and a close match is observed.

Similar behavior is noticed for BC structure, but with 25% reduced current (at $V_{ds}$ = $V_{gs}$ = –3 V), as shown in Figure 4.4b. This is due to limited hole injection from the source to the semiconductor in bottom contact structure. Therefore, the effect of potential difference between the Fermi level of contact metal (gold) and HOMO level of OSC (pentacene) is more prominent due to smaller contact area than that of top contact equivalents. The electrical parameters, such as $\mu$, $V_t$, SS, $I_{on}/I_{off}$, and $I_{ds}$ of top contact structure are tabulated and verified with respect to the experimental results in Table 4.2. The top contact structure exhibits a high drain current of 5.1 µA at $V_{ds}$ = $V_{gs}$ = –3 V that reasonably matches to the experimental current with an error of 2%. Similarly, other parameters such as mobility, threshold voltage, and subthreshold slope are also close to the experimental results with an error of 1.2%, 9%, and 6%, respectively.

Figure 4.5a shows the combined plots of transfer characteristics (at $V_{ds}$ = –1.5 V) in logarithmic scale for both TC and BC structures; while, the ratio of TC to BC drain current at different gate voltages is illustrated in Figure 4.5b. This current ratio is unity at lower gate voltage ($V_{gs}$ = –1.5 V); however, increment in $V_{gs}$ is liable to increase this ratio that shows performance difference owing to their structural difference. OTFT in bottom contact structure shows a reduction of 30% in mobility in comparison to the top contact. Besides this, it shows 6% increment in SS due to a higher trap density. On the other hand, BC device exhibits an improvement in $V_t$ by 10% because of closer proximity of

---

**TABLE 4.2    Validation of Simulated Parameters with Experimental Results for TC Structure**

| Performance Parameter | Simulation | Experiment[a] |
|---|---|---|
| $\mu$ (cm²/Vs) | 0.395 | 0.4 |
| $V_t$ (V) | −1.0 | −1.1 |
| SS (mV/dec) | 94 | 100 |
| $I_{on}/I_{off}$ | $2.5 \times 10^7$ | $10^7$ |
| $I_{ds}$ (μA) at $V_{ds} = V_{gs} = -3$ V | −5.1 | −5.0 |

[a] Data from Kim, J. B.; Hernandez, C. F.; Hwang, D. K.; Tiwari, S. P.; Potscavage Jr., W. J.; Kippelen, B., "Vertically stacked complementary inverters with solution processed organic semiconductors," *Org. Electron.* 2011, 12(7), 1132–1136.

---

contacts to the OSC–dielectric interface [12] in comparison to the top contact that might be helpful to build the channel at a lower gate voltage.

To observe the impact of semiconductor thickness, top and bottom contact structures of OTFT are analyzed at different $t_{osc}$, while $V_{gs}$ and $t_{ox}$ are kept constant at −3 V and 5.7 nm, respectively. Subsequently, a slight increase in saturation current (at $V_{ds} > -1.5$ V) is obtained for the top contact, whereas a monotonic increase in $I_{ds}$ is observed for bottom contact structure.

Figure 4.6 illustrates the output characteristics as a function of $t_{osc}$ for top and bottom contact structures. The small dependence of TC current against the pentacene thickness can be understood with the current density profile. The current density is observed at the center of conduction region across the OSC layer (*y*-distance) from the interface, as shown in Figure 4.7 for TC and BC structures. The maximum current density is found near the OSC–dielectric interface that reflects the formation of the conducting channel within the first few monolayers of the organic semiconductor.

A very thin accumulation region (about 5 nm) constitutes the major current flow as shown in Figure 4.8 for TC and BC structures. However, the rest of the semiconductor contributes a little to the overall drain current [1] that can be further increased by choosing a thicker pentacene layer. The top contact structure exhibits higher current density; whereas, it is lower in the BC structure and reduces with reduction in $t_{osc}$. On the other hand, the conduction current increases with reduction in $t_{ox}$ (with a constant $t_{osc} = 30$ nm) for both TC and BC devices.

This is due to generation of a high electric field in the thin dielectric layer that enables the accumulation of more charge carriers in the channel near the interface. Although a similar behavior is noticed for both devices, a higher magnitude of current is observed for top contact structure, particularly at thin layer of dielectric (20 nm), as shown in Figure 4.9 for TC and BC structures. Figure 4.10a illustrates the combined plots for $I_{ds}$ of top and bottom contact structures

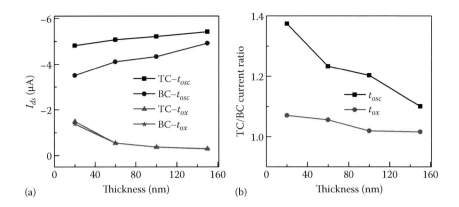

**FIGURE 4.10**    (a) Drain current of TC and BC devices and (b) TC/BC current ratio as a function of $t_{osc}$ and $t_{ox}$.

with respect to $t_{osc}$ and $t_{ox}$ (at $V_{ds} = V_{ds} = -3$ V); whereas the top to bottom current ratio at different $t_{osc}$ and $t_{ox}$ is presented in Figure 4.10b. This current ratio reduces by 22% while increasing $t_{osc}$ from 20 to 150 nm and approaches unity with higher $t_{osc}$.

This observation demonstrates almost similar current in both devices at higher $t_{osc}$; however, at lower thickness the behavior of two devices is quite different. This can be attributed to the prominent impact of the charge carrier injection barrier in the BC structure in comparison to its TC counterpart. In fact, this potential barrier degrades the performance of BC structure more significantly at small $t_{osc}$ due to fewer free charge carriers in the channel. However, at large OSC thickness the existence of more charge compensates for the deterioration caused by less charge injection from the source, leading to high current conduction in BC, similar to the TC structure. On the other hand, with a rise in $t_{ox}$ from 20 to 150 nm, almost similar changes are observed in the current of both structures, thereby resulting in a ratio closer to unity.

## 4.4  IMPACT OF $t_{osc}$ AND $t_{ox}$ ON PERFORMANCE PARAMETERS OF TOP CONTACT AND BOTTOM CONTACT ORGANIC THIN-FILM TRANSISTORS

This section demonstrates the variation in performance parameters of top and bottom contact structures with respect to $t_{osc}$ and $t_{ox}$. To specify the quantitative basis shift in parameters, these parameters, such as $\mu$, $I_{on}/I_{off}$, $V_t$, and $SS$, are extracted from simulated $I_{ds}-V_{gs}$ (0 to $-3$ V) characteristics at $V_{ds} = -1.5$ V. The mobility behavior of both the structures against the thickness ($t_{osc}$ and $t_{ox}$) is plotted in Figure 4.11.

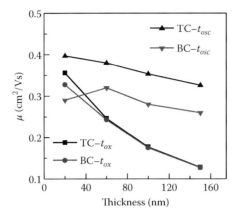

**FIGURE 4.11**    Mobility of TC and BC structure as a function of $t_{osc}$ and $t_{ox}$.

In the saturation region, mobility ($\mu_{sat}$) is extracted by using the expression

$$\mu_{sat} = \frac{2L}{WC_{ox}} \left( \frac{\partial\sqrt{I_{dsat}}}{\partial V_{gs}} \right)^2 \tag{4.4}$$

Mobility dependence on $t_{osc}$ is a function of electric field strength between the S and D electrodes. Maximum mobility for top contact structure is achieved at $t_{osc}$ of 20 nm that reduces with higher thickness of the OSC layer. It is due to a reduction in the average value of charge with an increment in active layer thickness, as shown in Figure 4.12. However, maximum mobility for bottom

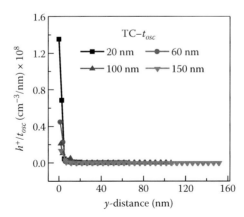

**FIGURE 4.12**    Hole concentration ($h^+/t_{osc}$) along $y$-distance at different $t_{osc}$ in TC structure.

contact is achieved at 60 nm that declines with positive or negative change in $t_{osc}$. The reason for reduction in mobility at lower $t_{osc}$ (20 nm) can be attributed to the interlayer surface potential difference in the semiconductor film that significantly affects the charge transport. A potential difference exists between the first and the second monolayer of OSC, forcing the charge carriers toward the OSC–dielectric interface more prominently in the thinner OSC due to higher electric field. This increases the charge localization in the vicinity of the dielectric surface, thereby resulting in inferior mobility with a thinner semiconductor. Besides this, the mobility also reduces due to a significant carrier trapping effect caused by the existence of dipoles in the OSC–dielectric interface [14]. Kirova and Bussac [15] and Hulea *et al.* [16] reported the interaction of charge carriers with the surface phonons of polar dielectrics that led to the self-trapping of carriers at the OSC–dielectric interface. This tends to reduce the mobility of carriers near the dielectric surface due to increased midgap states.

Upon increasing $t_{osc}$ from 20 to 60 nm, the initial increase in mobility occurs due to a reduction in the vertical electric field, thereby forcing fewer charge carriers to pass though the self-trapping effect at the dielectric surface. At a thicker OSC layer (60 nm), a similar magnitude of vertical electric field (*VE*) is observed as that of 20 nm but at a higher distance from the OSC–dielectric interface which shows the formation of the conducting channel slightly away from the dielectric surface, as shown in Figure 4.13a. Consequently, the carrier transport process experiences a weaker carrier trapping effect, resulting in improved device performance compared to a thinner semiconductor layer. Furthermore, thicknesses above 60 nm lead to a reduction in the electric field strength per unit thickness that deteriorates the device performance again, as illustrated in Figure 4.13b.

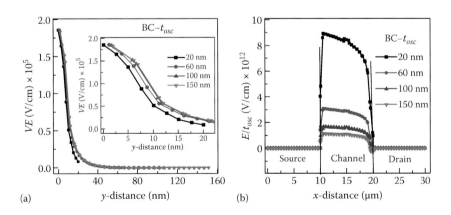

**FIGURE 4.13**   Characteristic plots of (a) vertical electric field (*VE*) along *y*-distance (inset shows a close view of *VE*) and (b) electric field profile ($E/t_{osc}$) along *x*-distance (at 1 nm above the OSC–dielectric interface) for BC structure.

The mobility is enhanced with an increase in overdrive voltage ($V_{gs} - V_t$). The carrier mobility can be improved even at constant $V_{gs}$ by thinning the dielectric layer that enables a high electric field, as shown in Figure 4.14a. Subsequently, at reduced $t_{ox}$, more surface charge carriers would be available to fill the traps, as shown in Figure 4.14b. Therefore, a rapid increase in mobility is observed by reducing $t_{ox}$ (Figure 4.11). Similar mobility behavior is noticed for both top and bottom contact devices at varying $t_{ox}$, while OSC thickness is kept constant.

The ratio of current in the accumulation mode to the depletion mode is termed as the *on/off* current ratio. Lowering the thickness of the semiconductor and dielectric layers is the key factor in producing a large difference between the *on-* and *off*-current of OTFTs. It can be analyzed with the expression

$$\frac{I_{on}}{I_{off}} = \frac{C_{ox}\,\mu(V_{gs} - V_t)^2}{t_{osc}\sigma V_{ds}}; \text{ where } I_{off} = \frac{W}{L}t_{osc}\sigma V_{ds} \quad (4.5)$$

where, $\sigma$ is the channel conductivity. The $I_{on}/I_{off}$ ratio is enhanced with either a rise in $I_{on}$ or a fall in $I_{off}$ due to a decrement in $t_{ox}$ and $t_{osc}$, respectively.

*Off*-current increases for a higher thickness of semiconductor due to an increment in the portion of the layer that allows the flow of current [2,3]. For the top and bottom contact devices, a sharp fall in this current ratio is observed for increasing $t_{osc}$ compared to $t_{ox}$, as shown in Figure 4.15a. Therefore, a faster increase in *off*-current can be considered as the dominant factor for reduction in the $I_{on}/I_{off}$ ratio. A reduction of 70% in *on*-current is observed for BC structure upon scaling $t_{ox}$ from 20 to 150 nm; whereas, *off*-current shows quite a large increment from 8 pA to 0.4 µA for the similar rise in $t_{osc}$, as shown in Figure 4.15b. A comparable response is observed for the top contact structure that

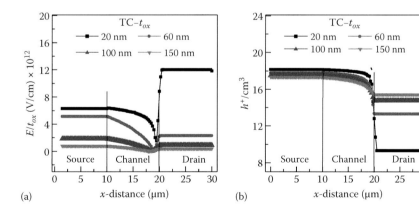

**FIGURE 4.14**   Characteristic plots of (a) electric field profile ($E/t_{ox}$) along *x*-distance at 1 nm below the OSC–dielectric interface and (b) hole concentration ($h^+$) along *x*-distance at 1 nm above the OSC–dielectric interface for TC structure.

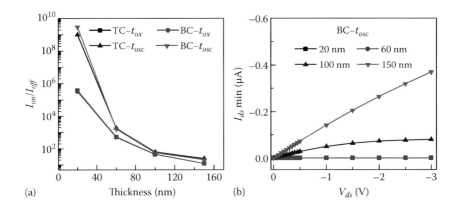

**FIGURE 4.15** Characteristic curves for (a) $I_{on}/I_{off}$ as a function of $t_{osc}$ and $t_{ox}$ for TC and BC and (b) $I_{ds}$ min versus $V_{ds}$ as a function of $t_{osc}$ (at $V_{gs} = 0$ V) for BC structure.

justifies the dependence of current *on/off* ratio more on the thickness of OSC than the dielectric [17] for both structures.

Threshold voltage is another parameter affected by $t_{osc}$ and $t_{ox}$. It is the minimum gate voltage required to accumulate the charge carriers at the OSC–dielectric interface. For portable devices, power consumption should be less than what can be achieved by operating the devices comparatively at lower bias. Obtaining lower $V_t$ is a key factor for reducing overall operating voltages. While decreasing $t_{ox}$ from 150 to 20 nm, $V_t$ is reduced to half for both top and bottom contact structures, as shown in Figure 4.16a. Contradictorily, a reduction in $V_t$ by 20% for TC and 8% for BC is observed with $t_{osc}$ scaling up from 20 to 150 nm. This improvement in $V_t$ is attributed to the enhancement in free charge carriers with higher active layer thickness, resulting in reduced $V_t$ values.

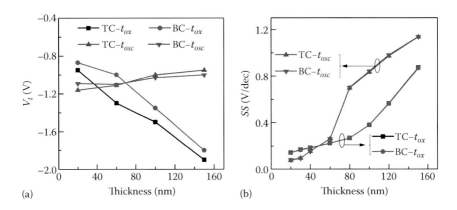

**FIGURE 4.16** Characteristic plots of (a) threshold voltage and (b) subthreshold slope as function of $t_{osc}$ and $t_{ox}$ for both TC and BC structures.

Subthreshold slope (*SS*) is a metric to measure the switching behavior of TFTs. Indirectly, it is a measure of impurity concentration, interface states, and trap density that mainly affect the switching behavior of a transistor [12]. At a higher impurity concentration, dipoles at the surface of the dielectric and trap density increases the *SS*. This slope is defined as the ratio of change in gate biasing to the change in the drain current in logarithmic scale that can be expressed as

$$SS = \frac{\partial V_{gs}}{\partial(\log_{10} I_{ds})} = \frac{k_B T}{q} \ln_{10}\left(1 + \frac{qN_{it}}{C_{ox}}\right) \tag{4.6}$$

where, $q$ and $N_{it}$ represent the electronic charge and the density of trap states at the OSC–dielectric interface, respectively. The subthreshold slope increases with thicker OSC and dielectric layers. This slope is observed higher by 14 and 6 times on scaling up $t_{osc}$ and $t_{ox}$ (20–150 nm), respectively. A similar variation in *SS* is observed for both TC and BC structures as demonstrated in Figure 4.16b. It increases with an increase in $t_{ox}$ due to lower capacitance. Besides this, it is observed that at a higher value of $t_{osc}$, the access resistance increases that in turn deteriorates the bulk current and thus the device current too. This leads to a gentle slope, thereby reducing the switching behavior. Increasing the drain and gate bias also results in a steeper subthreshold slope. An increase in $V_{gs}$ raises the number of charge carriers in the channel, whereas a higher $V_{ds}$ supports rapid movement of the carriers from source to drain, thereby improving the switching behavior of OTFTs.

Table 4.3 summarizes the variation in parameters $\mu$, $V_t$, *SS*, and $I_{on}/I_{off}$ of top and bottom contact devices with scaling down of $t_{osc}$ and $t_{ox}$ from 150 to 20 nm. An overall improvement in the performance is observed for thinner film of OSC and dielectric material. The current *on/off* ratio is found to be more dependent on $t_{osc}$ due to a dominant impact of the *off*-current over the *on*-current. When scaling down $t_{osc}$ from 150 to 20 nm, the magnitude of $I_{on}/I_{off}$ increases by $4 \times 10^7$ and $10^8$ times for TC and BC structures, respectively. However, it increases by $3 \times 10^4$ times for both structures after reducing $t_{ox}$ from 150 to 20 nm, which is

**TABLE 4.3    Improvement in the Performance Parameters of the TC and BC OTFTs with $t_{osc}$ and $t_{ox}$ Variations**

| Thickness (Scaling Down 150 to 20 nm) | Variation in Parameters | | | |
|---|---|---|---|---|
| | $\mu$ | $V_t$ | *SS* | $I_{on}/I_{off}$ |
| TC–$t_{osc}$ | 22% ↑ | 22% ↑ | 92% ↓ | $4 \times 10^7$ times ↑ |
| BC–$t_{osc}$ | 10% ↑ (20–60 nm) 18% ↓ (60–150 nm) | 10% ↑ | 90% ↓ | $10^8$ times ↑ |
| TC–$t_{ox}$ | 2.7 times ↑ | 50% ↓ | 84% ↓ | $3 \times 10^4$ times ↑ |
| BC–$t_{ox}$ | 2.4 times ↑ | 52% ↓ | 85% ↓ | $3 \times 10^4$ times ↑ |

significantly lower in comparison to scaling down of semiconductor thickness. Contradictorily, a large improvement in the mobility and threshold voltage is observed for similar reduction in the dielectric thickness as compared to OSC thickness. This can be attributed to the achievement of higher capacitance at lower $t_{ox}$ that in turn enables a higher density of accumulated charge at the OSC–dielectric interface.

## 4.5  CONTACT RESISTANCE EXTRACTION

The performance of an OTFT usually deviates from the ideal behavior due to higher contact resistance. A large difference between the Fermi level of contact metal and the HOMO/LUMO level of a semiconductor results in a potential barrier, leading to an insufficient carrier injection. This leads to a high contact resistance that drops a significant amount of the applied bias, thereby reducing the effective voltage across the channel responsible for current conduction. Top and bottom contact structures exhibit different device resistance ($R_C + R_{CH}$) due to their structural differences. Charge carrier injection through a large area results in low contact resistance for a top contact structure. However, the bottom contact structure displays a higher contact resistance due to smaller contact area. The drain current in the linear region can be expressed in terms of $R_{Total}$ as

$$I_{ds} = \frac{V_{ds}}{R_{Total}} = \frac{V_{ds}}{R_C + \left\{ W \mu C_{ox}(V_{gs} - V_t)/L \right\}^{-1}} \tag{4.7}$$

$$R_{Total} = R_{CH} + R_C = \frac{L}{W \mu C_{ox}(V_{gs} - V_t)} + R_C \tag{4.8}$$

To evaluate the contact resistance, various models have been proposed mainly based on the transmission line method, wherein to extract the $R_C$, the product of the total device resistance and $W$ is plotted against the channel length. Extrapolation of straight line to $L = 0$ provides the value of $R_C.W$, since the channel resistance disappears at zero channel length. This method is employed for extracting and analyzing the contact resistance for top and bottom contact structures for different channel lengths ranging from 10 to 40 μm. Furthermore, the dependence of contact resistance on $t_{osc}$ and $t_{ox}$ is also analyzed at different gate voltages ranging from −1.5 to −15 V for both TC and BC structures.

### 4.5.1  CONTACT RESISTANCE ANALYSIS IN TOP CONTACT STRUCTURE

The top contact device usually exhibits low contact resistance because of the large effective area for injecting charge carriers into the OSC channel from the metal contact. To extract the magnitude of $R_C$ for a particular semiconductor or dielectric thickness, the total device resistance is estimated at several channel lengths (10, 20, 30, and 40 µm).

The intercept of the straight line fit of $R_{Total} W$ versus $L$ curve at $L = 0$ provides the value of $R_C$, as shown in Figure 4.17 for different $t_{osc}$ and $t_{ox}$. The contact resistance dependence on $t_{osc}$ at different $V_{gs}$ is shown in Figure 4.18. At low $V_{gs}$ (−1.5 V), $R_C$ reduces by 42% as the $t_{osc}$ is increased from 20 to 150 nm. This is due to an increase in free charge carrier density in the channel at higher $t_{osc}$. On the contrary, $R_C$ increases with an increase in $t_{osc}$ at higher $V_{gs}$ (−15 V) due to reduction in free carrier density in the bulk OSC.

This is due to accumulation of nearly all the charge carriers at the OSC–dielectric interface under a higher electric field. Thus, the trap states in the bulk OSC significantly dominate the charge transport in the access region,

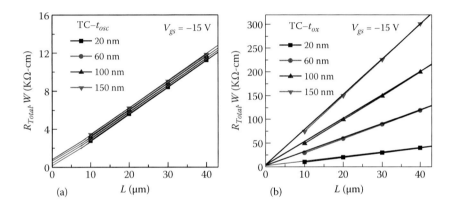

**FIGURE 4.17**    $R_{Total}.W$ versus $L$ plots as a function of (a) $t_{osc}$ and (b) $t_{ox}$ for TC OTFT.

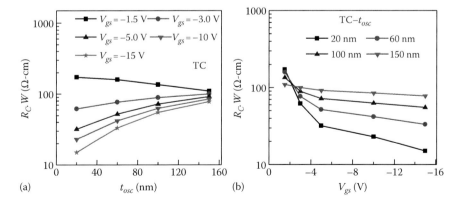

**FIGURE 4.18**    Characteristic plots of (a) $R_C.W$ versus $t_{osc}$ as a function of $V_{gs}$ and (b) $R_C.W$ versus $V_{gs}$ as a function of $t_{osc}$ for TC structure.

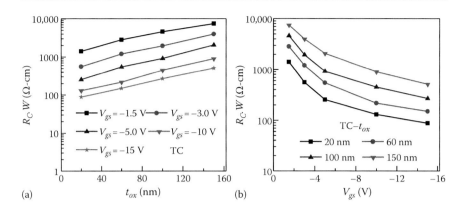

**FIGURE 4.19**   Characteristic plots of (a) $R_C.W$ versus $t_{ox}$ as a function of $V_{gs}$ and (b) $R_C.W$ versus $V_{gs}$ as a function of $t_{ox}$ for TC structure.

causing $R_{bulk}$ to increase. The access resistance also increases at higher $t_{osc}$ due to elongation of the path traversed by the charge carriers before reaching the channel. This results in a corresponding increase in $R_C$ on increasing $t_{osc}$ that is predominantly observed at higher $V_{gs}$. At lower $V_{gs}$, the device performance is found to be more affected by the contact resistance due to proportionately higher potential drop across the source contact, thereby reducing the net gate voltage ($V_{gs} - I_{ds}.R_C/2$) across the channel that is responsible for charge accumulation. However, this potential drop is not as pronounced at higher $V_{gs}$ due to the availability of higher net gate voltage that accumulates enough charge carriers for stronger currents. Furthermore, when scaling down $t_{ox}$ from 150 to 20 nm, a proportionate reduction is observed in the contact resistance for different $V_{gs}$, as shown in Figure 4.19.

### 4.5.2 CONTACT RESISTANCE ANALYSIS IN BOTTOM CONTACT STRUCTURE

The BC structure exhibit higher contact resistance in comparison to the TC structure, due to a smaller carrier injection area into the channel. The $R_C$ maximizes at $t_{osc}$ of 20 nm and reduces significantly at 60 nm, as illustrated in Figure 4.20a. However, total device resistance is very high for $t_{ox} = 150$ nm that is almost independent of channel length, as shown in Figure 4.20b. At higher $t_{ox}$, the impact of contact resistance substantially dominates over the channel resistance, which, therefore, keeps it almost independent of channel length.

Contact resistance as a function of $t_{osc}$ at different $V_{gs}$ is shown in Figure 4.21 for a bottom contact structure. The contact resistance is lowest at $t_{osc}$ of 60 nm that increases with positive or negative change in $t_{osc}$. At $t_{osc} < 60$ nm, the interlayer surface potential difference between the monolayer of OSC forces the charge carriers more toward the dielectric surface, thereby reducing free charge carriers in the bulk of semiconductor. This reduction of charge carriers causes

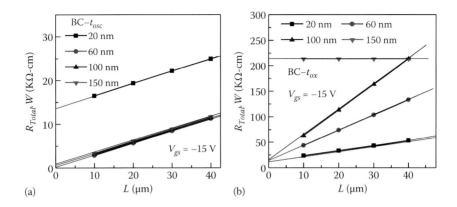

**FIGURE 4.20**  $R_{Total}.W$ versus $L$ plots as a function of (a) $t_{osc}$ and (b) $t_{ox}$ for BC structure.

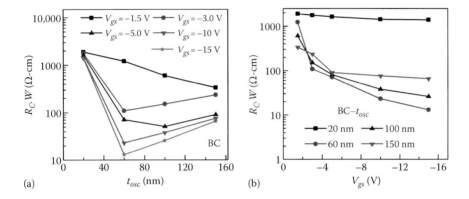

**FIGURE 4.21**  Characteristic plots of (a) $R_C.W$ versus $t_{osc}$ as a function of $V_{gs}$ and (b) $R_C.W$ versus $V_{gs}$ as a function of $t_{osc}$ for BC structure.

an increment in the bulk resistance, thereby increasing the contact resistance. Therefore, at $t_{osc}$ of 20 nm the device exhibits only a small change in already high $R_C$ values on varying $V_{gs}$.

At 60 nm the carrier transport experiences a weaker carrier trapping effect that results in improved device performance, and, therefore, the gate voltage shows better control on the conducting channel. However, at $t_{osc} > 60$ nm the device performance again declines due to high access resistance, thereby enhancing the contact resistance. At a particular $t_{ox}$, the resistance decreases with an increase of $V_{gs}$ as shown in Figure 4.22. At $t_{ox} = 20$ nm, the device conducts strongly due to accumulation of a larger number of charges at the interface. The device performance is less affected by variation in $V_{gs}$, producing a contact resistance change by only 2 times on reducing $V_{gs}$ from –15 to –1.5 V. However, this change is 24 times for $t_{ox} = 150$ nm, while $V_{gs}$ is varied from –15

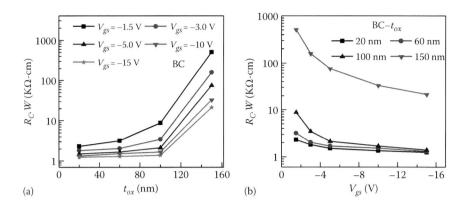

(a)  (b)

**FIGURE 4.22** Characteristic plots of (a) $R_C.W$ versus $t_{ox}$ as a function of $V_{gs}$ and (b) $R_C.W$ versus $V_{gs}$ as a function of $t_{ox}$ for BC structure.

to −1.5 V. The variation in the magnitude of $R_C$ is more prominent on varying $V_{gs}$ for larger values of $t_{ox}$.

## 4.6 CONCLUDING REMARKS

This chapter emphasizes the analyzed performances of top and bottom contact OTFTs based on semiconductor and dielectric thicknesses, each ranging from 20 to 150 nm. The results indicate higher mobility for TC structure as compared to BC but illustrate a different scenario when varying $t_{osc}$. An overall improvement in the performance parameters of both top and bottom contact structures is observed for thinner OSC and dielectric films. It is worth noting a large deviation in mobility and $V_t$ for the $t_{ox}$ variation as compared to $t_{osc}$. $V_t$ is observed to be 2 times higher for both TC and BC structures when increasing $t_{ox}$ from 20 to 150 nm. However, a reduction of 20% and 8% in $V_t$ is observed for TC and BC structures, respectively, with a similar increment in $t_{osc}$. Besides this, an increment in trap states is observed at higher $t_{osc}$ that leads to an increase in $SS$ by 14 times, thereby reducing the switching behavior. Furthermore, the results demonstrate that the structural difference plays a vital role in predicting the performance of OTFTs based on the thicknesses of the semiconducting and dielectric layers. The OTFTs are often realized as a top contact structure since it demonstrates higher mobility and lower contact resistance. However, the bottom contact structure is preferred for flexible displays due to its fabrication feasibility using low-cost printing methods. Rigorous improvements in the OTFT performance are required for upcoming applications like RFID tags, DNA sensors, flexible memory devices, and integrated circuits. Optimum thickness scaling of OSC and dielectrics is one of the crucial controlling parameters for device performance optimization and its employability in high-end applications.

## PROBLEMS
## MULTIPLE CHOICE

1. Which simulator is used to analyze OTFT devices?
   a. ModelSim
   b. Atlas Silvaco 2-D simulator
   c. SPICE
   d. All of the above
2. Which mobility model is used to analyze the behavior of OTFT devices and circuits?
   a. Low field mobility model
   b. Poole-Frenkel mobility model
   c. Charge drift model
   d. All of the above
3. A mesh is defined by
   a. A series of horizontal and vertical lines and the spacing between them
   b. A series of horizontal lines
   c. A series of vertical lines
   d. None of the above
4. The *off*-current ($I_{off}$) of an OTFT depends on
   a. Structural parameters ($W, L, t_{ox}$)
   b. Drain–source voltage ($V_{ds}$)
   c. Channel conductivity
   d. All of the above
5. The *on-off* current ratio ($I_{on}/I_{off}$) can be improved with
   a. Increment in $t_{ox}$ and $t_{osc}$
   b. Decrement in $t_{ox}$ and $t_{osc}$
   c. Increment in $t_{ox}$ and decrement in $t_{osc}$
   d. None of the above
6. What is the origin of contact resistance in an OTFT?
   a. Large contact area
   b. Variation in $t_{osc}$ and $t_{ox}$
   c. Potential barrier due to large difference between Fermi level of contact metal and HOMO/LUMO level of OSC layer
   d. All of the above
7. Total resistance in OTFT ($R_{Total}$) is equal to
   a. Contact resistance ($R_c$)
   b. Channel resistance ($R_{ch}$)
   c. Sum of contact resistance and channel resistance
   d. None of the above
8. Which method is used to evaluate contact resistance in OTFT?
   a. Transmission line method (TLM)
   b. Four-probe method

    c.  Differential method

    d.  All of the above

  9. Top contact OTFT structure has low contact resistance because of

    a.  High carrier traps density

    b.  Thick OSC layer

    c.  Larger carrier injection area

    d.  None of the above

10. *Off*-current rises with

    a.  An increase in $t_{osc}$ due to an increment in the bulk current

    b.  A decrease in $t_{osc}$ due to a decrement in the bulk current

    c.  An increase in $t_{osc}$ due to an decrement in the bulk current

    d.  None of the above

## ANSWER KEY

1. b; 2. b; 3. a; 4. d; 5. b; 6. c; 7. c; 8. d; 9. c; 10. a

## SHORT ANSWER

1. Describe the Atlas simulation process in brief with the help of an example of OTFT structure.
2. Why does the conduction current increase with the reduction in $t_{ox}$ for both TC and BC devices?
3. Define carrier trapping effect. What are its causes?
4. Write short notes on the following OTFT performance parameters:
   a. *On-off* current ratio ($I_{on}/I_{off}$)
   b. Threshold voltage ($V_{th}$)
   c. Subthreshold slope (*SS*)
   d. Contact resistance ($R_c$)
5. How is the transmission line method (TLM) used to evaluate contact resistance?
6. How does OTFT performance get affected with variation in $t_{osc}$ for both TC and BC devices?
7. Differentiate BGBC and BGTC structures with the help of suitable diagrams.
8. Derive an expression for mobility in the saturation region.
9. How does subthreshold slope get affected with the variations in $t_{ox}$ and $t_{osc}$?
10. Explain the overall improvements in performance parameters of TC and BC OTFTs with variations in thickness of organic semiconductors ($t_{osc}$) and dielectric ($t_{ox}$) layers.
11. Explain and derive the relationship between various performance parameters of OTFTs such as drain current, transconductance, field effect mobility, *on-off* current ratio, threshold voltage and subthreshold, and device contact resistance with respect to variations in thicknesses of OSC and insulator layer.

12. Explain the differences, advantages, and applications of the following simulation tools: (a) electronic device automation (EDA) and (b) technology computer-aided design (TCAD).
13. Explain the steps and functions involved in TCAD device simulation through pictorial representation.
14. Explain the simulation principle of Atlas numerical device simulation on the basis of (a) structural dimensions and mesh specification, (b) physical models, and (c) material parameters and operational bias conditions.

## EXERCISES

1. Analyze the impact of variation in insulator thickness on drive current of the organic thin-film transistor for the given parameters and biasing conditions: top contact OTFT with drain-source voltage ($V_{ds}$) of −3 V, gate-source voltage ($V_{gs}$) of −3 V, channel length ($L$) of 10 µm, channel width ($W$) of 100 µm, mobility ($\mu$) = 0.4 cm²/Vs, threshold voltage ($V_t$) = −1.0 V, and dielectric constant ($\varepsilon_{SiO2}$) = 3.9. The insulator thickness ($t_{ox}$) varies from 20 nm to 40 nm.

2. Analyze the change in *off* current with variation in OSC thickness ($t_{osc}$) for the bottom contact OTFT fabricated using p-type organic semiconductor (OSC). The thickness of organic semiconductor ($t_{osc}$) is varied from 20 nm to 60 nm. The other parameters for the analysis are given as drain-source voltage ($V_{ds}$) = −3 V, gate-source voltage ($V_{gs}$) = −3 V, channel length ($L$) = 10 µm, channel width ($W$) = 100 µm, mobility ($\mu$) = 0.4 cm²/Vs, threshold voltage ($V_t$) = −1.0 V, conductivity ($\sigma$) = 2.4 pS, dielectric constant ($\varepsilon_{SiO2}$) = 3.9, and oxide thickness ($t_{ox}$) = 20 nm.

3. Consider the following parameters for the analysis of top contact OTFT device: drain-source voltage ($V_{ds}$) = −3 V, gate-source voltage ($V_{gs}$) = −3 V, channel length ($L$) = 10 µm, channel width ($W$) = 100 µm, mobility ($\mu$) = 0.4 cm²/Vs, threshold voltage ($V_t$) = −1.0 V, conductivity ($\sigma$) = 2.4 pS, dielectric constant ($\varepsilon_{SiO2}$) = 3.9, and oxide thickness ($t_{ox}$) = 20 nm. OSC thickness ($t_{osc}$) varies from 20 nm to 60 nm. Estimate the change in the *on/off* current ratio with change in the OSC thickness.

4. Analyze the impact of variation of insulator thickness (20 nm to 60 nm) on the current *on/off* ratio. Consider the bottom contact OTFT parameters for the device analysis: drain-source voltage ($V_{ds}$) = −3 V, gate-source voltage ($V_{gs}$) = −3 V, channel length ($L$) = 10 µm, channel width ($W$) = 100 µm, mobility ($\mu$) = 0.4 cm²/Vs, threshold voltage ($V_t$) = −1.0 V, conductivity ($\sigma$) = 2.4 pS, dielectric constant ($\varepsilon_{SiO2}$) = 3.9, and OSC thickness ($t_{osc}$) = 20 nm.

5. Consider the following top contact OTFT parameters: drain-source voltage ($V_{ds}$) = −3 V, gate-source voltage ($V_{gs}$) = −3 V, channel length ($L$) = 10 µm, channel width ($W$) = 100 µm, and acceptor concentration

$(N_A) = 4 \times 10^{17}$ cm$^{-3}$. Analyze the impact of OSC $(t_{osc})$ and insulator thickness $(t_{ox})$ variation on the device threshold voltage $(\Delta V_t)$ if
   a.  OSC thickness $(t_{osc})$ is varied from 20 nm to 60 nm keeping oxide thickness $(t_{ox}) = 20$ nm.
   b.  Oxide thickness $(t_{ox})$ is varied from 20 nm to 60 nm keeping OSC thickness $(t_{osc}) = 60$ nm.
6. Extract the value of subthreshold slope for (a) insulator thickness $(t_{ox}) = 20$ nm and (b) insulator thickness $(t_{ox}) = 60$ nm. Other parameters for the analysis of bottom contact OTFT device are drain-source voltage $(V_{ds}) = -3$ V, gate-source voltage $(V_{gs}) = -3$ V, channel length $(L) = 10$ μm, channel width $(W) = 100$ μm, and trap density $(N_{it}) = 4 \times 10^{17}$ cm$^{-3}$.

# REFERENCES

1. Gupta, D.; Hong, Y. "Understanding the effect of semiconductors thickness on device characteristics in organic thin film transistors by way of two-dimensional simulations," *Org. Electron.* **2010**, 11(1), 127–136.
2. Resendiz, L.; Estrada, M.; Cerdeira, A.; Iniguez, B.; Deen, M. J. "Effect of active layer thickness on the electrical characteristics of polymer thin film transistors," *Org. Electron.* **2010**, 11(9), 1920–1927.
3. Islam, M. N. "Impact of film thickness of organic thickness on off-state current of organic thin film transistors," *J. Appl. Phys.* **2011**, 110(11), 114906-1–114906-10.
4. Klauk, H.; Zschieschang, U.; Halik, M. "Low voltage organic thin film transistors with large transconductance," *J. Appl. Phys.* **2007**, 102(7), 074514-1–074514-7.
5. Kano, M.; Minari, T.; Tsukagoshi, K.; Maeda, H. "Control of device parameters by active layer thickness in organic thin film transistors," *App. Phy. Lett.* **2011**, 98(7), 073307-1–073307-3.
6. *ATLAS and ATHENA User's Manual: Process and Device Simulation Software*, Santa Clara, Silvaco International, **2012**.
7. Shim, C. H.; Maruoka, F.; Hattori, R. "Structural analysis on organic thin film transistor with device simulation," *IEEE Trans. Electron Devices* **2010**, 57(1), 195–200.
8. Kim, J. B.; Hernandez, C. F.; Hwang, D. K.; Tiwari, S. P.; Potscavage Jr., W. J.; Kippelen, B. "Vertically stacked complementary inverters with solution processed organic semiconductors," *Org. Electron.* **2011**, 12(7), 1132–1136.
9. Horowitz, G. "Organic field-effect transistors," *Adv. Mater.* **1998**, 10(5), 365–377.
10. Gupta, D.; Jeon, N.; Yoo, S. "Modeling the electrical characteristics of TIPS-pentacene thin film transistors: Effect of contact barrier, field-dependent mobility and traps," *Org. Electron.* **2008**, 9(6), 1026–1031.
11. Kang, G. W.; Park, K. M.; Song, J. H.; Lee, C. H.; Hwang, D. H. "The electrical characteristics of pentacene-based organic field-effect transistors with polymer gate insulators," *Curr. Appl. Phys.* **2005**, 5(4), 297–301.
12. Ha, T. J.; Sparrowe, D.; Dodabalapur, A. "Device architectures for improved amorphous polymer semiconductor thin film transistors," *Org. Electron.* **2011**, 12(11), 1846–1851.
13. Gupta, D.; Katiyar, M.; Gupta, D. "An analysis of the difference in behavior of top and bottom contact organic thin film transistors using device simulation," *Org. Electron.* **2009**, 10(5), 775–784.

14. Pernstich, K. P.; Haas, S.; Oberhoff, D.; Goldmann, C.; Gundlach, D. J.; Batlogg, B.; Rashid, A. N.; Schitter, G. "Threshold voltage shift in organic field effect transistors by dipole monolayers on the gate insulator," *J. Appl. Phys.* **2004**, 96(11), 6431-1–6431-8.
15. Kirova, N.; Bussac, M. N. "Self-trapping of electrons at the field-effect junction of a molecular crystal," *Phys. Rev. B* **2003**, 68(23), 235312-1–235312-6.
16. Hulea, I. N.; Fratini, S.; Xie, H.; Mulder, C. L.; Iossad, N. N.; Rastelli, G.; Ciuchi, S.; Morpurgo, A. F. "Tunable Frohlich polarons in organic single-crystal transistors," *Nat. Mater.* **2006**, 5, 982–986.
17. Jia, H.; Gowrisanker, S.; Pant, G. K.; Wallace, R. M.; Gnade, B. E. "Effect of poly (3-hexylthiophene) film thickness on organic thin film transistor properties," *J. Vac. Sci. Technol. A* **2006**, 24(4), 1228–1232.

# Organic Light-Emitting Transistors

5

## 5.1 INTRODUCTION

Organic semiconductor materials, such as conjugated polymers and oligomers, have become materials of commercial interest owing to the less expensive and direct fabrication techniques used in formation of electronic and optoelectronic devices. Currently, the major application area of organic electronics is found in large-area display devices having the advantages of mechanical flexibility and relatively low cost. The complete circuitry of an organic active matrix display consists of an array of organic light-emitting diodes (OLEDs) with switching devices, that is, organic thin-film transistors (OTFTs). Nowadays, OLEDs have gained significant attention from the perspective of flexibility, lightweight, and low cost; however, circuit complexity is one of the major issues for large-area display applications. In order to reduce the intricacy, a novel concept is proposed to combine both optical and electrical functionality in a single device. An organic light-emitting transistor (OLET) is a bifunctional device with organic material that has light emission capability along with the switching characteristics. At present, OLETs are promising devices exhibiting some fascinating characteristics of established electroluminescent devices. Futuristic applications find OLETs as potential candidates in integrated circuits processing both electrical and optical signals. Recently, significant research efforts have been employed for the realization of electrically pumped organic lasers. Furthermore, OLETs can be favored candidates for fundamental studies in terms of charge transport, injection, and recombination to establish them as mature devices.

This chapter discusses the underlying physics and operational mechanisms of OLETs considering different materials and structures. A detailed description about unipolar and ambipolar OLETs is presented with necessary classifications and various properties of OLETs. Furthermore, a comparative analysis of OLETs and OLEDs is presented, based on the device geometry, emission efficiency, and device lifetime.

This chapter is organized in five sections including the current introductory Section 5.1. The working principle of OLETs is described in Section 5.2, and different organic materials used for OLET fabrication are presented in Section 5.3. The classification of OLETs is given in Section 5.4. Section 5.5 discusses different optical, electrical, and thermal properties of OLETs. Finally, this chapter concludes with a brief comparison between OLETs and OLEDs in Section 5.6.

## 5.2  WORKING PRINCIPLE

The basic operation of a unipolar OLET having an organic semiconductor layer of tetracene in a bottom contact/bottom gate (BC/BG) structure is presented in Figure 5.1. In the device, the channel can conduct only one carrier type (holes in this case), while another type of carrier (electrons) only appears in the channel by tunneling from the drain electrode under the influence of a high electric field near the drain electrode. In this context, light emission occurs by the excitons, that is, pairs of holes and electrons in a highly excited state that are formed near the drain electrode and engender radiative recombination.

As mentioned earlier, the channel contains majority of the holes as charge carriers, and hence, they have to travel through the entire channel so as to recombine with electrons at the drain end. The exciton formation takes place near the drain electrode, therefore, there is a high probability that excitons may decay by nonradiation. It is due to the escape of one of the charge carriers into the corresponding metal contact, serving as an incomplete recombination of holes and electrons.

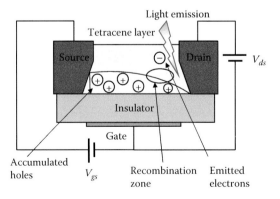

**FIGURE 5.1**   Schematic of light emission from tetracene.

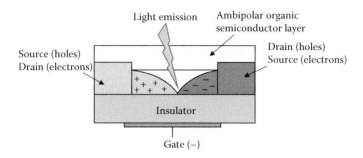

**FIGURE 5.2**    Schematic of bottom contact ambipolar OLETs.

This effect, also known as exciton quenching, results in weak light emission. To increase the emission efficiency, it is required to move the recombination region to the middle of the channel or far from the electrodes. This device structure is called ambipolar OLET. Accordingly, the basic structure of an OLET is modified by inserting two different organic semiconducting materials between the source and the drain having work functions suitable for injecting holes and electrons, respectively. The working of ambipolar OLETs is similar to a simple organic field-effect transistor (FET) with the major difference being both electrodes injecting charge carriers into the channel region, as depicted in Figure 5.2.

In the ambipolar OLET configuration, the source and drain are interchangeable for either of the charge carriers, that is, the holes and electrons. On the application of the electric field between the source and drain, the holes and electrons are driven to the highest occupied molecular orbital (HOMO) and the lowest unoccupied molecular orbital (LUMO), respectively. When opposite charges are combined to form an exciton, radiative recombination takes place and the exciton annihilates in the form of light photon and the rest in heat. Good electrical properties and intense light emission of OLETs depends on different structure-related concerns, processing and characterization conditions, device architecture, and choice of electrode materials.

## 5.3  MATERIALS FOR ORGANIC LIGHT-EMITTING TRANSISTOR (OLET) LAYERS

The OLET performance is strongly affected by the selection of different materials at various OLET layers as discussed in the following sections.

### 5.3.1  ORGANIC ACTIVE MATERIALS

The organic active materials in OLETs should possess the charge transport and luminescent properties. The elucidations for organic molecules are twofold: (1) polymers and (2) small molecules. A single unit is defined as monomer and classified as an oligomer if maximally five repeating units of molecules are present, whereas combinations of more than five monomers are defined as polymer. Till now, several small molecules, oligomers, and polymers have been used as an active material for the fabrication of OLETs.

### 5.3.1.1 SMALL MOLECULES AND OLIGOMERS

Previously, tetracene was the most widely used small molecule, single crystal material for fabrication of OLETs [1–4]. In particular, Rost *et al.* introduced a transistor using coevaporated thin film of two perylene derivatives named α-quinquethiophene (α-5T) as the hole transport material and N,N′-ditridecylperylene-3,4,9,10-tetracarboxylic diimide (Pl3) as the electron transport material [5]. The authors demonstrated an ambipolar operation of OLET with the slightly different approach of using bulk heterostructure [6]. Moreover, they manifested the thermal evaporation of pentacene as the hole transport material and *N,N′*-ditridecylperylene-3,4,9,10-tetracarboxylic diimide (PTCDI-C13H27) as the electron transport material. Pentacene is, however, not the most suitable material for light emission and hence, heterojunction structures using pentacene are strongly affected by exciton quenching [2]. Cicoira *et al.* in 2006, introduced the first solution-processed rigid core thiophene oligomer-based OLET using DTT7Me active organic material [7]. In other studies, small molecule based OLETs have been fabricated using two-component layered structures of *p*-type material (α,ω-dihexyl-quaterthiophene, DH4T) and *n*-type films of PTCDI-C13H27 (P13) [8]. Additionally, some other oligomers terfluorene [9], α,ω-bis(biphenyl-4-yl)-terthiophene (BP3T) [10], and 4,-4′-bis(styryl)biphenyl (BSBP) [11] were introduced in various OLETs. However, all the aforementioned transistors were only applicable in vacuum until Cicoira *et al.* used a synthesized tetracene derivative, namely, 2-(4-pentylstyryl) tetracene (PST) to operate in ambient air [12]. Owing to the benefits of organic materials, organic single crystals, namely, tetracene and rubrene, were introduced in the light-emitting devices to find a method for an organic laser [13]. Moreover, vapor-deposited conjugated oligomers α,ω-dihexylcarbonylquaterthiophene (DHCO4T) [14] and 2,5-bis(4-biphenylyl) thiophene (BP1T) [15] were used for stronger luminescent emission. Since BP3T has been reported as a unipolar material, for the first time Bisri *et al.* [16] reported the material as an ambipolar material using SiO$_2$ gate dielectric with PMMA buffer layers. Several new small molecules such as ditetracene [17], thiophene/phenylene co-oligomer, single crystals AC5-1CF$_3$-12OMe [18], and 4,4′-diphenyl-vinylene-anthracene (DPVA) [19] provided an advanced approach for analyzing the performance of OLETs. Various small molecules and oligomers are depicted in Figure 5.3.

### 5.3.1.2 POLYMERS

The development of OLETs inclined more toward the use of organic polymer materials rather than small molecules due to their robust characteristics under

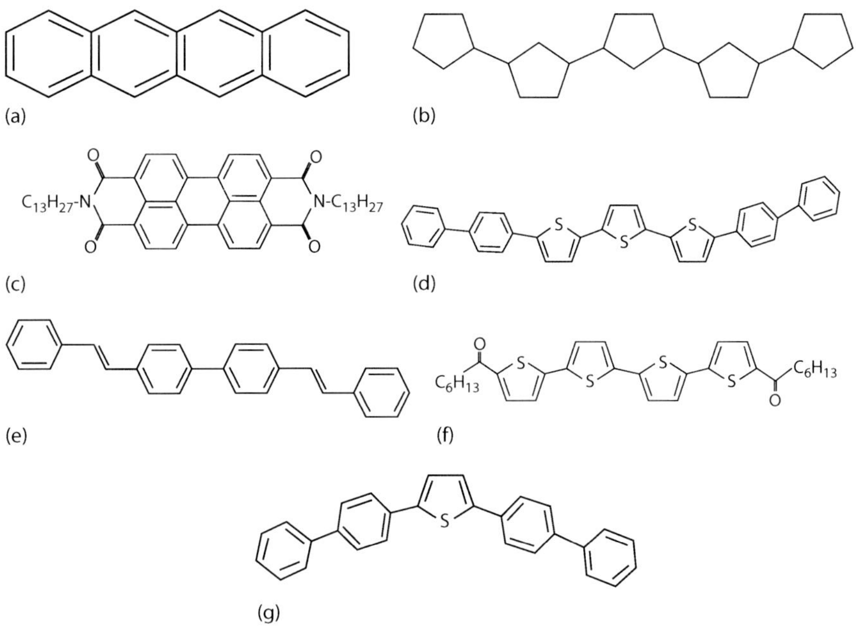

**FIGURE 5.3**    Molecular structures of small molecules and oligomers. (a) Teracene, (b) α-5T, (c) P13, (d) BP3T, (e) BSBP, (f) DHCO4T, and (g) BPIT.

ambient conditions. Until now, the uses of several polymers have been reported as active material in the OLETs [1,20]. In 2004, Ahles *et al.* [21] reported the first polymeric material, poly[9,9-di(ethylhexyl) fluorine] (PF2/6) by spin coating on the highly doped silicon substrate. While Sakanoue *et al.* [22] introduced [2-methoxy, 5-(2′-ethyl-hexoxy)-1,4-phenylenevinylene] (MEH-PPV) polymer material in a unipolar OLET by solution processing. An orange light emission was obtained from the device when operated in vacuum.

In general, the organic polymers result in sufficient emission efficiency but have low carrier mobilities, whereas single crystals have high mobilities with comparatively low luminescence. However, both the high emission efficiency and high carrier mobilities are essential for achieving a controlled light emission from the transistor. Accordingly, a conjugated polymer "Super Yellow" (SY) was introduced with quite simple fabrication steps and improved efficiency [23]. Another ambipolar transistor was reported by Zaumseil *et al.* [24] with a polyfluorene conjugated polymer material named poly(9,9-di-*n*-octyl-fluorene-*alt*-benzothiadiazole) (F8BT). Generally, the Fermi level of F8BT lies in the middle of both HOMO and LUMO and also near the work function of the gold (source/drain) material. Hence, ambipolar charge transport in the channel was obtained with holes and electron saturation mobilities of 7.5 × $10^{-4}$ cm$^2$V$^{-1}$s$^{-1}$ and 8.5 × $10^{-4}$ cm$^2$V$^{-1}$s$^{-1}$, respectively, and a maximum external quantum efficiency of 0.75%. Other reported polymeric luminescent materials

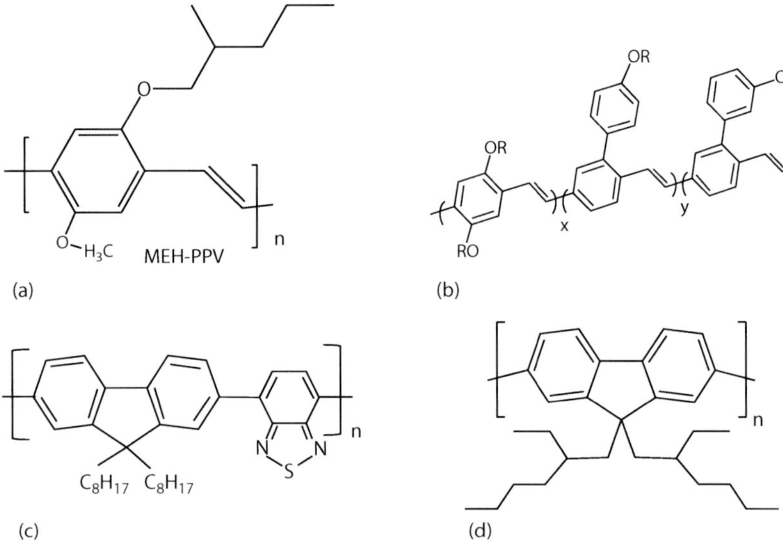

**FIGURE 5.4**  Molecular structures of different polymers. (a) MEH-PPV, (b) Super yellow, (c) F8BT, and (d) PF2/6.

are BBTDPP1 [25], PBTTT-C14 [26], and PBTTT [27]. For changing the color of emission by changing the thickness of employed layers, Oyamada *et al.* [20] doped a single crystal material rubrene into the polymer host material tetraphenylpyrene (TPPy) in the multilayer structure. In this device, the visible range of the emission spectrum can be altered by the upright position of the doped rubrene layer. Color tuning, however, was not possible by gate voltage, still the device obtained an external quantum efficiency of approximately 0.8%. This unipolar transistor resulted in significant improvements to photoluminescence (PL) efficiency; however, the mobility is still low.

Moreover, some electrolytic materials mixed with organic luminescent polymers have been demonstrated by Yumusak and Sariciftci [28]. In this approach, the researchers mixed Li triflate electrolyte with a conjugated polymer MDMO-PPV to produce an organic electrochemical LET. Kajii *et al.* [29] obtained blue emission from the transistor by using a liquid-crystalline semiconducting polymer, poly(9,9-dioctylfluorene) (F8), with doping of a red phosphorescent material Ir(piq)$_3$. The organic materials with high luminescence efficiency and good carrier mobilities are intensely required for sound growth of OLETs for future applications. Figure 5.4 presents the molecular structures of different organic polymers used in OLETs.

## 5.3.2  GATE DIELECTRICS

Primarily, the OLETs were designed and fabricated using inorganic gate dielectric SiO$_2$ as the insulating material between the gate and the active organic semiconductor layer [2,5–6,21–23]. Subsequently, silicon nitride was another inorganic dielectric material used for the fabrication of the OLETs [26]. Since, inorganic dielectrics have poor adhesion with the metal electrodes, an adhesion layer is required on the dielectric interface. Hence, the use of a thin film of metal as an adhesion layer is the most general method, however, it introduces another organic metal interface in the device and affects the characteristics. Consequently, Santato *et al.* employed self-assembled monolayer (SAM) to decrease the surface roughness of the $SiO_2$ dielectric [30]. In this study, Mylar foil was expended for fabrication of the first fully flexible organic LETs. Figure 5.5 presents an output characteristics of the OLET using Mylar [3] as a gate dielectric.

Generally, when emission occurs only at the drain electrode the device shows only unipolar behavior, and hence, is called a unipolar device. The gate dielectric materials and the OSC–dielectric interface play a crucial role in charge carrier transport in OLETs, since the hydroxyl groups (OH⁻) induce the trapping of electrons at the OSC–$SiO_2$–dielectric interface, as observed by Chua *et al.* [31]. To achieve electron transport in the channel, several treatments have been reported at the OSC–dielectric interface [14–18]. Earlier, the first ambipolar OLET based on a single active material for both the electron and the hole transport was reported. This device was fabricated through passivation of the gate dielectric $SiN_x$ with a thin film of polypropylene-co-1-butene (PPcB) [23]. To reduce the surface roughness of silicon nitride, a nonpolar polymer material, PPcB was used [32]. In contrast, Kawaguchi *et al.* reported ambipolar action of OLET controlled by introducing the hydroxyl (OH⁻) in the insulator layer [33]. Additionally, the researchers proposed a transistor having hydroxyl in the insulating layer consisting of lower threshold value, however, with quite low mobility. In this scenario, most of the researchers accepted the justification for polarity of the charge and trap states in insulators given by Chua *et al.* [31].

Several polymers are also applicable as a gate dielectric in OLETs as they are nonpolar and do not contain electron trapping groups. Noting the concept, several

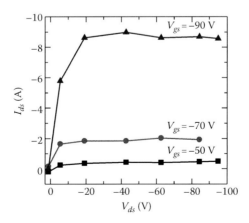

**FIGURE 5.5**    Output characteristics of OLETs using Mylar as the gate dielectric.

ambipolar OLETs with poly(methyl methacrylate) (PMMA) as the gate dielectric [24,34–36] were reported. While Burgi *et al.* [25] determined the highest mobility in ambipolar OLET with octyltrichlorosilane-treated $SiO_2$ dielectric in comparison to polymer dielectrics. The dielectric constant of insulating material also plays an important role in determining the field effect mobility and hereafter the device performance. Veres *et al.* [37] reported an increment in the charge carrier mobility in organic semiconductors by using gate dielectrics with a low dielectric constant; however, the charge density reduces simultaneously. To achieve higher recombination current, Naber *et al.* [38] used two dielectric layers of two different dielectric materials having low and high dielectric constants. Using the high dielectric constant material, a high electric field can be applied onto the device. This novel structure of OLET provides higher current than a device consisting of low-*k* single dielectric with comparable external quantum efficiency.

Water-soluble synthetic polymers have also been introduced as OLET gate dielectrics. In this context, a different approach for the dielectrics was reported by Capelli *et al.* [39] in 2011. The authors employed silk fibroin as the organic dielectric. The idea behind the use of silk protein as a dielectric is more effective in the *n*-type transistor than the *p*-type transistor in terms of threshold voltage, and *on/off* current ratio. In this approach, the gate buffer layer is essential for inhibiting the electron traps and to reduce the surface roughness. Henceforth, polymer dielectrics are the most widely used materials with $SiO_2$ for surface treatment at the $SiO_2$–OSC interface.

### 5.3.3 ELECTRODES

Typically, the basic OLET structure consists of three electrodes: source, drain, and gate. The optical and electrical characteristics of OLETs are strongly dependent on the charge carrier injection into the channel. Primarily, gold (Au) was used as the source and drain electrodes of OLET-based devices with heavily doped silicon as gate material on a bottom gate top contact approach. Previously, it was considered that the emission of light was produced due to the underetching of the source and drain electrodes. Subsequently, Rost *et al.* [6] demonstrated asymmetric contacts (source and drain contacts are of different materials) for an ambipolar heterostructure OLET; however, the light emission mechanism was not distinct. The energy band diagram and device structure of an ambipolar asymmetric OLET [6] are depicted in Figure 5.6.

In 2004, an accurate model was proposed for light emission assuming that the electrons are tunneled from the drain into the LUMO of OSC [2]. It was occurring due to the voltage drop between the gate and drain terminal that causes a distortion of the HOMO–LUMO levels of the organic material. Further, researchers reported that light can be produced without using underetched electrodes. In 2005, Oyamada *et al.* introduced a symmetrical S-D configuration having five electrodes of different materials [20]. This structure showed increased carrier injection, and, hence, high electroluminescence (EL)

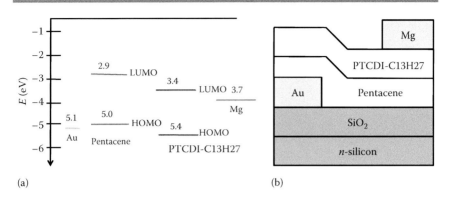

(a)                                      (b)

**FIGURE 5.6** (a) Energy band diagram and (b) device structure of ambipolar asymmetric OLET.

efficiency due to the slicing structure of S-D electrodes. The OLET device structure, output characteristics, and maximum external EL quantum efficiency on different S-D electrode material [20] are depicted in Figure 5.7.

Typically, the injection of holes and electrons in the channel is mainly dependent on the relative positions of the HOMO and LUMO level of organic material corresponding to the metal work functions [23]. In order to minimize the

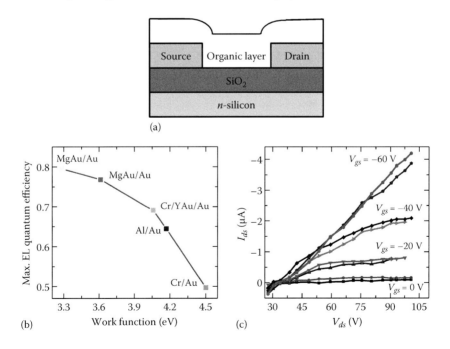

**FIGURE 5.7** (a) Device structure, (b) output characteristics, and (c) maximum external EL quantum efficiency on different S-D electrode material of OLETs.

luminescence quenching at the metal electrodes, the recombination or emission zone should be far from the metal electrodes. Therefore, the emission zone should be in the middle of the channel. To obtain an ambipolar operation of a transistor, one electrode (either source or drain) should inject holes while the other should inject electrons into the channel. The ambipolar operation is twofold:

1. An asymmetric contact, wherein, the electron injecting contact is a low work function metal and the hole injecting contact is a large work function metal. The work function of one electrode should be near the HOMO and the work function of other electrode should be near the LUMO of the OSC.
2. A symmetric metal electrode can be chosen to place the work function of the electrodes in the middle of the HOMO and LUMO level of the OSC. In this context, both holes and electrons are injected in the channel. The carrier injection efficiency is, however, less due to the high contact resistance caused by the difference between HOMO or LUMO and the work function. It is, henceforth, essential that the metal electrode and the OSC have well-matched energy levels to obtain a high current density of the charge carriers.

In the process of increasing electron density to perform the lasing action, Gwinner *et al.* tried to reduce the contact resistance for electrodes by patterning the *n*-type metal oxide (ZnO) over the gold electrode at the drain rather than using asymmetric contacts [40].

In the ambipolar regime, this technique increased the electron current about 20 times and significantly decreased the threshold voltage; however, it did not attain the advantage of flexible electronics, as it was difficult and not very efficient to achieve the processing temperatures to be compatible with plastic substrates. ZnO processed at lower temperatures did not show such good characteristics of *n*-type material. Second, mobility of the device was too small to attain any optical lasing action for futuristic applications.

Earlier, the gate electrodes of OLETs were fabricated using highly doped silicon in the bottom gate approach, whereas several researchers used gold as a gate material. Subsequently, Suganuma *et al.* [41] presented a novel structure of OLETs using two gates, wherein one gate was used to control the electrons independently, while the other gate was used to control the holes in the corresponding channels that formed a heterojunction. The typical characteristics are shown in Figure 5.8. Although, this device was suitable for *n*-type driving polarity, the gate used to control the *p*-channel was not independent in nature. Still, the *on/off* ratio of the device was improved with the luminance of about 100 cd/m² and the turn-*on* voltage of less than 10 V.

Most of the research work has already been carried out to improve the carrier injection through the source and drain electrodes in the channel. However, for further improvement in the performance and a better control of OLETs, it is required to focus on the gate electrodes.

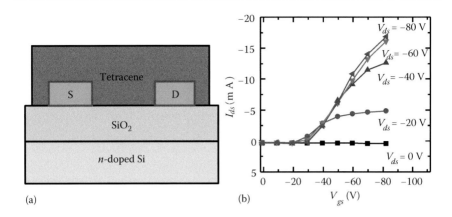

**FIGURE 5.8**    (a) Structure and (b) transfer characteristics of OLETs.

## 5.4 CLASSIFICATION OF OLETs

To obtain a detailed analysis of low cost, less complex, and flexible light-emitting transistors, it is required to categorize OLET structures. OLETs can be classified in three categories based on organic active materials, applied gate bias, and device structures, as discussed in the following sections.

### 5.4.1 CLASSIFICATION BASED ON CHARGE CARRIERS

Based on the charge carriers the OLETs can be divided in two categories: unipolar OLETs and ambipolar OLETs.

#### 5.4.1.1 UNIPOLAR OLETs

Unipolar OLETs are the materials in the OLET structure in which only one carrier type is transported and the emission occurs at the electrodes. Formerly, organic semiconductor materials were considered to be intrinsic in nature; however, they showed $p$-type or $n$-type behavior. Owing to the benefits of organic materials, several unipolar organic materials have been investigated in the formation of light-emitting FETs.

Primarily, single organic luminescent materials have been used for the fabrication of OLETs and light emission in these devices occurred near the electrodes. Earlier, it was demonstrated that light emission occurs due to the underetching of source and drain electrodes.

The first OLET was a unipolar OLET having an organic semiconductor layer of tetracene emitting green color light near the drain electrode proposed by Hepp *et al.* in 2003 [1]. The reported hole mobility ($\mu_h$) and the threshold voltage of this device was 0.05 cm²V⁻¹sec⁻¹ and –25 V, respectively. OLET device structure and transfer characteristics are presented in Figure 5.8. Later, it was proposed that the voltage drop near the metal–organic interface causes a distortion of the HOMO–LUMO levels of the organic material [2]. Figure 5.9 shows the OLET structure with current directions, electrical output characteristics, electroluminescence intensity, and transient characteristics.

Consequently, Ahles *et al.* demonstrated the first polymeric unipolar OLET [21]. It was a *p*-type OTFT in which electron injection took place due to the high electric field at the drain electrode and the luminance was controlled by the gate bias. Different than the previous devices, Sakanoue *et al.* introduced an MEH-PPV–based unipolar OLET [22]. This device was fabricated using heterostructure electrodes and gave almost the same justification of light emission as before; however, it was not capable of operating without vacuum. The output characteristics with Al/Au electrodes of MEH-PPV and photo current versus drain current for Cr/Au and Al/Au electrodes of OLET are shown in Figure 5.10.

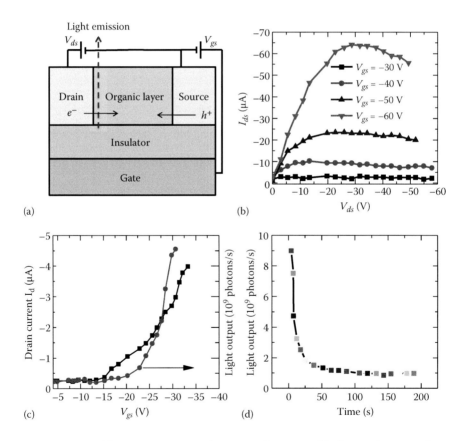

**FIGURE 5.9**    (a) OLET structure with current directions, (b) electrical output characteristics, (c) electroluminescence intensity and external quantum efficiency, and (d) transient characteristics.

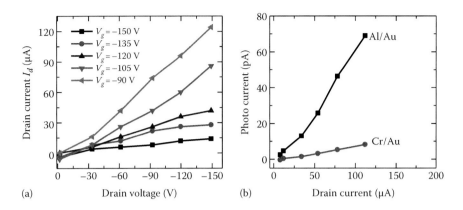

**FIGURE 5.10**    (a) Output characteristics with Al/Au electrodes of MEH-PPV and (b) photo current versus drain current for Cr/Au and Al/Au electrodes of OLETs.

Several multilayer structures were also fabricated in the unipolar OLETs. However, the purpose of realization of these structures is color tuning and increasing emission efficiency. In the similar manner, Oyamada *et al.* introduced a local doping method of highly luminescent guest material rubrene in the host layer TPPy [42]. Using this method, the characteristics of OLET are improved, also the width of the electroluminescence spectrum can be measured. Conversely, the authors reported that emission spectrum width is independent of gate bias and the external quantum efficiency of the device is decreased by increasing gate voltage as the host layer carriers are quenched by high gate voltage. To measure the electric field intensity in the emission spectrum, electroluminescence was used to relate it with the microstructure of the polymer channel. In another multilayer approach, Seo *et al.* introduced an emissive layer between the top layer of conjugated polyelectrolyte and the bottom layer of a hole transporting material [43]. This structure transformed the work function of electrodes and hence carrier injection. However, the device could not achieve the ambipolar action, and, hence, the emission spectrum position is independent of gate bias. Later, the researchers fabricated the multilayer OLET using different emissive layers to emit multicolors. This device achieved a highest brightness of ~650 Cd/m² for the super yellow emissive layer; however, efficiency was still low (0.005 cd/A) [44]. Table 5.1 summarizes the semiconductor materials and processes used to fabricate unipolar OLETs.

### 5.4.1.2  AMBIPOLAR OLETs

In order to increase emission efficiency, it is necessary to fabricate an OLET device having both holes and electrons in the channel. While emission occurs between the channels, the radiative recombination can be increased with reduced electron quenching near the drain electrode. In this context, a single material ambipolar transistor using a conjugated polymer "Super Yellow" (SY)

**TABLE 5.1    Materials and Performance of Unipolar OLETs**

| Material and Processing | Performance Parameters | Reference |
|---|---|---|
| Single crystalline small molecule Tetracene; Vacuum deposition | $\mu_h = 5 \times 10^{-2}$ cm²V⁻¹s⁻¹ $I_{on}/I_{off} = 1 \times 10^6$ | 1 |
| PF2/6; Spin coating | $\mu_h = 4.4 \times 10^{-4}$ cm²V⁻¹s⁻¹ | 21 |
| MEH-PPV; Solution processing | $I_{on}/I_{off} = 1 \times 10^5$ in vacuum $I_{on}/I_{off} = 1 \times 10^3$ in air | 22 |
| Tetracene; Physical vapor deposition | $\mu_h = 5 \times 10^{-4}$ cm²V⁻¹s⁻¹ | 3 |
| Rubrene doped TPPy; Vacuum deposition | $\mu_h = 1.7 \times 10^{-5}$ cm²V⁻¹s⁻¹ | 20 |
| TPPy with doping of 1 wt% rubrene; Vacuum deposition | PL efficiency ($\Phi_{PL}$) = 68 – 100% $\mu_h \sim 10^{-5}$ cm²V⁻¹s⁻¹ $\eta_{ext} = 0.01 - 0.5\%$ | 42 |
| Tetracene; Silylation by MPTMS | $\mu = 1 \times 10^{-2}$ cm²V⁻¹s⁻¹ (Au/MPTMS) $\mu = 1 \times 10^{-3}$ cm²V⁻¹s⁻¹ (Au/Cr) $I_{on}/I_{off} = 10^6$ (for both) | 30 |
| DTT7Me; Vacuum sublimation | $\mu = 2 \times 10^{-2}$ cm²V⁻¹s⁻¹ $I_{on}/I_{off} = 10^6$ | 7 |
| 2-(4-pentylstyryl) tetracene (PST); Vacuum sublimation | $\mu_h = 0.05$ to $0.15$ cm²V⁻¹s⁻¹ $I_{on}/I_{off} = 5 \times 10^6$ | 12 |
| MDMO-PPV mixed with polymer electrolyte poly(ethylene oxide) and Li triflate; Vacuum evaporation | $\mu = 2.2 \times 10^{-4}$ cm²V⁻¹s⁻¹ | 28 |
| PFN+BIm4− as the CPE, PBTTT-C$_{14}$ (p-type), Super Yellow as emissive layer; Solution processing | $\mu_h = 0.02$ cm²V⁻¹s⁻¹ $I_{on}/I_{off} = 1 \times 10^8$ Brightness = 520 cd/m² Efficiency = 0.08 cd/A | 43 |
| DH4T as p-type semiconductor, P13 as n-type semiconductor | $\mu_h = 1.3 \times 10^{-2}$ cm²V⁻¹s⁻¹ (DH4T) $\mu_n = 4 \times 10^{-2}$ cm²V⁻¹s⁻¹ (P13) | 39 |
| PBTTT as hole transporting polymer, Super Yellow as light-emitting polymer | $\mu_h = 1.5 \times 10^{-2}$ cm²V⁻¹s⁻¹ Brightness = 4500 cd/m² Efficiency = 0.9 cd/A | 27 |
| PFN+BIm4− as an electron injection layer, PBTTT as a p–type-semiconductor; Solution processing | $\mu = 1.5 \times 10^{-2}$ cm²V⁻¹s⁻¹ $I_{on}/I_{off} = 10^7$ | 44 |

using the passivation of the gate dielectric $SiN_x$ with 14 wt% 1-butene PPcB has been reported [23]. In the proposed device, the emission zone was in the middle of the channel and its location was moved by changing the gate bias [23], as presented in Figure 5.11a, while Figure 5.11b shows the optical characteristics.

Subsequently, Rost *et al.* used ambipolar bulk heterostructure that is fabricated with a coevaporated thin film of α-quinquethiophene (α-5T) and N,N′-ditridecylperylene-3,4,9,10-tetracarboxylic diimide (P13) as the hole and electron transport material, respectively [5]. It was the first ambipolar OLET that demonstrated the general concept of adjusting electron and hole mobilities

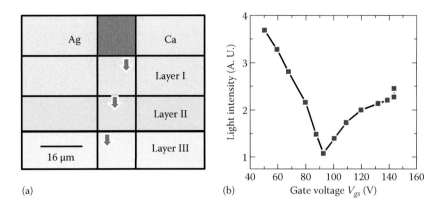

(a)    (b)    Gate voltage $V_{gs}$ (V)

**FIGURE 5.11**    (a) Moving emission pattern at different gate bias and (b) optical characteristics.

by coevaporation of two different organic semiconductors. Thereafter, the researcher presented an asymmetric bulk heterostructure ambipolar transistor using asymmetric electrodes. Subsequently, several efforts have been made for the multilayer and heterostructure approaches for ambipolar operation [26,45]. The typical characteristics by Namdas *et al.* [26] are shown in Figure 5.12.

Multilayer and heterostructure approaches are, however, less efficient for light-emitting FETs from ease of fabrication and processing perspectives. Therefore, a single material, that is, ambipolar material, is desired to fabricate an efficient OLET. Chua *et al.* determined that electron transport can be achieved in conjugated polymers [31]. It is achieved using the passivation of the $SiO_2$ with pure nonpolar polymer dielectrics, since the hydroxyl groups serve as traps for electrons at the polymer–$SiO_2$ interface. Subsequently, single ambipolar organic materials have been reported, constituting both electron and

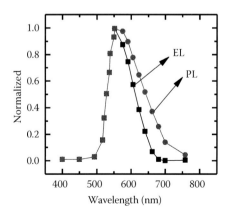

**FIGURE 5.12**    EL and PL spectrum of ambipolar OLET.

hole currents depending on the bias arrangements. Table 5.2 summarizes the materials and processes used for the fabrication and performance parameters extracted from ambipolar OLETs.

## 5.4.2 CLASSIFICATION BASED ON BIASING CONDITIONS

The basic structure and working principle of OLETs was described in Section 5.2. Further, the detailed study of OLETs can be performed on the basis of gate bias, that is, direct current gated OLETs and alternating current gated OLETs.

### 5.4.2.1 DIRECT CURRENT (DC) GATED OLETs

The detailed study of an ambipolar OLET with direct current (DC) gate bias can be based on the basic structures as described in Figure 5.1 and Figure 5.2. Generally, when a DC gate bias is applied to an OLET, it operates in two operating regimes: unipolar and ambipolar. Light emission occurs in the ambipolar regime from one electrode to another electrode based on the potential difference between gate and both electrodes. Hence, the emission zone can be moved by varying the gate voltage, and luminescence can be altered by varying the drain-source voltage ($V_{ds}$).

The operation can be understood when positive voltages are applied on the gate and drain electrodes. The injection of holes is obtained for $V_d - V_g > |V_{t,h}|$, or when drain terminal is more positive than gate. The working of DC gated OLETs is shown in Figure 5.13 and the operational consequences based on the gate bias applied are discussed next.

1. Initially, when $V_g = 0$ V, holes will accumulate in the channel. Until the gate voltage is less than the threshold voltage of electrons, no electrons are injected in the channel from the source terminal. At the same time, no recombination takes place and the current is hole dominated as shown in Figure 5.13.
2. When the gate voltage $V_g$ is further increased, that is, $V_g > V_{t,e}$, electron injection starts from the source into the channel and the transistor enters into the ambipolar regime. Recombination and emission occur near the electrode. In the ambipolar regime, the current is reduced due to the recombination.
3. As $V_G$ increases further ($V_d - V_g < |V_{t,h}|$), more electrons are tunneled into the channel. The voltage drop between the drain and the gate reduces, and, hence, the number of holes accumulated in the channel tends to reduce. Thus, the emission zone moves from source to drain. At a certain gate voltage when the recombination zone reaches the drain electrode, no holes are injected into the channel and only an electron-dominated current exists. Henceforth, the transistor again enters into the unipolar regime.

## TABLE 5.2    Materials and Performance of Ambipolar OLETs

| Material and Processing | Performance | Reference |
|---|---|---|
| α-5T and P13; Coevaporated thin films | $\mu_h = 10^{-4}$ cm$^2$V$^{-1}$s$^{-1}$ <br> $\mu_e = 10^{-3}$ cm$^2$V$^{-1}$s$^{-1}$ | 5 |
| Pentacene and PTCDI-C$_{13}$H$_{27}$; Thermal evaporation (bulk heterostructures) | $\mu_h = 9.0 \times 10^{-2}$ cm$^2$V$^{-1}$s$^{-1}$ <br> $\mu_e = 9.3 \times 10^{-3}$ cm$^2$V$^{-1}$s$^{-1}$ | 6 |
| Super Yellow (SY); Shadow mask evaporation | $\mu_h = 3 \times 10^{-4}$ cm$^2$V$^{-1}$s$^{-1}$ <br> $\mu_e = 6 \times 10^{-5}$ cm$^2$V$^{-1}$s$^{-1}$ | 23 |
| F8BT; Solution processing | $\mu_h = 7.5 \times 10^{-4}$ cm$^2$V$^{-1}$s$^{-1}$ <br> $\mu_e = 8.5 \times 10^{-4}$ cm$^2$V$^{-1}$s$^{-1}$ <br> $\eta = 0.75\%$ | 24 |
| DH4T and PTCDIC$_{13}$H$_{27}$, P13 | $\mu_h = \mu_e = 3 \times 10^{-2}$ cm$^2$V$^{-1}$s$^{-1}$ | 8 |
| Ter(9,9-diarylfluorene) (Terfluorene) | $\eta_{ext} = 0.60\%$ | 11 |
| BP3T with an electron injection layer of pentacene; Vapor deposition | $\mu_h = 9.13 \times 10^{-3}$ cm$^2$V$^{-1}$s$^{-1}$ <br> $\mu_e = 2.94 \times 10^{-2}$ cm$^2$V$^{-1}$s$^{-1}$ | 10 |
| 4,-4'-bis(styryl)biphenyl (BSBP); Vacuum deposition | $\mu_h = 1.5 \times 10^{-3}$ cm$^2$V$^{-1}$s$^{-1}$ <br> $\mu_e = 2.5 \times 10^{-5}$ cm$^2$V$^{-1}$s$^{-1}$ <br> Photoluminescence efficiency = 20% | 9 |
| Super Yellow (SY) | $\mu_h = 2.1 \times 10^{-5}$ cm$^2$V$^{-1}$s$^{-1}$ <br> $\mu_e = 1.6 \times 10^{-5}$ cm$^2$V$^{-1}$s$^{-1}$ | 32 |
| Tetracene and rubrene single-crystal; Spin coated | $\mu_h = 2.3$ cm$^2$V$^{-1}$s$^{-1}$ <br> $\mu_e = 0.12$ cm$^2$V$^{-1}$s$^{-1}$ | 13 |
| F8BT; Spin coating | $\mu_h = 5 \times 10^{-4}$ cm$^2$V$^{-1}$s$^{-1}$ <br> $\mu_e = 2.5 \times 10^{-4}$ cm$^2$V$^{-1}$s$^{-1}$ | 46 |
| Copper–phthalocyanine | Various mobilities for different dielectrics | 33 |
| Poly[3,6-bis-(4'-dodecyl-[2,2']bithiophenyl-5-yl)-2,5-bis-(2-hexyl-decyl)-2,5-dihydro-pyrrolo[3,4] pyrrole-1,4-dione] (BBTDPP1); Solution processing | $\mu_h = 0.1$ cm$^2$V$^{-1}$s$^{-1}$ <br> $\mu_e = 0.09$ cm$^2$V$^{-1}$s$^{-1}$ <br> (For SiO$_2$, Ba top contact) | 25 |
| Poly(2,5-bis(3-tetradecylthiophen-2-yl) thieno[3,2-*b*] thiophene) (PBTTT-C$_{14}$) and Super Yellow; Solution processing | $\mu_h = 3 \times 10^{-2}$ cm$^2$V$^{-1}$s$^{-1}$ <br> $\mu_e = 0.12$ cm$^2$V$^{-1}$s$^{-1}$ <br> Brightness = 1500 cd/m$^2$ <br> $\eta_{ext} = 7.5 \times 10^{-2}\%$ | 26 |
| F8BT; Solution processing | $\mu_h = \mu_e = 0.01$ cm$^2$V$^{-1}$s$^{-1}$ <br> $\eta_{ext} = 7.5 \times 10^{-2}\%$ | 38 |
| R,ω-dihexylcarbonylquaterthiophene (DHCO4T); Vacuum evaporation | $\mu_h = 10^{-3}$ cm$^2$V$^{-1}$s$^{-1}$ <br> $\mu_e = 10^{-1}$ cm$^2$V$^{-1}$s$^{-1}$ | 14 |
| 2,5-bis(4-biphenylyl)thiophene (BP1T) | $\eta_{ext} = 1.3 \times 10^{-2}\%$ | 15 |
| α,ω-bis(biphenylyl) terthiophene (BP3T) and 1,3,6,8-tetraphenylpyrene (TPPy) | $\mu_h = 1.18 \times 10^{-3}$ cm$^2$V$^{-1}$s$^{-1}$ <br> $\mu_e = 4.59 \times 10^{-3}$ cm$^2$V$^{-1}$s$^{-1}$ | 16 |
| Tetracene/ditetracene double stack | $\mu_h = 0.126$ cm$^2$V$^{-1}$s$^{-1}$ <br> $\mu_e = 0.033$ cm$^2$V$^{-1}$s$^{-1}$ <br> $\eta_{ext} = 0.012\%$ | 17 |

(*Continued*)

**TABLE 5.2 (CONTINUED)    Materials and Performance of Ambipolar OLETs**

| Material and Processing | Performance | Reference |
|---|---|---|
| 5,5″-bis(biphenylyl)-2,2′:5′,2″-terthiophene (BP3T); Spin coating | Current density = 12.3 kA/cm² | 45 |
| Poly(9,9-dioctylfluorene-*alt*-benzothiadiazole) (F8BT) and poly(9,9-dioctylfluorene) (F8); Spray pyrolysis and spin coating | $\mu_h = 1.4 \times 10^{-3}$ cm²V⁻¹s⁻¹ $\mu_e = 6 - 7 \times 10^{-4}$ cm²V⁻¹s⁻¹ | 40 |
| AC5-1CF$_3$-12OMe | $\mu_h = 6.0 \times 10^{-3}$ cm²V⁻¹s⁻¹ $\mu_e = 1.1 \times 10^{-4}$ cm²V⁻¹s⁻¹ | 18 |
| F8BT | $\mu_h = 1.7 \times 10^{-4}$ cm²V⁻¹s⁻¹ $\mu_e = 3.5 \times 10^{-4}$ cm²V⁻¹s⁻¹ | 34 |
| F8BT | Power efficiency = 1.37 lm/W Luminance efficiency = 42.87 cd/A | 35 |

At a constant drain-to-source voltage the emission intensity remains constant in the ambipolar regime; however, at the electrodes it reduces due to metal quenching. Similar to the above explanation, the working of OLET can be described for opposite charge carriers using negative drain and gate bias voltages.

### 5.4.2.2  ALTERNATING CURRENT (AC) GATED AMBIPOLAR OLETs

For the first time, a novel operating OLET is proposed by Yamao *et al.* using alternating current (AC) voltages to the gate electrode [15]. The working principle of this device is slightly different from the DC gated ambipolar OLET. In the circuit, an AC voltage of frequency *f* and amplitude $V_g$ is applied to the gate terminal. The drain terminal and the source terminal are kept at a constant DC voltage. The basic circuitry is shown in Figure 5.14. To understand the basic working principle, we first assume that the gate voltage initialized from a negative half cycle of AC signal.

For a negative initial gate voltage, the positive charge carriers are occupied in the channel and the transistor enters into the saturation regime, as shown in Figure 5.15a. Furthermore, gate voltage starts to increase and becomes equal to $V_s$ (at $V_g = V_s$). Simultaneously, the hole density close to the interface between the source terminal and the active layer would be zero and the channel reaches the pinch-off point (from PL1 to PL2) as shown in Figure 5.15b. The corresponding gated AC signal with operating points PL1 and PL2 is plotted in Figure 5.15c.

Another AC driven OLET was demonstrated by Liu *et al.*, wherein light is emitted from both sides of the channel [47]. In particular, the holes are injected and accumulated into the channel from both the source and drain (symmetric electrode) for the negative half cycle of AC voltage. Again for the positive half cycle, the electrons are injected from both electrodes, as shown in Figure 5.16.

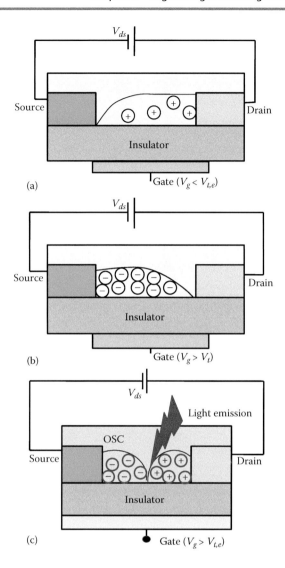

**FIGURE 5.13**   Working of DC gated OLET at different gate and drain biases: (a) $V_d - V_g > |V_{t,h}|$ and $V_g < V_{t,e}$, (b) $V_d - V_g < |V_{t,h}|$ and $V_g > V_{t,e}$, and (c) $V_d - V_g > |V_{t,h}|$ and $V_g > V_{t,e}$.

For high frequency, the speed of shifting the AC voltage is fast enough that the holes still stay in the semiconductor. The holes recombine with the subsequently injected electrons and light emission occurs at both electrodes. In the same study, the authors reported that a major cause for the electron tunneling is the space-charge field not the gate voltage. The characteristics by Liu *et al.* [47] can be seen in Figure 5.17 in terms of temperature dependence of PL and EL intensity and the EL spectrum.

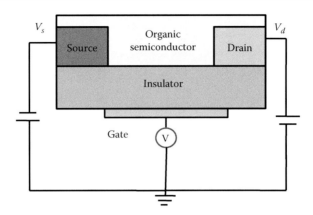

**FIGURE 5.14** Schematic of AC gated OLETs.

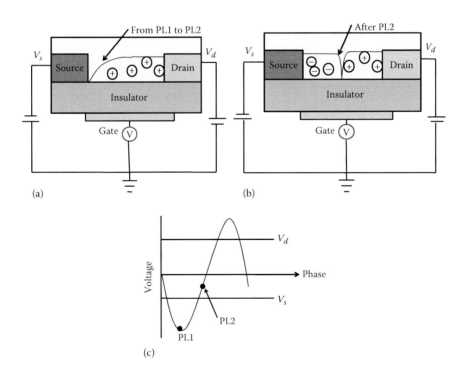

**FIGURE 5.15** Illustration of an AC gated OLET operating in (a) saturation regime (from PL1 to PL2) and (b) pinch-off regime (from PL2) with (c) indication of operating points (PL1 and PL2) on an AC signal.

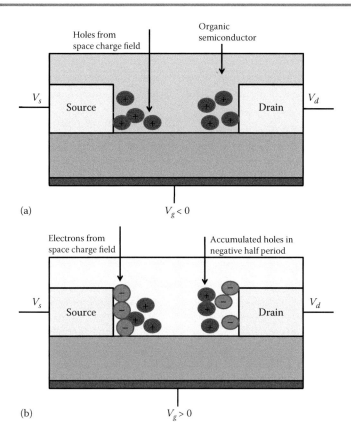

**FIGURE 5.16**    Illustration of operating principle of another AC driven OLET with symmetric contacts: (a) for negative half cycle, (b) for positive half cycle.

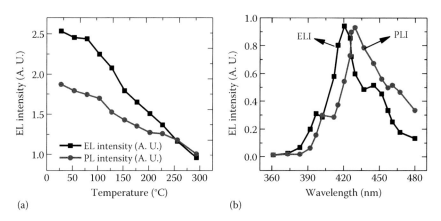

**FIGURE 5.17**    (a) Temperature dependence of PL and EL intensity, and (b) EL spectrum.

## 5.4.3 CLASSIFICATION BASED ON ARCHITECTURE

Extensive research has been carried out for organic transistors having electroluminescent properties. This technology has the potential to replace organic LEDs. On the basis of structure, OLETs can be classified into two categories: planar OLETs and vertical OLETs.

### 5.4.3.1 PLANAR OLETs

The planar structure or lateral structure of OLETs is based on the basic structure of an (FET), that is, carriers flow parallel to the channel from source to drain. OLETs having planar structure were described in Sections 5.2 and 5.4.2.

### 5.4.3.2 VERTICAL OLETs

Generally, the vertical transistor is a static induction transistor type organic transistor having a vertical driving element. The basic structure of a vertical OLET is same as the structure of an OLED, that is, the source and drain are on the top and bottom, and carriers flow perpendicular to the channel [48]. Like OLEDs, one electrode, either source or drain, is of transparent material. In the vertical OLET structure shown in Figure 5.18, the top electrode is defined as the gate electrode (G), the bottom transparent electrode is the drain electrode (D), and the middle thin electrode is the source electrode (S). An organic luminescent material is sandwiched between source and drain electrode as active layer. The source electrode is chosen in a way that it can provide a large injection barrier between the active layer and source electrode. There is, hence, no injected electron from the source into the active layer at zero gate bias.

For light emission, the holes and electrons are radiatively recombined in the active layer, injected by the drain and source electrode, respectively. For

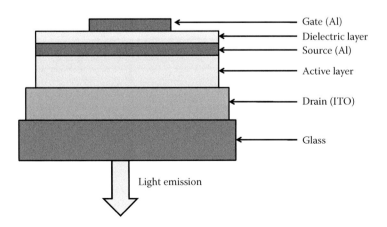

**FIGURE 5.18**    Schematic view of a vertical OLET.

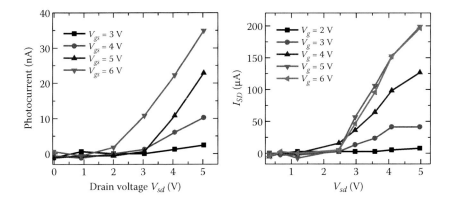

**FIGURE 5.19**   Output and optical characteristics cross-sectional images of a vertical OLET.

radiative recombination in the active layer, balanced and efficient electron and hole injection is required. Initially, when there is no gate voltage or $V_g = 0$ V, the electron injection is negligible due to the high injection barrier between the work function of the source electrode and LUMO of active layer. Electrons and holes are not balanced in the active layer, and, therefore, no emission takes place. When gate voltage of an appropriate polarity is applied, the electron injection barrier is reduced and a number of electrons are injected from the source into the channel. When a balanced carrier injection is achieved, excitons are formed in the active layer and they recombine radiatively. Strong light emission is obtained at efficient injection. The example of a vertical OLET in Figure 5.19 is demonstrated by Xu *et al.* [49].

Vertical OLETs provide relatively high currents and high speeds with low operational voltages. However, the structure is complex and it requires a better control of gate to attain a high *on/off* ratio.

## 5.5 STANDARD TERMS AND PROPERTIES OF LIGHT EMISSION IN ORGANIC SEMICONDUCTORS

This section includes the basic concepts, properties, and standard terms essential to understanding the light emission mechanism in OLETs. The basic terms and properties are explained as follows.

Host—Host is the molecules in the light-emitting device on which the charge is combined to form an exciton. Good transport property is required for a decent host material.

Guest—Guest is the molecule that accepts the exciton through energy transfer. In some systems, the guest molecule forms the exciton by trapping the charge.

Singlet and triplet excited states—The exciton formed can be assumed as two electrons and, therefore, the wave functions can be considered in four possible ways: (1) both spin up, (2) both spin down, (3) a symmetric linear combination of one spins up and the other spins down, and (4) an antisymmetric linear combination of one spins up and the other spins down [50]. The wave function equations are intuitively defined as

$$\psi = \uparrow(1)\uparrow(2) \tag{5.1}$$

$$\psi = \downarrow(1)\downarrow(2) \tag{5.2}$$

$$\psi = \frac{1}{\sqrt{2}}\{\uparrow(1)\downarrow(2)+\downarrow(1)\uparrow(2)\} \tag{5.3}$$

$$\psi = \frac{1}{\sqrt{2}}\{\uparrow(1)\downarrow(2)-\downarrow(1)\uparrow(2)\} \tag{5.4}$$

Equation 5.4 shows the antisymmetric wave function, wherein $S = 0$ and is known as the singlet excited state. Equations 5.1, 5.2, and 5.3 describe the triplet states in which $S = 1$ and are symmetric with respect to the particles 1 and 2 [50].

Fluorescence—When the singlet excited state decays to the ground state, radiation is emitted in the form of light called fluorescence [51].

Phosphorescence—When the triplet state decays to the ground, the radiation is called phosphorescence. Generally, it has a higher lifetime (slow) than that of fluorescence [51].

Spin-orbit coupling—Usually, the magnetic field generated within the material caused spin flips, which caused the intermixing of singlet and triplet states called spin-orbit coupling. The heavy metal atoms have more nuclear charge that caused strong spin-orbit coupling. Spin-orbit coupling is an important factor in phosphorescent devices.

Light emission in an organic molecule—Unlike the band theory in inorganic semiconductors, organic semiconductors have discrete vibrational and rotational levels caused by their amorphous nature [52]. A molecule in its excited state, first, relaxes by nonradiative transitions to other vibrational states and then goes to the ground state by fluorescence or phosphorescence. In a basic OLET structure (ambipolar), charges are injected from source and drain, which meet in the organic active layer under the influence of the electric field. In active layers, both charge carriers (i.e., holes and electrons) combine to form an excited state. When the emitter is phosphorescent or fluorescent, radiative recombination takes place to emit light. This phenomenon is called electroluminescence, as the electric field between the source and drain drives the carriers in the device to make them combine for light emission. The absorption and luminescence phenomenon are depicted in Figure 5.20a,b. To enhance the emission efficiency, several researchers used a host–guest system, wherein a host molecule transfers the exciton to a guest molecule.

Types of energy transfer processes—Typical energy transfer processes in the organic materials can be categorized in the following manner.

Cascade transfer—Cascade transfer is the simplest energy transfer phenomenon. In the cascade transfer mechanism, the exciton formed on the host undergoes radiative decay and the emitted light is absorbed by the guest molecule that further forms an exciton. This exciton decay radiatively emits light. The reabsorption occurs when there is an overlap in the fluorescence spectrum of the host and the absorption spectrum of the guest molecule. This effect of reabsorption and fluorescence increases the effective lifetime of the singlet fluorescence.

Förster transfer—Förster transfer is a long-range transfer process, wherein Coulombic exchange or induced dipole interaction takes place. In the Förster transfer mechanism, the relaxation of the host excited state is coupled with absorption in the guest, and

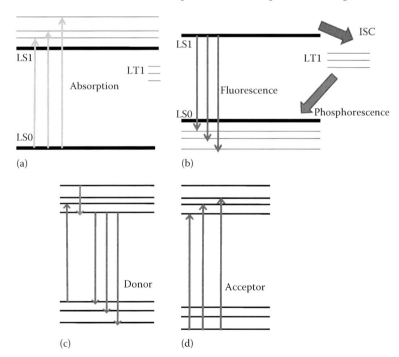

**FIGURE 5.20**   Mechanism of light emission in organic molecule: (a) absorption, (b) luminescence, and (c, d) schematics of Förster energy transfer.

fluorescence from the host does not occur unlike the cascade transfer. Therefore, coupling occurs between the host excited state and guest ground state electrons due to the Coulombic interaction. In the Förster mechanism, transfer from a singlet host to a triplet guest is impossible, as it requires spin conservation [53]. However, it also requires overlap between the luminescence spectrum of host and the absorption spectrum of the guest. Figure 5.20c,d depicts the energy transfer mechanism.

Dexter transfer—Dexter transfer is a short-range transfer mechanism that occurs through the physical transfer of electrons between the host and the guest molecules. The electron transfer is possible when two molecules are so close to each other that their electron clouds overlap. The conservation of the total spin is a constraint on this type of transfer, therefore only triplet-to-triplet transfer is feasible with this process. The mechanism is illustrated in Figure 5.21. Here D* denotes the excited state of the host and A* denotes the excited state of the guest molecule [54].

Triplet energy transfer—Triplet energy transfer process occurs through a Dexter transfer process. Generally, it can be summarized as

$$3D^* + 1A \rightarrow 1D + 3A^* \tag{5.5}$$

Singlet energy transfer—Singlet energy transfer can occur by both the Dexter and Förster transfer processes. However, Förster transfer having lower guest concentration dominates due to its long-range nature. It can be summarized as

$$1D^* + 1A \rightarrow 1D + 1A^* \tag{5.6}$$

Triplet–singlet energy transfer—Triplet state to singlet state energy transfer is called sensitized fluorescence and it is possible by the Förster transfer process. It can be summarized as

$$3D^* + 1A \rightarrow 1D + 1A^* \tag{5.7}$$

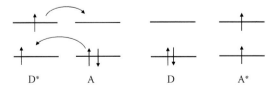

$$D^* \qquad A \qquad D \qquad A^*$$

**FIGURE 5.21**   Schematic of Dexter transfer. D* denotes the excited state of the host and A* denotes the excited state of the guest molecule.

Intersystem crossing—For a good luminescence it is required that excitons transferred to singlet states also get transferred to triplet states. Intersystem crossing (ISC) is a process whereby an electron is exchanged between a singlet and a triplet excited state. This process is caused by spin-orbit coupling and flipping of spin is obtained. In internal conversion processes, that is, singlet-to-singlet or triplet-to-triplet transfer, no spin inversion takes place. Thus, both these processes are nonradiative and completely differ from intersystem crossing.

## 5.6 COMPARISON OF ORGANIC LIGHT-EMITTING TRANSISTORS WITH CONVENTIONAL ORGANIC LIGHT-EMITTING DIODES

For the fabrication of low-cost and large-area display applications, OLEDs have gained lots of research attention from the perspective of flexibility, lightweight, brightness, and low power consumption. To the counterpart of these advantages, OLEDs are quite unstable, exhibit effective shorter lifetime, and are easily affected by the ambient conditions. However, OLETs have high potential to overcome these issues with a capability to integrate nanoscale electroluminescent devices. There are a number of physical and technical differences between OLEDs and OLETs.

1. In general, OLEDs have vertical structure while the structures of OLETs are mostly planar (except vertical OLETs). Therefore, in OLEDs, the charge carriers flow perpendicular to the channel; however, in OLETs it is parallel to the channel.
2. OLEDs are used in active matrix displays with an array of switching devices (OTFTs) to complete the functionality. Therefore, it adds to the complexity when integrated in large-area display applications. Whereas in OLETs, it is easy to obtain a highly integrated circuit as these devices exhibit both optical and electrical functionality.
3. In a typical OLED structure, the lifetime of electrons and holes are of few tens of nanometers, whereas in OLETs both carriers must travel typically tens of micrometers in length. Therefore, the charge transport properties can be analyzed in OLETs with more ease. Secondly, in OLETs it is possible to visualize the emission (recombination) zone and hence fundamental studies of charge injection, transport, and recombination can be performed.
4. In OLETs, the control at the recombination zone provides the facility to move the emission zone and achieves a balanced charge carrier current throughout the channel. Therefore, the virtually higher electroluminescence quantum efficiency can be obtained in comparison to OLEDs by reducing the exciton quenching.

5. The carrier mobility in OLETs can be about 4 orders of magnitude higher than in OLEDs that result in high current density and high emission efficiency.
6. In OLEDs, the transparent electrode has high refractive index that causes the maximum (typically 80%) emitted light to be trapped in waveguided modes in the substrate. It makes light coupling structures complex, as the transparent electrode cannot be removed. On the contrary, there is no transparent electrode in OLETs, and, hence, light coupling is naturally easier than OLEDs.
7. OLETs have a wide scope of improvement in the performance, as they allow the use of high mobility organic semiconductors, whereas OLEDs support low-mobility organic semiconductors to maintain efficient recombination.

Taking cognizance of these facts, the OLET has become the most promising candidate in the field of electroluminescent device evolution in the current research scenario.

## PROBLEMS
## MULTIPLE CHOICE

1. OLED has gained significant attention due to its
   a. Flexibility
   b. Light weight
   c. Low cost
   d. All of the above
2. In OLET the light emission occurs due to
   a. Electrons
   b. Holes
   c. Excitons
   d. Both (a) and (c)
3. Excitons quenching results in
   a. Strong light emission
   b. Weak light emission
   c. Blue light emission
   d. Green light emission
4. The organic active materials in OLETs should have
   a. Charge transport property
   b. Luminescent properties
   c. Both (a) and (b)
   d. Only (b)
5. The ambipolar action of OLET can be controlled by introducing
   a. Ketone group
   b. Silicon nitride

   c.  Double layer of active material
   d.  Hydroxyl group
6. The first OLET was a
   a.  Unipolar OLET
   b.  Ambipolar OLET
   c.  DC-gated OLET
   d.  AC-gated OLET
7. The radiation when the triplet state decays to the ground is called
   a.  Fluorescence
   b.  Phosphorescence
   c.  Spin-orbit coupling
   d.  Both (a) and (c)
8. It is possible to visualize the emission zone in
   a.  OLEDs
   b.  OLETs
   c.  LEDs
   d.  LCDs
9. In which energy transfer process do Coulombic interaction occur Cascode transfer
   a.  Dexter transfer
   b.  Förster transfer
   c.  Triplet energy transfer
   d.  None of the above
10. Light out-coupling structure is complex in OLEDs
   a.  OLETs
   b.  LEDs
   c.  LCDs
   d.  All of the above

## ANSWER KEY

1. d; 2. c; 3. b; 4. c; 5. d; 6. a; 7. b; 8. b; 9. c; 10. a

## SHORT ANSWER

1. Describe the structure and working principle of OLET.
2. Explain the term excitons quenching.
3. Draw the structure of the following P13.
   a.  DHCO4T
   b.  BPIT
   c.  F8BT
   d.  Super yellow
4. Discuss the different energy transfer processes.
5. Explain the mechanism of light emission in organic molecules.
6. Describe the working of DC-gated OLET at different gate bias.

7. What is the key difference between the fluorescence and phosphorescence?
8. Explain the phenomenon of electroluminescence.
9. Give the classification of OLET.
10. Explain in brief the singlet and triplet excited state.

## REFERENCES

1. Hepp, A.; Heil, H.; Weise, W.; Ahles, M.; Schmechel, R.; von Seggern, H. "Light-emitting field-effect transistor based on a Tetracene thin film," *Phys. Rev. Lett.* **2003**, 91, 157406.
2. Santato, C.; Capelli, R.; Loi, M. A.; Murgia, M.; Cicoira, F.; Roy, V. A. L.; Stallinga, P.; Zamboni, R.; Rost, C.; Karg, S.F.; Muccini, M. "Tetracene-based organic light-emitting transistors: Optoelectronic properties and electron injection mechanism," *Synth. Met.* **2004**, 146, 329–334.
3. Santato, C.; Manunza, I.; Bonfiglio, A.; Cicoira, F.; Cosseddu, P.; Zamboni, R.; Muccini, M. "Tetracene light-emitting transistors on flexible plastic substrates," *Appl. Phys. Lett.* **2005**, 86, 141106.
4. Reynaert, J.; Cheyns, D.; Janssen, D.; Müller, R.; Arkhipov, V. I.; Genoe, J.; Borghs, G.; Heremans, P. "Ambipolar injection in a submicron-channel light-emitting tetracene transistor with distinct source and drain contacts," *J. Appl. Phys.* **2005**, 97, 114501.
5. Rost, C.; Karg, S.; Riess, W.; Loi, M. A.; Murgia, M.; Muccini, M. "Ambipolar light-emitting organic field-effect transistor," *Appl. Phys. Lett.* **2004**, 85(9), 1613–1615.
6. Rost, C.; Karg, S.; Riess, W.; Loi, M. A.; Murgia, M.; Muccini, M. "Light-emitting ambipolar organic heterostructure field-effect transistor," *Synth. Met.* **2004**, 146, 237–241.
7. Cicoira, F.; Santato, C.; Melucci, M.; Favaretto, L.; Gazzano, M.; Muccini, M.; Barbarella, G. "Organic light-emitting transistors based on solution-cast and vacuum-sublimed films of a rigid core thiophene oligomer," *Adv. Mater.* **2006**, 18, 169–174.
8. Dinelli, F.; Capelli, R.; Loi, M. A.; Murgia, M.; Muccini, M.; Facchetti, A.; Marks, T. J. "High-mobility ambipolar transport in organic light-emitting transistors," *Adv. Mater.* **2006**, 18, 1416–1420.
9. Oyamada, T.; Chang, C.-H.; Chao, T.-C.; Fang, F.-C.; Wu, C.-C.; Wong, K.-T.; Sasabe, H.; Adachi, C. "Optical properties of oligo(9,9-diarylfluorene) derivatives in thin films and their application for organic light-emitting field-effect transistors," *J. Phys. Chem. C* **2007**, 111, 108–115.
10. Yamane, K.; Yanagi, H.; Sawamoto, A.; Hotta, S. "Ambipolar organic light emitting field effect transistors with modified asymmetric electrodes," *Appl. Phys. Lett.* **2007**, 90, 162108.
11. Sakanoue, T.; Yahiro, M.; Adachi, C.; Uchiuzou, H.; Takahashi, T.; Toshimitsu, A. "Ambipolar light-emitting organic field-effect transistors using a wide-band-gap blue-emitting small molecule," *Appl. Phys. Lett.* **2007**, 90, 171118.
12. Cicoira, F.; Santato, C.; Dadvand, A.; Harnagea, C.; Pignolet, A.; Bellutti, P.; Xiang, Z.; Rosei, F.; Meng, H.; Perepichka, D. F. "Environmentally stable light emitting field effect transistors based on 2-(4-pentylstyryl)tetracene," *J. Mater. Chem.* **2008**, 18, 158–161.

13. Takenobu, T.; Bisri, S. Z.; Takahashi, T.; Yahiro, M.; Adachi, C.; Iwasa, Y. "High current density in light-emitting transistors of organic single crystals," *Phys. Rev. Lett.* **2008**, 100, 066601.
14. Capelli, R., Dinelli, F.; Toffanin, S.; Todescato, F.; Murgia, M.; Muccini, M.; Facchetti, A.; Marks, T. J. "Investigation of the optoelectronic properties of organic light-emitting transistors based on an intrinsically ambipolar material," *J. Phys. Chem. C* **2008**, 112, 12993–12999.
15. Yamao, T.; Shimizu, Y.; Terasaki, K.; Hotta, S. "Organic light emitting field-effect transistors operated by alternating-current gate voltages," *Adv. Mater.* **2008**, 20, 4109–4112.
16. Bisri, S. Z.; Takenobu, T.; Yomogida, Y.; Yamao, T.; Yahiro, M.; Hotta, S.; Adachi, C.; Iwasa, Y. "Fabrication of ambipolar light-emitting transistor using high-photoluminescent organic single crystal," *Proc. of SPIE* **2008**, 6999.
17. Feldmeier, E. J.; Schidleja, M.; Melzer, C.; von Seggern, H. "A color-tuneable organic light-emitting transistor," *Adv. Mater.* **2010**, 22, 3568–3572.
18. Katagiri, T.; Shimizu, Y.; Terasaki, K.; Yamao, T.; Hottam, S. "Light-emitting field-effect transistors made of single crystals of an ambipolar thiophene/phenylene co-oligomer," *Org. Electron.* **2011**, 12, 8–14.
19. Bisri, S. Z.; Takenobu, T.; Sawabe, K.; Tsuda, S.; Yomogida, Y.; Yamao, T.; Hotta, S.; Adachi, C.; Iwasa, Y. "P-I-N homojunction in organic light-emitting transistors," *Adv. Mater.* **2011**, 23, 2753–2758.
20. Oyamada, T.; Uchiuzou, H.; Akiyama, S.; Oku, Y.; Shimoji, N.; Matsushige, K.; Sasabe, H.; Adachi, C. "Lateral organic light-emitting diode with field-effect transistor characteristics," *J. Appl. Phys.* **2005**, 98, 074506.
21. Ahles, M.; Hepp, A.; Schmechel, R.; von Seggern, H. "Light emission from a polymer transistor," *Appl. Phys. Lett.* **2004**, 84(3), 428–430.
22. Sakanoue, T.; Fujiwara, E.; Yamada, R.; Tada, H. "Visible light emission from polymer-based field-effect transistors," *Appl. Phys. Lett.* **2004**, 84(16), 3037–3039.
23. Swensen, J. S.; Soci, C.; Heeger, A. J. "Light emission from an ambipolar semiconducting polymer field-effect transistor," *Appl. Phys. Lett.* **2005**, 87, 253511.
24. Zaumseil, J.; Donley, C. L.; Kim, J.-S.; Friend, R. H.; Sirringhaus, H. "Efficient top-gate, ambipolar, light-emitting field-effect transistors based on a green-light-emitting polyfluorene," *Adv. Mater.* **2006**, 18, 2708–2712.
25. Bürgi, L.; Turbiez, M.; Pfeiffer, R.; Bienewald, F.; Kirner, H.-J.; Winnewisser, C. "High-mobility ambipolar near-infrared light-emitting polymer field-effect transistors," *Adv. Mater.* **2008**, 2217–2224.
26. Namdas, E. B.; Ledochowitsch, P.; Yuen, J. D.; Moses, D.; Heeger, A. J. "High performance light emitting transistors," *Appl. Phys. Lett.* **2008**, 92, 183304.
27. Namdas, E. B.; Hsu, B. B. Y.; Yuen, J. D.; Samuel, I. D. W.; Heeger, A. J. "Optoelectronic gate dielectrics for high brightness and high-efficiency light-emitting transistors," *Adv. Mater.* **2011**, 23, 2353–2356.
28. Yumusak, C.; Sariciftci, N. S. "Organic electrochemical light emitting field effect transistors," *Appl. Phys. Lett.* **2010**, 97, 033302.
29. Kajii, H.; Kusumoto, Y.; Ikezoe, I.; Ohmori, Y. "Top-gate type, ambipolar, phosphorescent light-emitting transistors utilizing liquid-crystalline semiconducting polymers by the thermal diffusion method," *Org. Electron.* **2012**, 13, 2358–2364.
30. Santato, C.; Cicoira, F.; Cosseddu, P.; Bonfiglio, A.; Bellutti, P.; Muccini, M.; Zamboni, R.; Rosei, F.; Mantoux, A.; Doppelt, P. "Organic light-emitting transistors using concentric source/drain electrodes on a molecular adhesion layer," *Appl. Phys. Lett.* **2006**, 88, 163511.

31. Chua, L. L.; Zaumseil, J.; Chang, J.-F.; Ou, E. C.-W.; Ho, P. K.-H.; Sirringhaus, H.; Friend, R. H. "General observation of n-type field-effect behavior in organic semiconductors," *Nature* **2005**, 434, 194–199.

32. Swensen, J. S.; Yuen, J.; Gargas, D.; Buratto, S. K.; Heeger, A. J. "Light emission from an ambipolar semiconducting polymer field effect transistor: Analysis of the device physics," *J. Appl. Phys.* **2007**, 102, 013103.

33. Kawaguchi, H.; Taniguchi, M.; Kawai, T. "Control of device characteristics of ambipolar organic field-effect transistors using the hydroxyl in organic insulator," *Synth. Met.* **2008**, 158, 355–358.

34. Ohshima, Y.; Lim, E.; Manaka, T.; Iwamoto, M.; Sirringhaus, H. "Observation of electron behavior in ambipolar polymer-based light-emitting transistor by optical second harmonic generation," *J. Appl. Phys.* **2011**, 110, 013715.

35. Gwinner, M. C.; Kabra, D.; Roberts, M.; Brenner, T. J. K.; Wallikewitz, B. H.; McNeill, C. R.; Friend, R. H.; Sirringhaus, H. "Highly efficient single-layer polymer ambipolar light-emitting field-effect transistors," *Adv. Mater.* **2012**, 24, 2728–2734.

36. Feldmeier, E. J.; Melzer, C. "Multiple colour emission from an organic light-emitting transistor," *Org. Electron.* **2011**, 12, 1166–1169.

37. Veres, J.; Ogier, S.; Lloyd, G.; de Leeuw, D. M. "Gate insulators in organic field-effect transistors," *Chem. Mater.* **2004**, 16, 4543–4555.

38. Naber, R. C. G.; Bird, M.; Sirringhaus, H. "A gate dielectric that enables high ambipolar mobilities in polymer light-emitting field-effect transistors," *Appl. Phys. Lett.* **2008,** 93, 23301.

39. Capelli, R.; Amsden, J. J.; Generali, G.; Toffanin, S.; Benfenati, V.; Muccini, M.; Kaplan, D. L.; Omenetto, F. G.; Zamboni, R. "Integration of silk protein in organic and light-emitting transistors," *Org. Electron.* **2011**, 12, 1146–1151.

40. Gwinner, M. C.; Vaynzof, Y.; Banger, K. K.; Ho, P. K. H.; Friend, R. H.; Sirringhaus, H. "Solution-processed zinc oxide as high performance air-stable electron injector in organic ambipolar light-emitting field-effect transistors," *Adv. Funct. Mater.* **2010**, 20, 3457–3465.

41. Suganuma, N.; Shimoji, N.; Oku, Y.; Okuyama, S.; Matsushige, K. "Organic light-emitting transistors with split-gate structure and PN-hetero-boundary carrier recombination sites," *Org. Electron.* **2008**, 9, 834–838.

42. Oyamada, T.; Sasabe, H.; Oku, Y.; Shimoji, N.; Adachi, C. "Estimation of carrier recombination and electroluminescence emission regions in organic light-emitting field-effect transistors using local doping method," *Appl. Phys. Lett.* **2006**, 88, 093514.

43. Seo, J. H.; Namdas, E. B.; Gutacker, A.; Heeger, A. J.; Bazan, G. C. "Conjugated polyelectrolytes for organic light emitting transistors," *Appl. Phys. Lett.* **2010**, 97, 043303.

44. Seo, J. H.; Namdas, E. B.; Gutacker, A.; Heeger, A. J.; Bazan, G. C. "Solution-processed organic light-emitting transistors incorporating conjugated polyelectrolytes," *Adv. Funct. Matter.* **2011**, 21, 3667–3672.

45. Sawabe, K.; Takenobu, T.; Bisri, S. Z.; Yamao, T.; Hotta, S.; Iwasa, Y. "High current densities in a highly photoluminescent organic single-crystal light-emitting transistor," *Appl. Phys. Lett.* **2010**, 97, 043307.

46. Zaumseil, J.; Kline, R. J.; Sirringhaus, H. "Electroluminescence imaging and microstructure of organic light-emitting field-effect transistors," *Appl. Phys. Lett.* **2008**, 92, 073304.

47. Liu, X.; Kjelstrup-Hansen, J.; Boudinov, H.; Rubahn, H.-G. "Charge-carrier injection assisted by space-charge field in AC-driven organic light-emitting transistors," *Org. Electron.* **2011**, 12, 1724–1730.

48. Park, B.; Takezoe, H. "Enhanced luminescence in top-gate-type organic light-emitting transistors," *Appl. Phys. Lett.* **2004**, 85, 1280.

49. Xu, Z.; Li, S.; Ma, L.; Li, G.; Yang, Y. "Vertical organic light emitting transistor," *Appl. Phys. Lett.* **2007**, 91, 092911.

50. Baldo, M.; Segal, M. "Phosphorescence as a probe of exciton formation and energy transfer in organic light emitting diodes," *Phys. Stat. Sol.* **2004**, 201, 1205.

51. Segal, M.; Singh, M.; Rivoire, K.; Diey, S.; Voorhis, T. V.; Baldo, M. A. "Extrafluorescent electroluminescence in organic light-emitting devices," *Nat. Mater.* **2007**, 6, 374.

52. Dodabalapur, A. "Highlights in condensed matter physics and materials science," *Solid State Commun.* **1997**, 102, 259.

53. Förster, T. *Discussions of Faraday Society* **1959**, 27, 7.

54. Turro, N. *Modern Molecular Photochemistry*, University Science Books, **1991**.

# Organic Device
# Applications

# Static and Dynamic Analysis of Organic All-*p*, Organic Complementary, and Hybrid Complementary Inverter Circuits

6

## 6.1 INTRODUCTION

Organic transistors find extensive applicability in low-cost large-area flexible displays, ring oscillators, integrated circuits, memory and radio-frequency identification (RFID) tags; wherein, an inverter is used as the most basic digital circuit element. To ensure proper operation of an inverter circuit, some stability factors should be considered; especially, the noise margin and propagation delay. A complementary inverter is beneficial in terms of low static power consumption, high noise margin, high gain, and operational robustness. For such a complementary organic inverter, the mobility and threshold voltage of a $p$-type transistor should be comparable to its $n$-type counterpart.

The mobility of $n$-type organic semiconductors (OSCs) is often lower in comparison to the $p$-type due to their large band gaps. Difficulty in obtaining a high mobility $n$-type organic transistor forced researchers to propose an all $p$-type organic inverter circuit [1,2]. However, an all $p$-type inverter also faces challenges such as low voltage swing, poor balance between pull-up and pull-down operation, high power dissipation, and low noise margins. Taking these limitations into account, hybrid complementary circuits have been proposed. Dodabalapur *et al.* first suggested hybrid complementary technology by replacing an $n$-type organic transistor with an inorganic hydrogenated amorphous silicon thin-film transistor (a-Si:H TFT) [3]. Development of organic and hybrid circuits, using simple and low-cost process techniques, permits the integration of complete digital circuits.

This chapter presents the realization of different inverter configurations, such as organic all-$p$, organic complementary, and hybrid complementary with an assortment of organic/inorganic TFTs. This chapter presents a comparison of the performance of inverter circuits based on different materials for driver and load TFTs, starting with independent designs of different TFTs. Simulated electrical characteristics and parameters of all $p/n$ transistors are verified through experimental results. Furthermore, this chapter investigates the static and dynamic behavior of different inverters using CuPc–$F_{16}$CuPc, pentacene–$C_{60}$, pentacene–ZnO, pentacene–a-Si:H, and pentacene–pentacene combinations to address the effect of various materials and configurations. Besides this, the imitated propagation delay and switching threshold voltage of various combinations of inverter designs are validated with the analytical results.

This chapter is arranged in five sections, including the current introductory Section 6.1. Independent designs of different TFTs along with their performance validation through experimental results are presented in Section 6.2. The performance of organic/hybrid complementary and all-$p$ organic inverter circuits with different TFT combinations are discussed in Section 6.3, and Section 6.4 presents a comparison of the static and dynamic performance of the proposed inverter configurations. Finally, Section 6.5 concludes the chapter with important outcomes.

## 6.2  ANALYSIS OF DIFFERENT $p$- AND $n$-TYPE DEVICES

Circuit performance depends on electrical characteristics and performance parameters of particular devices. This section presents an analysis of electrical characteristics and operational parameters of individual $p$- and $n$-type TFTs to be used in inverter circuit design. Table 6.1 summarizes the dimensional parameters and the materials combinations for different TFTs. Device and circuit analysis is performed using the finite-element-based Atlas 2-D numerical device simulator.

**TABLE 6.1    Dimensional Parameters and Materials of Different TFTs Used for Inverters**

| Parameter | Organic p-Type | | Organic n-Type | | Inorganic n-Type | |
|---|---|---|---|---|---|---|
| | TFT-1 [4] | TFT-2 [5] | TFT-3 [5] | TFT-4 [6] | TFT-5 [7] | TFT-6 [4] |
| $L$ (μm) | 90 | 50 | 50 | 100 | 40 | 90 |
| $W$ (μm) | 500 | 1000 | 1000 | 1000 | 80 | 500 |
| $t_{ox}$ (nm) | 12.5 ($Al_2O_3$+$SiO_2$) | 300 ($SiO_2$) | 300 ($SiO_2$) | 144 ($SiO_2$, $TiSiO_2$) | 300 ($SiN_x$) | 12.5 ($Al_2O_3$+$SiO_2$) |
| $t_{osc}$ (nm) | 50 (Pentacene) | 10 (CuPc) | 10 ($F_{16}$CuPc) | 60 ($C_{60}$) | 250 (a-Si:H) | 60 (ZnO) |
| $t_s/t_d$ (nm) | 10 (Al) | 10 (Au) | 10 (Au) | 101 (LiF+Al) | 30 (Silicide) | 10 (Al) |
| $t_g$ (nm) | 10 ($n^+$ Si) | 10 ($n^+$ Si) | 10 (p-Si) | 30 (p-Si) | 80 (p-Si) | 10 (p-Si) |

## 6.2.1  p-TYPE ORGANIC THIN-FILM TRANSISTORS (OTFTs)

Organic materials offer strong assurance in terms of properties, processing, and cost effectiveness. The majority of OTFTs that have been reported in literature are p-type due to their higher mobility and better stability in comparison to their n-type counterparts. In this section, the electrical behavior of p-type devices based on pentacene and CuPc active materials is analyzed.

### 6.2.1.1  PENTACENE-BASED p-TYPE OTFT

Pentacene has proven to be the most widely used p-type OSC due to higher hole mobility that is a result of orbital overlapping among the molecules in the crystal lattice. In addition to high mobility, it shows good chemical stability even in adverse environmental conditions, orderly formation in thin film structure, and good interface with gold and aluminum, commonly used electrode metals. Figure 6.1a shows the device structure of a pentacene-based OTFT, with a

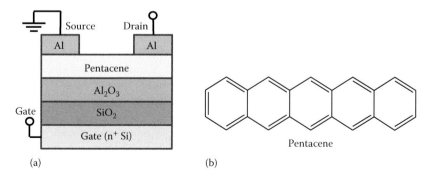

(a)    (b)

**FIGURE 6.1**    (a) Device structure of pentacene-based OTFT and (b) chemical structure of pentacene.

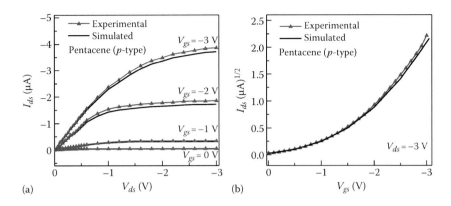

**FIGURE 6.2**    (a) $I_{ds}-V_{ds}$ and (b) $\sqrt{I_{ds}}-V_{gs}$ characteristics of a pentacene-based OTFT.

$W$ and $L$ of 500 and 90 μm, respectively. The gate electrode is of heavily doped silicon with a thickness of 10 nm.

A bilayer of dielectric materials, $Al_2O_3$ (10 nm) and $SiO_2$ (2.5 nm), results in a higher capacitance value of 350 nF/cm². Furthermore, the thicknesses of the pentacene semiconducting layer and aluminum source/drain (S/D) contacts are 50 and 10 nm, respectively. Effective density of states in the conduction and valence band are assumed to be of the order of $10^{21}$ cm$^{-3}$ and energy band gap of pentacene is taken as 2.2 eV [8]. Poole-Frenkel model parameters $\Delta_h$ and $\beta_h$ are $1.792 \times 10^{-2}$ eV and $7.758 \times 10^{-5}$ eV (cm/V)$^{0.5}$, respectively [9]. Zero field mobility and relative permittivity of pentacene are 0.5 cm²/Vs and 4, respectively.

The chemical structure of a pentacene semiconductor is shown in Figure 6.1b that consists of a chain of five benzene rings. Figure 6.2 illustrates the plots of simulated/experimental $I_{ds}-V_{ds}$ and $\sqrt{I_{ds}}-V_{gs}$ characteristics, respectively. The device demonstrates a reasonably strong current of 3.7 μA at a low $V_{ds}$ and $V_{gs}$ of −3 V due to high capacitance produced by the thin layer (12.5 nm) of dielectric. The mobility, threshold voltage, and *on/off* current ratio are 1.0 cm²/Vs, −0.65 V, and $1.3 \times 10^4$, respectively.

### 6.2.1.2 CuPc-BASED p-TYPE OTFT

Copper phythalocyanine OSC demonstrates an excellent chemical and thermal stability in environmental conditions. Therefore, it is considered as the most promising *p*-type OSC for modern optoelectronic devices such as organic light-emitting diodes (OLEDs), solar cells, gas sensors, and OTFTs. Besides this, it can be easily sublimed, which results in high purity thin films without decomposition. However, its mobility is quite low in comparison to pentacene due to higher band gap. The structure of a CuPc-based *p*-type device is shown in Figure 6.3a, and the chemical structure of CuPc material is presented in Figure 6.3b.

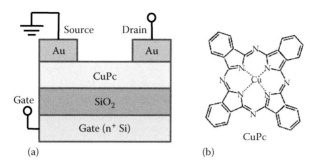

(a)                                                  (b)

**FIGURE 6.3**    (a) Device structure of a CuPc-based OTFT and (b) chemical structure of a CuPc.

A heavily doped donor type, silicon wafer with 10 nm thickness is used as the gate electrode. The thicknesses of CuPc (OSC) and thermally grown $SiO_2$ (dielectric) are 10 and 300 nm, respectively. Furthermore, the S and D electrodes are of gold with 10 nm each, and $L$ and $W$ are 50 and 1000 μm, respectively. The band gap and hole mobility of CuPc material is 3.1 eV and $5 \times 10^{-4}$cm²/Vs, which is inferior in comparison to pentacene, therefore it requires higher operating voltage. The density of states in both the conduction and valence band is $10^{21}$ cm$^{-3}$. Besides this, the work function of silicon and gold is considered as 3.9 and 5.1 [10], respectively.

The electrical characteristics of CuPc-based OTFTs are illustrated in Figure 6.4. The device shows a current of 18 μA at a quite high $V_{ds} = V_{gs} = -100$ V. The mobility and threshold voltage extracted from the $\sqrt{I_{ds}}-V_{gs}$ curve at $V_{ds}$ of $-100$ V are 0.02 cm²/Vs and $-5$ V, respectively. The device exhibits a low capacitance of 12 nF/cm². This is due to a thicker layer of dielectric (300 nm) that in turn results in a high threshold voltage, thereby requiring enormously high voltage to enable a sufficiently large drain current.

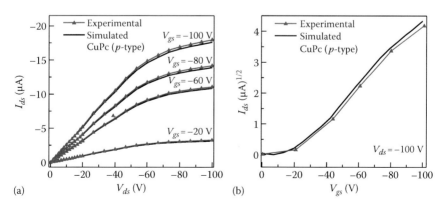

(a)                                                  (b)

**FIGURE 6.4**    (a) $I_{ds}-V_{ds}$ and (b) $\sqrt{I_{ds}}-V_{gs}$ characteristics of a CuPc-based OTFT.

## 6.2.2 *n*-TYPE OTFTs

Continuous improvement in the performance of *n*-type organic transistors is attributed to the synthesis of novel OSC materials, surface treatment of the gate dielectric, using a self-assembled monolayer (SAM) of dielectric on the substrate and the optimization of fabrication techniques. Although *n*-type organic TFTs exhibit lower mobility and poor stability in comparison to the *p*-type, both types (*p*- and *n*-) of TFTs are required for an organic complementary circuit that derives benefits in terms of lower power consumption, higher noise margin, and better stability as compared to all *p*-type organic circuits. To design an *n*-type OTFT, the semiconductor must be utilized that can allow the injection of electrons into its lowest unoccupied molecular orbital (LUMO). This section analyzes *n*-type OTFTs based on $F_{16}CuPc$ and $C_{60}$.

### 6.2.2.1 $F_{16}CuPc$-BASED *n*-TYPE OTFT

The high performance *n*-type OSCs can be synthesized by adding –Cl, –CN, and –F groups to the outermost orbital of molecules, since these groups are strongly capable of withdrawing the electrons. With the same thought, several researchers have demonstrated an example of fabricating *n*-type material, $F_{16}CuPc$, by adding a –F group to the *p*-type material, CuPc. Figure 6.5 demonstrates the device structure of $F_{16}CuPc$-based OTFT and chemical structure of the active material. Similar to CuPc-based OTFTs discussed earlier, this device also comprises a highly doped silicon wafer (10 nm) with a thick layer of thermally grown $SiO_2$ gate dielectric (300 nm) and gold S/D contacts (10 nm).

The thickness of active layer material, $F_{16}CuPc$, is 10 nm, and the channel length and width are 50 and 1000 μm, respectively. The electron affinity for $F_{16}CuPc$ is 4.2 eV. The majority of *n*-type organic materials exhibit an electron affinity near 4.0 eV that is lower than the work function of highly optimized gold electrode metal (5.1 eV). Therefore, the carrier injection in *n*-type OSCs is limited by an energy barrier height of approximately 1.1 eV, resulting in a lower

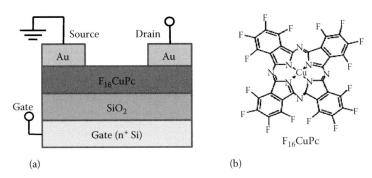

(a)                                  (b)

**FIGURE 6.5**   (a) Device structure of an $F_{16}CuPc$-based OTFT and (b) chemical structure of an $F_{16}CuPc$.

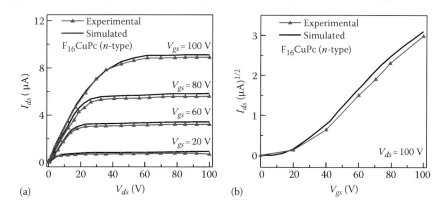

**FIGURE 6.6** (a) $I_{ds}$–$V_{ds}$ and (b) $\sqrt{I_{ds}}$–$V_{gs}$ characteristics of an $F_{16}$CuPc-based OTFT.

performance. The electrical characteristics shown in Figure 6.6 demonstrate an output current of 9 μA at high drain and gate voltages of 100 V, due to lower mobility and lower device capacitance. Besides this, the mobility and threshold voltage are 0.011 cm²/Vs and 7 V, respectively. Due to lower mobility, $F_{16}$CuPc-based $n$-type OTFTs are limited to low-speed applications.

### 6.2.2.2 $C_{60}$-BASED $n$-TYPE OTFT

Fullerene has been considered a promising $n$-type material due to its highest mobility among other OSCs. A high-quality thin film of fullerene organized under an ultra-high vacuum exhibits the electron mobility comparable to that of amorphous silicon. To obtain a $C_{60}$ film from amorphous to highly crystalline structure, Faimen *et al.* reported the range of substrate temperature from room temperature to 250°C [11]. This material is well suited to being deposited on the plastic substrate, thereby preferring to obtain a high-performance, flexible, $n$-type OTFT. A number of high-performance soluble fullerene derivatives have been investigated during the last decade. Figure 6.7a shows the device structure of a $C_{60}$-based $n$-type OTFT, and Figure 6.7b presents the chemical structure of a $C_{60}$ semiconductor that consists of a soccer ball-like simple structure.

A titanium silicon-dioxide (TiSiO₂) layer of 132 nm is sandwiched between 8 and 4 nm SiO₂ dielectric layers. Upper and lower layers of SiO₂ are employed to stabilize device performance by suppressing gate leakage. Furthermore, the thicknesses of silicon gate and $C_{60}$ active material are 30 and 60 nm, respectively. The S/D top contacts made up of a bilayer of lithium fluoride (LiF) and aluminum are 1 and 100 nm thick, respectively.

The channel length and width are 100 and 1000 μm, respectively. In addition, the electron affinity for $C_{60}$ active material is 3.7 eV. The electrical characteristics are illustrated in Figure 6.8. This device demonstrates a current of 5.2 μA at $V_{ds} = V_{gs} = 5$ V, owing to the high dielectric constant of 15.7 for multilayers of dielectric that produce a capacitance of 96.5 nF/cm². The device

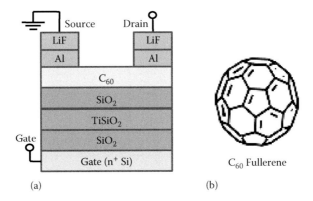

**FIGURE 6.7** (a) Device structure of a $C_{60}$-based OTFT and (b) chemical structure of a $C_{60}$.

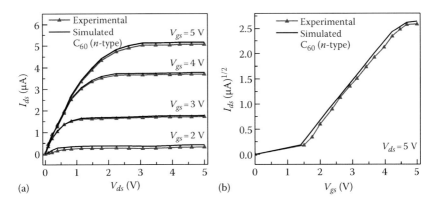

**FIGURE 6.8** (a) $I_{ds}$–$V_{ds}$ and (b) $\sqrt{I_{ds}}$–$V_{gs}$ characteristics of a $C_{60}$-based OTFT.

shows reasonably good performance in terms of mobility, threshold voltage and *on/off* current ratio, which are obtained as 0.92 cm²/Vs, 1.3 V, and $0.84 \times 10^5$, respectively.

### 6.2.3 *n*-TYPE INORGANIC TFTs

Generally, difficulties experienced during the designing of organic complementary inverter circuits are due to lower mobility of *n*-type OTFTs as compared to their *p*-type counterparts. Furthermore, these are known to be more susceptible to atmospheric decay; however, *p*-type materials are quite stable in the air. Therefore, a hybrid complementary design is also proposed, wherein an *n*-type organic TFT is replaced by an inorganic TFT. This section analyzes the behavior of *n*-type devices based on a-Si:H and ZnO semiconductors.

### 6.2.3.1 a-Si:H–BASED *n*-TYPE INORGANIC TFT

Inorganic *n*-type TFTs show a higher mobility and better electrical characteristics as compared to organic *n*-type TFTs. However, the fabrication cost and flexibility are obvious constraints. Figure 6.9 shows the device structure of an a-Si:H transistor, with an $L$ and $W$ of 40 and 80 μm, respectively. The active layer is of a-Si:H material with a thickness of 250 nm. Furthermore, the thickness of SiN$_x$ gate dielectric is 300 nm; whereas, the Ni-silicide material with 30 nm thickness is used for the S/D contacts.

Simulation parameters for a-Si:H material, such as band gap, electron affinity, permittivity, and bulk carrier mobility, are 1.9 eV, 4 eV, 11, and 20 cm²/Vs, respectively [7]. In addition, the density of states, $N_C$ and $N_V$, are $10^{20}$ cm$^{-3}$. The $I_{ds}$–$V_{ds}$ and $\sqrt{I_{ds}}$–$V_{gs}$ characteristics of a-Si:H–based TFT are plotted in Figure 6.10, respectively. The mobility and threshold voltage obtained from the $\sqrt{I_{ds}}$–$V_{gs}$ curve at $V_{ds}$ = 10 V, are 1.1 cm²/Vs and 2.4 V, respectively.

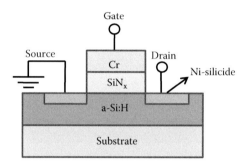

**FIGURE 6.9**   Device structure of an a-Si:H–based thin-film transistor.

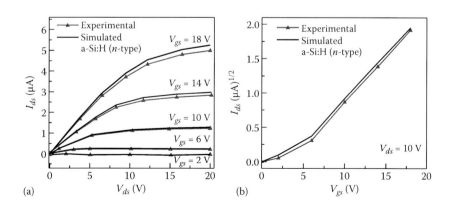

**FIGURE 6.10**   (a) $I_{ds}$–$V_{ds}$ and (b) $\sqrt{I_{ds}}$–$V_{gs}$ characteristics of an a-Si:H–based thin-film transistor.

### 6.2.3.2 ZnO-BASED *n*-TYPE INORGANIC TFT

Zinc oxide (ZnO)-based TFTs show substantially improved electrical characteristics. The device performance is stable when exposed to air with impressively good mobility. For flexible display applications, it can be proved to be a suitable replacement of an a-Si:H material due to its lower fabrication temperature (<100°C) and capability to be integrated on a glass or plastic substrate.

The device structure for a ZnO-based TFT is shown in Figure 6.11 that has $W$ and $L$ dimensions of 500 and 90 µm, respectively. Heavily doped $p$-type silicon, ZnO, and aluminum of thicknesses 10, 60, and 10 nm are used as the gate electrode, active layer, and S/D contacts, respectively. The drain current of magnitude 2.7 µA is achieved at $V_{ds} = V_{gs} = 3$ V, as depicted in Figure 6.12a. The $\sqrt{I_{ds}}-V_{gs}$ characteristic is shown in Figure 6.12b. The device exhibits mobility, threshold voltage, and *on/off* current ratio as 1.06 cm²/Vs, 1.1 V, and $0.7 \times 10^4$, respectively.

Among all the $p/n$ devices, the capacitance is highest for pentacene- and ZnO-based devices due to significantly lower dielectric layer thickness (12.5 nm); whereas it is lowest for CuPc- and $F_{16}$CuPc-based OTFTs. It is due to a comparatively

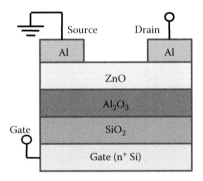

**FIGURE 6.11**    Device structure of a ZnO-based thin-film transistor.

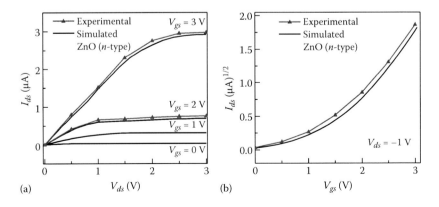

**FIGURE 6.12**    (a) $I_{ds}-V_{ds}$ and (b) $\sqrt{I_{ds}}-V_{gs}$ characteristics of a ZnO-based TFT.

**TABLE 6.2    Comparison between Experimental and Simulated Parameters for All *p*- and *n*-Type Devices**

| Device | Semiconductor | $C_{ox}$ (nF/cm²) | $V_t$ (V) Sim. | $V_t$ (V) Exp. | $\mu$ (cm²/Vs) Sim. | $\mu$ (cm²/Vs) Exp. | $I_{on}/I_{off}$ Sim. | $I_{on}/I_{off}$ Exp. |
|---|---|---|---|---|---|---|---|---|
| TFT-1 [4] | Pentacene | 350 | −0.65 | −0.6 | 1.0 | 1.03 | $1.34 \times 10^4$ | $2.1 \times 10^4$ |
| TFT-2 [5] | CuPc | 12 | −5.0 | −6.0 | 0.02 | 0.018 | $2 \times 10^3$ | NA |
| TFT-3 [5] | $F_{16}$CuPc | 12 | 7.0 | 8.2 | 0.011 | 0.01 | $1 \times 10^3$ | NA |
| TFT-4 [6] | $C_{60}$ | 96.5 | 1.3 | 1.13 | 0.92 | 1.0 | $0.84 \times 10^5$ | $1 \times 10^5$ |
| TFT-5 [7] | a-Si:H | 22 | 2.4 | 2.7 | 1.1 | 1.0 | $2.7 \times 10^6$ | $2 \times 10^6$ |
| TFT-6 [4] | ZnO | 350 | 1.1 | 1.05 | 1.06 | 1.12 | $0.7 \times 10^4$ | $1.2 \times 10^4$ |

higher thickness of the $SiO_2$ dielectric layer (300 nm) as well as the lower permittivity (3.9), thereby resulting in enormously high threshold voltages. Besides this, the mobility is also very low for both CuPc and $F_{16}$CuPc OTFTs; however, all other TFTs exhibit comparable mobility. Simulated performance parameters of all the devices are compared with reported experimental results and summarized in Table 6.2. Simulated results are in agreement with the experimental results with an average error of 1.8% and 4.0% for threshold voltage and mobility, respectively. Based on the performance parameters of all the devices, five different inverters with TFTs combinations, namely, pentacene–$C_{60}$, CuPc–$F_{16}$CuPc, pentacene–a-Si:H, pentacene–ZnO, and pentacene–pentacene, are analyzed in the next section.

## 6.3  ANALYSIS OF INVERTER CIRCUITS WITH DIFFERENT TFT COMBINATIONS

An inverter is considered the most basic circuit element in CMOS (complementary metal-oxide semiconductor) technology. This section analyzes the performance of various inverter circuits consisting of different device combinations for driver and load. For an appropriate comparison on a common platform, all the inverter configurations are analyzed with *W/L* of 2000/90 and 500/90 for driver and load, respectively. Inverter configurations are simulated under mixed-mode that includes elements simulated using device simulation and compact circuit models. It incorporates advanced numerical algorithms that effectively analyze DC, transient, small signal AC, and small signal networks. Under mixed-mode, each input file is split in two parts: one describes the circuit net list/analysis and the other illustrates device simulation and model parameters. The circuit description includes circuit topology, electrical models, and parameters of circuit components.

The voltage transfer characteristic (VTC) of a typical inverter is depicted in Figure 6.13. Low noise margin ($NM_L$) and high noise margin ($NM_H$) are related

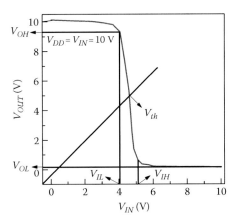

**FIGURE 6.13**   Voltage transfer characteristics of a typical inverter.

to reliable and robust operation of a logic circuit and can be obtained from different transition levels of output voltage as

$$NM_L = V_{IL} - V_{OL} \tag{6.1}$$

$$NM_H = V_{OH} - V_{IH} \tag{6.2}$$

where, $V_{OH}$ and $V_{OL}$ are maximum and minimum output voltage levels, respectively. The transition point $V_{IH}$ is defined as the minimum input voltage that can be treated as logic 1; whereas, $V_{IL}$ is the maximum input voltage to be interpreted as logic 0. In an ideal transfer curve, switching from high to low output level occurs exactly at half of the supply voltage, which shows good balancing between pull-up and pull-down inverter transistors. It can be adjusted by proper selection of transconductance ratio ($k_R$) of the driver-to-load transistor. Furthermore, the switching threshold of the inverter ($V_{th}$) can be defined as

$$V_{th} = \frac{V_{t,n} + \sqrt{\dfrac{1}{k_R}}\left(V_{DD} - |V_{t,p}|\right)}{\left(1 + \sqrt{\dfrac{1}{k_R}}\right)} \tag{6.3}$$

where, $V_{t,p}$ and $V_{t,n}$ are the threshold voltages of $p$- and $n$-type transistors, respectively.

Propagation delays, $\tau_{PLH}$ and $\tau_{PHL}$, determine input-to-output signal delay during low-to-high and high-to-low transitions of output, respectively, as

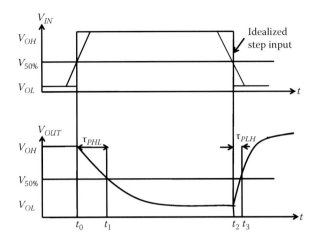

**FIGURE 6.14**    Input and output voltage waveforms of a typical inverter.

shown in Figure 6.14. These delay times, considering step input with zero rise ($\tau_r$) and fall ($\tau_f$) times, can be analytically obtained as

$$\tau_{PLH}(0) = \frac{C_{TL}}{k_p\left(V_{OH} - V_{OL} - |V_{t,p}|\right)}\left[\frac{2|V_{t,p}|}{\left(V_{OH} - V_{OL} - |V_{t,p}|\right)} + \ln\left(\frac{4\left(V_{OH} - V_{OL} - |V_{t,p}|\right)}{V_{OH} - V_{OL}} - 1\right)\right]$$

$$(6.4)$$

$$\tau_{PHL}(0) = \frac{C_{TL}}{k_n(V_{OH} - V_{t,n})}\left[\frac{2V_{t,n}}{(V_{OH} - V_{t,n})} + \ln\left(\frac{4(V_{OH} - V_{t,n})}{V_{OH} + V_{OL}} - 1\right)\right] \quad (6.5)$$

where, $\tau_{PLH}(0)$ and $\tau_{PHL}(0)$ are the propagation delays with ideal step input, $C_{TL}$ is the total load capacitance, and $k_p/k_n$ are the respective transconductance of $p/n$ transistors. However, for input pulse with some finite rise and fall times, these propagation delays can be obtained as

$$\tau_{PLH}(1) = \sqrt{\tau_{PLH}^2(0) + \left(\frac{\tau_f}{2}\right)^2} \quad (6.6)$$

$$\tau_{PHL}(1) = \sqrt{\tau_{PHL}^2(0) + \left(\frac{\tau_r}{2}\right)^2} \quad (6.7)$$

where, $\tau_{PLH}(1)$ and $\tau_{PHL}(1)$ are the actual propagation delays obtained at the step input pulse with some finite rise and fall time. The propagation delay $\tau_{PLH}$ and

$\tau_{PHL}$ can be evaluated from input and output waveforms of a typical inverter circuit by using the following expressions:

$$\tau_{PLH} = t_3 - t_2 \qquad (6.8)$$

$$\tau_{PHL} = t_1 - t_0 \qquad (6.9)$$

$$\tau_p = \frac{\tau_{PLH} + \tau_{PHL}}{2} \qquad (6.10)$$

The time $t_0$, $t_1$, $t_2$, and $t_3$ are represented in Figure 6.14. The average propagation delay time ($\tau_p$) is the time required for the input signal to propagate through the inverter circuit. Actual propagation delays take into account the charge up and charge down time of internal capacitors of the devices. For evaluating an approximate value of load capacitance, a Fan-Out-1 circuit is considered. To analyze the driving capability of each inverter, an identical inverter configuration ($P1 = P2$ and $N1 = N2$) is considered as the load and individual load capacitances are calculated for a dual-stage inverter circuit as shown in Figure 6.15.

Total load capacitance ($C_{TL}$) can be expressed as

$$C_{TL} = (C_{gd})_{P1} + (C_{gd})_{N1} + C_g \qquad (6.11)$$

where, $(C_{gd})_{P1}$ and $(C_{gd})_{N1}$ are the effective gate-drain capacitance of $p$- and $n$-type transistors, respectively, at the output terminal. $C_g$ is the gate capacitance of inverter 2 and can be expressed as

$$C_g = C_{ox}(WL)_{P2} + C_{ox}(WL)_{N2} \qquad (6.12)$$

Load capacitance for each individual device combination is summarized in Table 6.3.

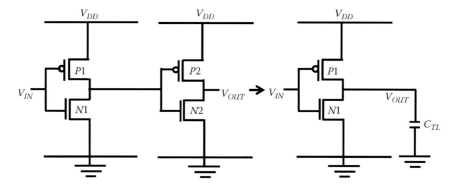

**FIGURE 6.15** Schematic of dual-stage inverter and single-stage inverter with equivalent $C_{TL}$.

**TABLE 6.3  Load Capacitance of Different Device Combinations for Inverter Circuits**

| Inverter Configuration | Device Combination | $C_{TL}$ (F) |
|---|---|---|
| Fully organic complementary | Pentacene–$C_{60}$ | $0.7 \times 10^{-9}$ |
| | CuPc–$F_{16}$CuPc | $3 \times 10^{-12}$ |
| Hybrid complementary | Pentacene–a-Si:H | $0.4 \times 10^{-9}$ |
| | Pentacene–ZnO | $1.6 \times 10^{-9}$ |
| Organic all $p$-type | Pentacene–Pentacene | $1.6 \times 10^{-9}$ |

To verify the load capacitance, simulations are performed for dual-stage inverters and single-stage inverters with $C_{TL}$ at output. Similar transient responses are obtained for both circuits that validate the analytical value of $C_{TL}$.

### 6.3.1  FULLY ORGANIC COMPLEMENTARY INVERTER CIRCUITS

In this section, fully organic complementary inverters based on two different TFT combinations are analyzed. Foremost, the circuit is designed by integrating pentacene and $C_{60}$ devices and then it is compared with the CuPc–$F_{16}$CuPc combination. These OTFT combinations are chosen because of their comparable performance. The schematic, VTC, and transient response of a pentacene–$C_{60}$ inverter are shown in Figure 6.16.

To analyze the response of an inverter circuit, input ($V_{IN}$) with high and low transition of 10 and 0 V is applied at the input terminal. In the complementary inverter, at low $V_{IN}$, the $p$-type transistor acts as a closed switch; whereas, an $n$-type transistor acts as an open switch. Hence, a $p$-device pulls up the voltage to supply voltage ($V_{DD}$) through its source. However, the case is reversed at high $V_{IN}$ and thus an $n$-device pulls down the output close to 0 V.

Considering device dimensions and load, all the inverter configurations are analyzed for an input pulse with 0.15 ms rise/fall time. It has been found sufficient to drive the load in all the cases except for a CuPc–$F_{16}$CuPc organic complementary inverter circuit. The schematic and VTC of a CuPc–$F_{16}$CuPc inverter are shown in Figure 6.17.

Analytical delay times at $W/L$ ratio of 500/90 (CuPc) and 2000/90 ($F_{16}$CuPc) are more than 30 ms due to very low mobility of both CuPc- and $F_{16}$CuPc-based OTFTs. Therefore, this OTFT combination is not responding properly to the input signal with 0.15 ms rise/fall time, as shown in Figure 6.18a. Thus, a large $W/L$ ratio is required for both transistors, which are adjusted as 500/5 and 2000/5 for CuPc and $F_{16}$CuPc transistors, respectively, for obtaining a comparable performance to other inverters, as illustrated in Figure 6.18b.

A complementary inverter circuit with pentacene and $C_{60}$ OTFTs demonstrates better characteristics in comparison to the CuPc–$F_{16}$CuPc combination due to significantly higher mobility and lower threshold voltages of devices in the

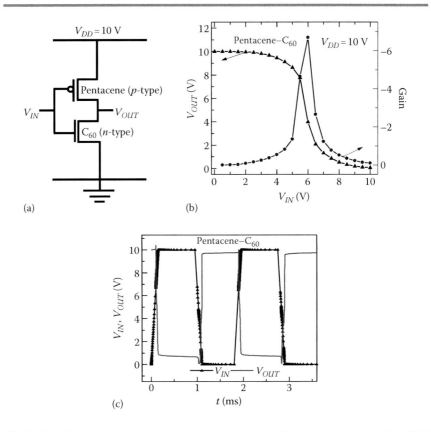

**FIGURE 6.16**    Fully organic complementary inverter (a) circuit schematic, (b) VTC/gain, and (c) transient response with pentacene and $C_{60}$ OTFTs.

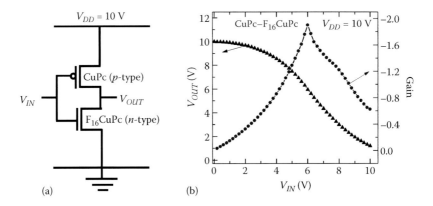

**FIGURE 6.17**    (a) Circuit schematic and (b) combined plots of VTC and gain for a $CuPc–F_{16}CuPc$ organic complementary inverter.

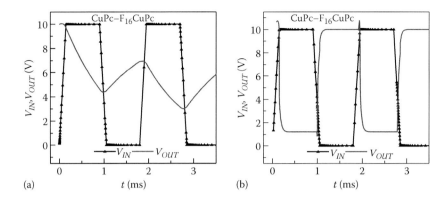

**FIGURE 6.18**   Transient response of organic complementary inverter with $W/L$ of (a) 500/90 and 2000/90, and (b) 500/5 and 2000/5 for CuPc and $F_{16}$CuPc OTFTs.

former configuration than the latter one. Critical voltages $V_{OH}$, $V_{IH}$, $V_{OL}$, and $V_{IL}$ are 8.8 (8.7), 6.3 (7.0), 0.12 (1.2), and 4.8 (4.6) V for pentacene–$C_{60}$ (CuPc–$F_{16}$CuPc) organic complementary inverters. Compared to a CuPc–$F_{16}$CuPc inverter, the voltage swing is higher by 16% for a pentacene–$C_{60}$ inverter. Similarly, the gain is also in the excess of 3.3 times. Besides this, an improvement of 39% in the propagation delay is observed for a pentacene–$C_{60}$ inverter. The mobility of an $F_{16}$CuPc OTFT is lower by 40% as compared to CuPc that, in turn, slows the pull-down action, thereby resulting in a higher delay even for higher $W/L$ ratios.

## 6.3.2  HYBRID COMPLEMENTARY INVERTER CIRCUITS

In this section, two hybrid complementary inverters are analyzed: (1) A combination of pentacene organic TFT with a-Si:H inorganic TFT, (2) a pentacene with ZnO TFT. The schematic diagram, VTC, and transient response of a pentacene–a-Si:H inverter are shown in Figure 6.19. The a-Si:H material has ambipolar characteristics and can be operated in both $p$- and $n$-channel mode, but a $p$-channel TFT shows significantly inferior characteristics as compared to its counterpart.

The $p$-type OSC exhibits comparable performance to $n$-channel a-Si:H, therefore, hybrid technology is incorporated to combine the advantages of $n$-channel a-Si:H and pentacene $p$-channel OSC. These circuits illustrate good transfer characteristics and are promising from a performance standpoint. It can be analyzed from the VTC and transient plots of two hybrid inverters that the transition from high to low output voltage is steeper for pentacene–ZnO combination as compared to pentacene–a-Si:H. It is due to a higher transconductance of ZnO transistor compared to a-Si:H, since the capacitance of the ZnO transistor is 16 times higher. Therefore, the $\tau_{PHL}$ of pentacene–ZnO is half of that of its counterpart, regardless of the same $W/L$ ratios for $p$- and $n$-type TFTs in both hybrid inverters. The schematic diagram, VTC/gain, and transient response of a pentacene–ZnO inverter are shown in Figure 6.20.

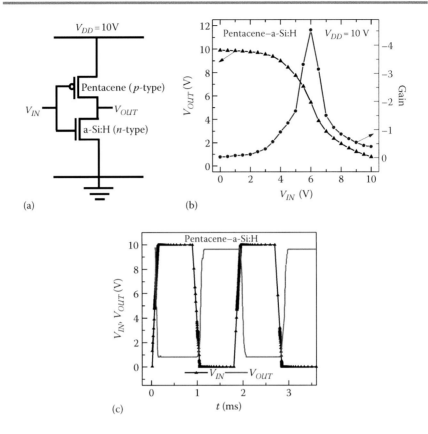

**FIGURE 6.19** Hybrid complementary inverter (a) circuit schematic, (b) VTC/gain, and (c) transient response for the pentacene–a-Si:H combination.

Critical voltages for a pentacene–ZnO (pentacene–a-Si:H) inverter are $V_{OH}$ = 9.3 (9.0), $V_{IH}$ = 3.8 (6.7), $V_{OL}$ = 0.02 (0.9), and $V_{IL}$ = 4.5 (4.0) V. Due to the large pull-down current of ZnO TFT, an increment of 15% in voltage swing and 29% in propagation delay is obtained, as compared to the pentacene–a-Si:H combination. Besides this, the gain is also improved by 2 times for a pentacene–ZnO hybrid inverter. In addition, the $NM_H$ and $NM_L$ are also 2.4 and 1.5 times higher for a pentacene and ZnO TFTs hybrid inverter in comparison to an inverter based on pentacene and a-Si:H TFT.

### 6.3.3  ALL $p$-TYPE ORGANIC INVERTER CIRCUIT

In this section, an all $p$-organic inverter configuration is designed using only a pentacene device for both the driver and load component. An all $p$-type inverter, shown in Figure 6.21a, is realized using zero-$V_{gs}$ load logic (ZVLL); wherein, the gate of load transistor is connected to its source, therefore it maintains a

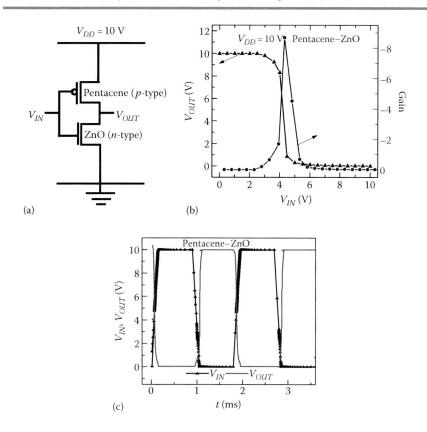

**FIGURE 6.20** Hybrid organic complementary inverter (a) circuit schematic, (b) VTC/gain, and (c) transient response of the pentacene–ZnO combination.

constant drain current through the load transistor. In ZVLL configuration, the load transistor, which is acting as a current source, would either be completely *off* or operate in the subthreshold regime. In both the cases, it only produces very low current. At low $V_{IN}$, the channel resistance of driver appears much lower than the load, since it turns into an *on* state. Therefore, $V_{OUT}$ is obtained closer to the $V_{DD}$ by means of a resistive divider between the driver and the load transistors. At high input, the driver has zero-$V_{gs}$ and thus a larger $W/L$ ratio is needed to increase the pull-down action. Therefore, the width of load transistor (2000 µm) is kept 4 times higher than the driver (500 µm).

The static and dynamic response of a $p$-type organic inverter is demonstrated in Figure 6.21b and c, respectively. The extracted static performance parameters include $V_{OH} = 8.8$, $V_{IH} = 7.4$, $V_{OL} = 0.5$, and $V_{IL} = 6.0$ V. The $V_{OL}$ approaches 0 V, but the $V_{OH}$ does not attain transition up to $V_{DD}$ due to the threshold voltage drop in the transistors. The noise margins $NM_H$ and $NM_L$ are 1.4 and 5.5, respectively. The $NM_L$ is reasonably high and close to the desired value, $V_{DD}/2$, but the $NM_H$ is quite low due to small pull-down current of the

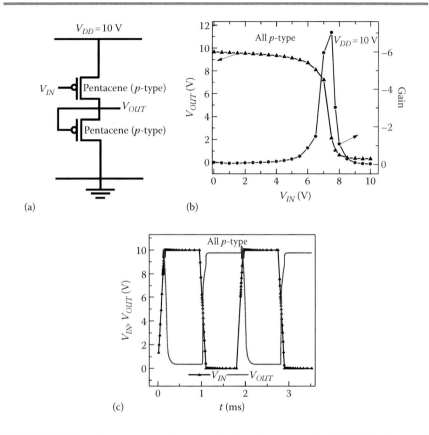

**FIGURE 6.21** All *p*-organic inverter (a) circuit schematic, (b) VTC/gain, and (c) transient response with pentacene *p*-type devices only.

load. This in turn increases the magnitude of $V_{IH}$, thereby resulting in a low $NM_H$ and high $V_{th}$ (6.8 V) for an all-*p* organic inverter. Furthermore, the voltage swing, gain, and propagation delay are 8.3 V, –7.0, and 92 μS, respectively, for an all-*p* organic inverter.

## 6.4 RESULTS AND DISCUSSION

This section summarizes and compares the performance of all proposed inverter circuits based on different organic/inorganic TFTs. Table 6.4 includes the threshold voltages of different inverters, obtained analytically (Equation 6.3) and through simulation that show a close match with an average error of 1.3%. The $V_{th}$ of pentacene–$C_{60}$ and pentacene–ZnO is the closest to $V_{DD}/2$. It is higher by 0.7 V in the former combination, whereas it is lower by 0.7 V in the later one. It is due to more than 3 times higher capacitance of a ZnO-TFT in comparison to a $C_{60}$-TFT, thereby resulting in a high-to-low transition before

**TABLE 6.4   Comparison of Simulated and Analytical $V_{th}$ of Different Inverter Circuits**

| Inverter Configuration | Device Combination | W/L | $V_{th}$ (V) Simulated | $V_{th}$ (V) Analytical |
|---|---|---|---|---|
| Fully organic complementary | $p$: Pentacene | 500/90 | 5.7 | 5.5 |
|  | $n$: C$_{60}$ | 2000/90 |  |  |
|  | $p$: CuPc | 500/5 | 6.0 | 6.2 |
|  | $n$: F$_{16}$CuPc | 2000/5 |  |  |
| Hybrid complementary | $p$: Pentacene | 500/90 | 5.9 | 6.2 |
|  | $n$: a-Si:H | 2000/90 |  |  |
|  | $p$: Pentacene | 500/90 | 4.3 | 4.0 |
|  | $n$: ZnO | 2000/90 |  |  |
| Organic all $p$-type | Pentacene-1 | 500/90 | 6.8 | 6.5 |
|  | Pentacene-2 | 2000/90 |  |  |

$V_{DD}/2$ due to a stronger pull-down operation, comparatively. On the other hand, in an all-$p$ organic inverter, the $V_{th}$ is higher by 36% from the desired value due to zero-$V_{gs}$ of load leading to a slow pull-down action.

The static performance of different inverter circuits mainly in terms of critical voltages, noise margin, and voltage swing and gain is summarized in Table 6.5. The output voltage logic level conservation is better for the inverter circuits based on pentacene–C$_{60}$ and pentacene–ZnO when compared to the other materials combinations. The noise margin is highest for the pentacene–ZnO inverter in terms of both $NM_H$ and $NM_L$. However, for other inverters such as pentacene–C$_{60}$, CuPc–F$_{16}$CuPc, and pentacene–a-Si:H, the magnitude of $NM_H$ is quite low due to a significantly lower capacitance of their respective $n$-type devices in comparison to a ZnO-based device. Besides this, the $NM_H$ is also inferior in all-$p$ organic inverters due to zero-$V_{gs}$ of the load transistor. The gain of the pentacene–ZnO inverter is enhanced by 1.3, 1.2, 1.9, and 4.3 times than the pentacene–C$_{60}$, all-$p$, pentacene–a-Si:H, and CuPc–F$_{16}$CuPc inverters, respectively.

**TABLE 6.5   Static Performance of Fully Organic Complementary, Hybrid Complementary, and All $p$-Organic Inverters**

| Device Combination | Output Voltages (V) $V_{OH}$ | $V_{IH}$ | $V_{OL}$ | $V_{IL}$ | Noise Margin (V) $NM_H$ | $NM_L$ | Voltage Swing (V) | Gain |
|---|---|---|---|---|---|---|---|---|
| Pentacene–C$_{60}$ | 8.8 | 6.3 | 0.12 | 4.8 | 2.5 | 4.7 | 8.7 | −6.7 |
| CuPc–F$_{16}$CuPc | 8.7 | 7.0 | 1.2 | 4.6 | 1.7 | 3.4 | 7.5 | −2.0 |
| Pentacene–a-Si:H | 9.0 | 6.7 | 0.9 | 4.0 | 2.3 | 3.1 | 8.1 | −4.5 |
| Pentacene–ZnO | 9.3 | 3.8 | 0.02 | 4.5 | 5.5 | 4.5 | 9.3 | −8.6 |
| All Pentacene | 8.8 | 7.4 | 0.5 | 6.0 | 1.4 | 5.5 | 8.3 | −7.0 |

**TABLE 6.6  Comparison of Simulated and Analytical Delay of Different Inverter Circuits**

| | Propagation Delay (μs) | | | | | |
| | Simulated | | | Analytical | | |
| **Device Combination** | $\tau_{PLH}(1)$ | $\tau_{PHL}(1)$ | $\tau_p$ | $\tau_{PLH}(1)$ | $\tau_{PHL}(1)$ | $\tau_p$ |
|---|---|---|---|---|---|---|
| Pentacene–$C_{60}$ | 59 | 65 | 62 | 71 | 75 | 73 |
| CuPc–$F_{16}$CuPc | 80 | 121 | 101 | 75 | 111 | 93 |
| Pentacene–a-Si:H | 111 | 120 | 116 | 120 | 127 | 124 |
| Pentacene–ZnO | 116 | 50 | 83 | 105 | 55 | 80 |
| All Pentacene | 64 | 120 | 92 | 79 | 108 | 94 |

Table 6.6 shows a comparison between the simulated and analytically obtained delay of different inverter circuits. A reasonable match is obtained that yielded an average error of 3% for average propagation delay. The CuPc–$F_{16}$CuPc combination shows a 63%, 22%, and 10% higher delay than the pentacene–$C_{60}$, pentacene–ZnO, and all-$p$ inverters, respectively. The speed of the CuPc–$F_{16}$CuPc inverter is low, even at 18 times higher $W/L$ ratio for both $p$- and $n$-type devices as compared to the devices in other inverter configurations. It is due to the lower mobility and higher threshold voltage of CuPc and $F_{16}$CuPc organic devices. On the other hand, the pentacene–$C_{60}$ combination shows the lowest delay with approximately equal $\tau_{PLH}$ and $\tau_{PHL}$, which demonstrates a good balance between pull-up and pull-down action performed by $p$- and $n$-type OTFTs, respectively.

## 6.5  CONCLUDING REMARKS

The analysis and comparison of static and dynamic responses of fully organic complementary, hybrid complementary, and all $p$-organic inverter circuits emphasize the effect of different materials and configurations. The performance of inverter circuits is explored mainly in terms of voltage swing, noise margin, gain, switching threshold voltage, and propagation delay. Individual devices based on pentacene, CuPc, $F_{16}$CuPc, $C_{60}$, a-Si:H, and ZnO semiconductors were discussed for the optimal combination of driver and load transistors. The observations show a reasonable match between the simulation and the experimental results with an average error of 1.8% and 4% for mobility and threshold voltage, respectively.

Circuit level analysis of pentacene–$C_{60}$ and pentacene–ZnO TFT-based inverters present the switching threshold voltage near to the desired value of $V_{DD}/2$. On the contrary, an all-$p$ organic inverter results in a higher switching threshold because gate and source terminals of the load transistor are tied together, leading to a slow pull-down action. The critical output voltage logic levels are reasonable for the inverters; wherein, the pentacene transistor is used with $C_{60}$- and ZnO-based transistors due to their comparable mobility and threshold voltages. The noise margins $NM_H$ and $NM_L$ of the pentacene–ZnO inverter are approximately equal to $V_{DD}/2$, which explains the better functioning of the inverter. Besides this, pentacene–$C_{60}$ and all-$p$ organic inverters exhibit a satisfactory $NM_L$, however, the $NM_H$ is quite inferior due to a lower strength of the pull-down transistor in both configurations.

The achieved gain for the pentacene–ZnO inverter is highest, lower for all-$p$ and pentacene–$C_{60}$ based inverters, and worst for pentacene–a-Si:H and CuPc–$F_{16}$CuPc TFT inverters. The poor gain of a-Si:H and $F_{16}$CuPc inverters is due to a lower capacitance and larger threshold voltage. As compared to the pentacene–ZnO, all-$p$, pentacene–a-Si:H, and CuPc–$F_{16}$CuPc inverters, faster switching speed is observed for the pentacene–$C_{60}$ inverter due to good balance between the pull-up and the pull-down operations in the inverter. These observations illustrate an overall good performance for pentacene–ZnO and pentacene–$C_{60}$ combinations compared to other inverter combinations making them useful for designing faster, low-cost and flexible organic integrated circuits and systems.

## PROBLEMS
## MULTIPLE CHOICE

1. A hybrid complementary circuit consists of
   a. Both $p$-type and $n$-type organic TFTs
   b. $p$-Type organic and $n$-type inorganic TFTs
   c. Both $p$-type and $n$-type inorganic TFTs
   d. None of these
2. Which is the most widely used $p$-type organic semiconductor material?
   a. CuPc
   b. Pentacene
   c. P3HT
   d. $C_{60}$
3. The commonly used energy band gap value of pentacene is
   a. 1.0 eV
   b. 5 eV
   c. 2.2 eV
   d. None of these
4. Which $n$-type material has the highest mobility?
   a. $F_{16}$CuPC
   b. $C_{60}$
   c. TCNQ
   d. PCBM
5. Among all $p/n$ devices, the capacitance is highest for which organic material-based OTFT?
   a. Pentacene
   b. CuPc

    c.  ZnO
    d.  Both a and c
6.  Which of the following inverter types has the highest threshold voltage?
    a.  Fully organic complementary
    b.  Hybrid complementary
    c.  All $p$-type organic
    d.  None of these
7.  Noise margin is highest in
    a.  Pentacene–ZnO
    b.  Pentacene–$C_{60}$
    c.  Pentacene–a-Si:H
    d.  All pentacene
8.  Which inverter combination has the lowest speed?
    a.  Pentacene–$C_{60}$
    b.  Pentacene–a-Si:H
    c.  Pentacene–ZnO
    d.  All pentacene
9.  Which inverter combination shows highest switching speed?
    a.  Pentacene–ZnO
    b.  All pentacene
    c.  Pentacene–a-Si:H
    d.  Pentacene–$C_{60}$
10. Gain of which inverter combination is the highest?
    a.  Pentacene–$C_{60}$
    b.  Pentacene–a-Si:H
    c.  Pentacene–ZnO
    d.  All Pentacene

**ANSWER KEY**

1. b; 2. b; 3. c; 4. b; 5. d; 6. c; 7. a; 8. b; 9. d; 10. c

**SHORT ANSWER**

1.  What are the imperative characteristics of a complementary inverter?
2.  Why is an all $p$-type organic inverter circuit proposed and what are the challenges experienced?
3.  Describe the electrical characteristics and operational parameters of individual $p$- and $n$-type organic and inorganic TFTs suitable for organic inverter designs.
4.  What are the advantages of a pentacene-based $p$-type OTFT over CuPc?
5.  Suggest important factors for the performance improvement of $n$-type OTFTs.
6.  Explain the types and behavior of hybrid inverter circuits.
7.  Briefly analyze the response of fully organic complementary inverter circuits.

8. Describe the extraction of performance parameters of complementary inverters.
9. Explain the static and dynamic response of an all $p$-type organic inverter.
10. Compare the performance of all the inverter circuits discussed in the chapter.
11. Describe the static and dynamic behavior of the following organic inverter circuits:
    a.  Fully organic inverter circuit
    b.  All $p$-type organic inverter circuit

## EXERCISES

1. Design and simulate, a $p$-type single gate organic thin-film transistor using a TCAD tool in bottom gate top contact (BGTC) structure with pentacene as an OSC material of the dimension $(t_{osc})$ 50 nm and the device parameters are given as: channel length $(L)$ = 90 μm, channel width $(W)$ = 500 μm, insulator thickness $(t_{ox})$ = 10 + 2.5 nm $(Al_2O_3 + SiO_2)$, thickness of source/drain $(t_s/t_d)$ = 10 nm (Gold), and gate thickness $(t_g)$ of 10 nm $(n^+si)$. Based on the Poole and Frenkel mobility model plot, the transfer curve and output curve of the device, for the transfer curve $V_{ds}$ is at −3 V and $V_{gs}$ varies from 0 to −3 V. Whereas for the output curve the $V_{gs}$ and $V_{ds}$ varies from 0 to −3 V. Based on these simulations calculate the performance parameters for the device such as threshold voltage, subthreshold slope, transconductance, mobility, drive current, and current *on/off* ratio.

2. Design and simulate an $n$-type single gate organic thin-film transistor, using a TCAD tool in bottom gate top contact (BGTC) structure with $C_{60}$ as an OSC material of the dimension $(t_{osc})$ 60 nm and the device parameters are channel length $(L)$ = 100 μm, channel width $(W)$ = 1000 μm, insulator thickness $(t_{ox})$ = 8 nm + 132 nm + 4 nm $(SiO_2 + TiSiO_2 + SiO_2)$, thickness of source/drain $(t_s/t_d)$ = 1 + 100 nm (LiF + Al), and gate thickness $(t_g)$ = 30 nm $(p$-si$)$.

    Based on the Poole and Frenkel mobility model plot the transfer curve and output characteristics of the device, for the transfer curve $V_{ds}$ is at 5 V and $V_{gs}$ varies from 0 to 5 V. Whereas for the output curve the $V_{gs}$ and $V_{ds}$ varies from 0 to 5 V. Based on these simulations calculate the performance parameters for the device such as threshold voltage, sub-threshold slope, transconductance, mobility, drive current and current *on/off* ratio.

3. Design and simulate a fully organic complementary inverter circuit using the devices from Q 1 and 2. The aspect ratio for $p$-type device is $W/L$ = 500/90 and for $n$-type device $W/L$ = 2000/90. The supply voltage is $V_{IN} = V_{DD}$ = 10 V. Estimate the low and high noise margin ($NM_L$ and $NM_H$).

4. Design an all $p$-type inverter circuit using the device in exercise 1. The aspect ratio for the $p$-type driver device is $W/L = 500/90$ and for $p$-type load is 2000/90. The supply voltage $V_{DD}$ is 10 V. Thus, calculate the low and high noise margin ($NM_L$ and $NM_H$).

5. Design and simulate a hybrid inverter circuit with the $p$-type device from exercise 1. The dimensions of the ZnO $n$-type device are given as thickness of OSC ($t_{osc}$) = 60 nm, channel length ($L$) = 90 μm, channel width ($W$) = 500 μm, bilayer insulator thickness ($t_{ox}$) = 12.5 nm (10 + 2.5 nm) ($SiO_2$ + $Al_2O_3$), thickness of source/drain ($t_s/t_d$) = 10 nm (Al), and gate electrode thickness ($t_g$) = 10 nm ($p$-Si). The aspect ratio for the $p$-type driver transistor is $W/L = 500/90$ and for the $n$-type load transistor is $W/L = 2000/90$. The supply voltage is $V_{DD}$ = 10 V. Calculate the low and high noise margin ($NM_L$ and $NM_H$).

6. Consider a hybrid inverter using a $p$-type pentacene transistor and $n$-type ZnO transistor with $(W/L)_n = (W/L)_p = 500/90$, $C_{ox,p} = C_{ox,n} = 350$ nF/cm$^2$, $\mu_p = 1.0$ cm$^2$/Vs, $\mu_n = 1.06$ cm$^2$/Vs, $V_{t,p} = -0.65$ V and $V_{t,n} = 1.3$ V, $V_{OH} = 9.7$ V, $V_{OL} = 0.1$ V, $V_{IH} = 5.0$ V, $V_{IL} = 3.0$ V. Consider $C_{T,L} = 1.6$ nF. Calculate $NM_H$, $NM_L$ and propagation delay ($\tau_{PHL}$ and $\tau_{PLH}$).

## REFERENCES

1. Raval, H. N.; Tiwari, S. P.; Navan, R. R.; Mhaisalkar, S. G.; Rao, V. R. "Solution processed bootstrapped organic inverters based on P3HT with a high-k gate dielectric material," *IEEE Electron Device Lett.* **2009**, 30(5), 484–486.
2. Resendiz, L.; Estrada, M.; Cerdeira, A.; Cabrera, V. "Analysis of the performance of an inverter circuit: Varying the thickness of the active layer in polymer thin film transistors with circuit simulation," *Jap. J. Appl. Phys.* **2012**, 51(4S), 04DK04-1–04DK04-6.
3. Dodabalapur, A.; Baumbach, J.; Baldwin, K.; Katz, H. E. "Hybrid organic/inorganic complementary circuits," *Appl. Phys. Lett.* **1996**, 68(16), 2246–2248.
4. Oh, M. S.; Hwang, D. K.; Lee, K.; Choi, W. J.; Kim, J. H.; Im, S.; Lee, S. "Pentacene and ZnO hybrid channels for complementary thin film transistor inverters operating at 2V," *J. Appl. Phys.* **2007**, 102(7), 076104–076104-3.
5. Neon, S.; Kanehira, D.; Yoshomoto, N.; Fages, F.; Ackermann, C. V. "Shelf-life time test of $p$ and $n$-channel organic thin film transistors using copper phthalocyanines," *Thin Solid Films* **2010**, 518(19), 5593–5598.
6. Na, J. H.; Kitamura, M.; Arakawa, Y. "High performance $n$-channel thin film transistors with an amorphous phase $C_{60}$ film on plastic substrate," *Appl. Phys. Lett.* **2007**, 91(19), 193501-1–193501-3.
7. Huang, Y.; Hekmatshoar, B.; Wanger, S.; Sturm, J. C. "Top-gate amorphous silicon TFT with self-aligned silicide source/drain and high mobility," *IEEE Electron Device Lett.* **2008**, 29(7), 737–739.
8. Gupta, D.; Katiyar, M.; Gupta, D. "An analysis of the difference in behavior of top and bottom contact organic thin film transistors using device simulation," *Org. Electron.* **2009**, 10(5), 775–784.
9. Shim, C. H.; Maruoka, F.; Hattori, R. "Structural analysis on organic thin film transistor with device simulation," *IEEE Trans. Electron Devices* **2010**, 57(1), 195–200.

10. Li, C.; Pan, F.; Wang, X.; Wang, L.; Wang, H.; Wang, H.; Yan, D. "Effect of the work function of gate electrode on hysteresis characteristics of organic thin-film transistors with $Ta_2O_5$/polymer as gate insulator," *Org. Electron.* **2009**, 10(5), 948–953.
11. Faiman, D.; Goren, S.; Katz, E. A.; Koltun, M.; Melnik, N.; Shames, A. "Structure and optical properties of $C_{60}$ thin films," *Thin Solid Films* **1997**, 295(1–2), 283–286.

# Robust Organic Inverters and NAND/NOR Logic Circuits Based on Single and Dual Gate OTFTs

**7**

## 7.1 INTRODUCTION

Integrated circuits are the backbone of the modern very large-scale integration (VLSI) system design that builds up mainly using different combinations of logic gates. The performance of conventional semiconductor-based logic circuits is unquestionably superior; however, the production cost and flexibility are obvious constraints. Organic logic circuits provide an ideal solution being inexpensive, flexible, and promising enough to achieve high performance. Inverter and universal logic gates are essentially required in several organic thin-film transistor (OTFT) applications, but only few logic designs based on organic TFTs have been reported to date [1–3].

To produce a high-performance logic circuit, a complementary design is beneficial. Nevertheless, for such designs both *p*- and *n*-type transistors should exhibit similar performance. It was previously discussed that *n*-type organic materials exhibit inferior performance compared to their *p*-type counterparts. This is due to a large barrier height in *n*-type OTFTs, as most of the commonly used contact metals exhibit work functions better suited to inject the holes into HOMO (highest occupied molecular orbital) than the electrons into LUMO (lowest unoccupied molecular orbital). Therefore, the majority of organic circuits make use of $p$-type designs only. Within available $p$-type materials, pentacene is the material of choice because of its higher mobility and good chemical stability compared to other organic small molecules [4].

Dual gate (DG) transistors are often preferred in silicon technology due to their tremendous superiority to single gate transistors. Similar to the improvements noticed in silicon, the characteristics of organic transistors can also be improved using a dual gate mechanism. Compared to a single gate (SG) transistor, the DG organic transistor exhibits improved performance due to charge carrier modulation [5] with biasing of both gates. Moreover, an increase in gate oxide charge density also helps in improving transistor performance. These transistors exhibit numerous advantages such as higher mobility, higher *on/off* current ratio, higher transconductance, steeper subthreshold slope, and, most important, a better control on the threshold voltage in comparison to the single gate. The characteristics of organic transistors can certainly be improved using double gate, thereby improving the performance of digital circuits as well.

This chapter analyzes and compares the performance of SG and DG OTFTs based all-$p$ inverter circuits in DLL (diode load logic) and ZVLL (zero-$V_{gs}$ load logic) configurations. Higher performance is observed for the ZVLL configuration in terms of voltage swing, noise margin, and gain; however, there is a trade-off in terms of speed. Besides this, back-gate biasing and bootstrapping techniques improve the performance of DG inverter circuits for their robust low-cost digital VLSI circuit applications.

This chapter is arranged in six sections along with the current introductory Section 7.1. Thereafter, Section 7.2 analyzes and compares the performance of pentacene-based SG and DG devices, and in Section 7.3, different load configurations for all-$p$ inverters are discussed. Section 7.4 analyzes the performance of all $p$-type organic inverter circuits in DLL and ZVLL configurations. Section 7.5 illustrates the performance of DG device-based inverters in DLL and ZVLL configurations using back-gate biasing and bootstrapping techniques. Organic NAND and NOR logic gate implantations are described in Sections 7.6 and 7.7, respectively. Finally, the major conclusions are carried out in Section 7.8.

## 7.2 SINGLE AND DUAL GATE DEVICE ANALYSIS

The single gate OTFT shown in Figure 7.1a consists of a thin film of organic semiconductor (OSC), usually fabricated as an inverted structure with a gate at the bottom. An additional dielectric and gate contact forming DG configuration improves the performance but at an extra fabrication cost. However, compared to SG OTFTs, the DG organic TFTs demonstrate various merits that include improvement in $\mu$, $I_{on}/I_{off}$, and $g_m$. Besides this, it exhibits steeper

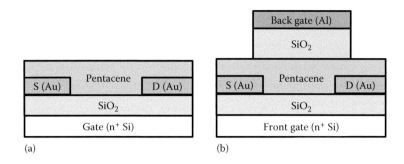

**FIGURE 7.1**    OTFT schematics of (a) SG and (b) DG devices.

subthreshold slope (*SS*) and better controlling on $V_t$. Although, $V_t$ determines the switching behavior of any device and is dependent on doping concentration, an external bias control on $V_t$ itself can bring about highly controlled operation of the device [6]. The DG OTFT shown in Figure 7.1b consists of an additional gate and dielectric at the top, while other thin films such as OSC, dielectric, G, and S/D metal contacts are similar to an SG transistor.

The back (or top) gate electrostatically modifies the charge carrier distribution in the channel, accumulated by the front (or bottom) gate. The front gate strongly drives the transistor, whereas the back gate weakly couples to the channel and linearly shifts the $V_t$ [2]. The back gate also works as a passivation layer that limits performance degradation under environmental conditions [7]. Based on the bias conditions, a DG transistor can be operated in front, back, and dual gate modes. In front gate mode, a voltage at the front gate is applied, while the back gate is kept at ground potential; however, the biasing is reversed in back gate mode. In single gate bias mode, the second gate has no effect on conduction, whereas both the gates play a vital role in accumulating the charge at the OSC–dielectric interface in dual gate mode. Total charge ($Q_{Total}$) produced by both the gates in dual gate mode is expressed as [8]

$$Q_{Total} = C_{ox,F} \, V_F + C_{ox,B} \, V_B \qquad (7.1)$$

where, $C_{ox,F}$, $V_F$ and $C_{ox,B}$, $V_B$ are the oxide capacitance and voltage of the front- and back-gate electrodes, respectively. The current in DG OTFT at a given $V_{ds}$ is determined by interplay between the biases at two gate electrodes. The electrical characteristics of both single and dual gate devices are investigated at identical dimensions as summarized in Table 7.1.

The SG and DG devices consist of $W$ and $L$ of 800 and 25 µm, respectively. The front- and back-gate electrodes are of heavily doped silicon and aluminum, respectively, each with a thickness of 150 nm. Furthermore, the $SiO_2$ of thicknesses 100 and 300 nm is used as the front- and back-gate dielectrics, respectively. The thicknesses of the pentacene semiconducting layer and gold S/D contacts are 200 and 80 nm, respectively. The $I_{ds}$–$V_{ds}$ characteristics of the SG

**TABLE 7.1    Structural Dimensions and Materials of Single and Dual Gate Devices**

| Device Parameter | Thickness | Material |
| --- | --- | --- |
| Front-gate electrode, $t_{g,F}$ | 150 nm | Silicon |
| Back-gate electrode, $t_{g,B}$ | 150 nm | Aluminum |
| Front-gate dielectric, $t_{ox,F}$ | 100 nm | $SiO_2$ |
| Back-gate dielectric, $t_{ox,B}$ | 300 nm | $SiO_2$ |
| Semiconductor, $t_{osc}$ | 200 nm | Pentacene |
| S/D contact, $t_s/t_d$ | 80 nm | Gold |

*Source:* Kim, J. B.; Hernandez, C. F.; Hwang, D. K.; Tiwari, S. P.; Potscavage Jr., W. J.; Kippelen, B., "Vertically stacked complementary inverters with solution processed organic semiconductors," *Org. Electron.* 2011, 12(7), 1132–1136.

transistor are shown in Figure 7.2a that exhibits a significant slope in saturation region, whereas it demonstrates almost constant saturation current in the DG organic transistor as illustrated in Figure 7.2b. The *on*-current of a DG device is almost double as compared to its counterpart at $V_{ds}$ and $V_{gs}$ of –10 V. The $I_{ds}$–$V_{gs}$ characteristics of SG and DG OTFTs are drawn in Figure 7.3.

In comparison to DG OTFT, the potential drop within the channel increases from S to D more steeply in SG, as shown in Figure 7.4a. The threshold voltage of a DG transistor is controlled by applying back-gate bias. A positive biasing at the back gate partially depletes the front accumulation channel that is compensated by an equal shift in front-gate biasing [1]. On the other hand, a negative back-gate bias produces an additional current by introducing a second channel at the top interface. The characteristics curves of $I_{ds}$ with respect to front-gate voltage ($V_{FG}$) are presented in Figure 7.4b with the change of biasing at the back gate ($V_{BG}$) in the step size of 20 V. The $V_t$ shifts toward more negative when the

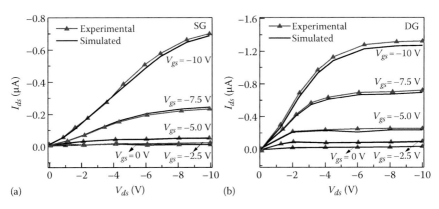

**FIGURE 7.2**    $I_{ds}$–$V_{ds}$ characteristics of (a) SG and (b) DG OTFTs as a function of $V_{gs}$.

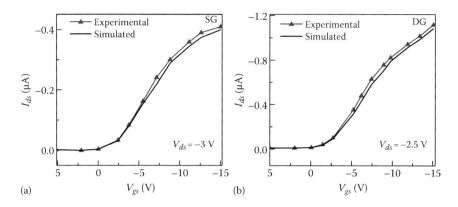

**FIGURE 7.3**  $I_{ds}-V_{gs}$ characteristics of (a) SG and (b) DG OTFTs.

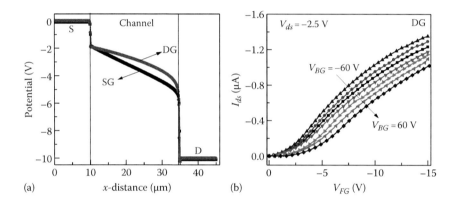

**FIGURE 7.4**    (a) Potential drop in SG and DG devices at the OSC–front dielectric interface and (b) $I_{ds}-V_{gs}$ characteristics of DG transistor with $V_{BG}$ swept from –60 to 60 V.

$V_{BG}$ is swept from negative to positive. This changes the transistor operation from depletion mode to enhancement mode. Thus, a large negative voltage is required at the back gate as compared to zero bias to accumulate the holes at the front gate.

The electrical characteristics and performance parameters obtained by simulation are verified with reported experimental results [6] for both SG and DG OTFTs. The drain current at $V_{ds} = V_{gs} = -10$ V is 0.7 µA for SG and 1.3 µA for DG that reasonably matches to the experimental current with an error of 2.8% and 3.5%, respectively. Similarly, other parameters such as $V_t$, $\mu$, SS, $I_{on}/I_{off}$ and $g_m$ are closely matched to the experimental results with an average error of 1%, 9%, 8%, 5%, and 1%, respectively, as summarized in Table 7.2. The mobility of the DG device is 5 times higher than the SG. Besides this, $I_{on}/I_{off}$ is 42% higher and SS is 41% lower proving a lower trap density in a DG device compared to

**TABLE 7.2    Simulated and Experimental Parameters of SG and DG OTFTs**

| | SG OTFT | | DG OTFT | |
|---|---|---|---|---|
| Parameters | Simulated | Experimental[a] | Simulated | Experimental[a] |
| $\mu$ (cm²/Vs) | 0.018 | 0.02 | 0.092 | 0.1 |
| $V_t$ (V) | −2.1 | −2.0 | −2.13 | −2.2 |
| SS (V/dec) | 2.3 | 2.0 | 1.35 | 1.3 |
| $I_{on}/I_{off}$ | $2.8 \times 10^3$ | $3.2 \times 10^3$ | $3.97 \times 10^3$ | $3.8 \times 10^3$ |
| $g_m$ (μS) | 0.040 | 0.044 | 0.13 | 0.12 |

[a]  Data from Kim, J. B.; Hernandez, C. F.; Hwang, D. K.; Tiwari, S. P.; Potscavage Jr., W. J.; Kippelen, B., "Vertically stacked complementary inverters with solution processed organic semiconductors," *Org. Electron.* 2011, 12(7), 1132–1136.

SG. As per transfer characteristics, the channel transconductance $\partial I_{ds}/\partial V_{gs}$ (at constant $V_{ds}$) is higher than the channel conductance $\partial I_{ds}/\partial V_{ds}$ (at constant $V_{gs}$) for both SG and DG OTFTs, which is the prerequisite for voltage amplification in a logic circuit [9].

It is well known that the mobility of OSCs varies in accordance with mobility enhancement factor, $\alpha$, that further depends on trap density, grain size of OSC, and subthreshold slope of the device [10]. Parameter $\alpha$ is associated with the conduction mechanism of the device [11,12]. It depends on doping density and dielectric permittivity of OSC material. The DG current in the accumulation region $I_{ds}^{acc}$ can be expressed in terms of alpha power law function as [13–15]

$$I_{ds}^{acc} = -\frac{\mu_0}{(2+\alpha)} \frac{W}{L} C_{ox} \left[ \left( \| V_t - V_{gs} \| \right)^{2+\alpha} - \left( \| -V_{gs} + V_t + V_{ds} \| \right)^{2+\alpha} \right] \qquad (7.2)$$

Furthermore, in the depletion region, the *off*-current in terms of *off*-resistance, $R_{off}$, is

$$I_{ds}^{off} = \frac{V_{ds}}{R_{off}} \qquad (7.3)$$

In the subthreshold region, the drain current, $I_{ds}^{st}$, is expected to increase exponentially with $V_{gs}$ and can be expressed as [15]

$$I_{ds}^{st} = \Delta \exp \left[ \frac{\ln 10}{ISS} \left( \| V_{gs} - V_t \| \right) \right] \qquad (7.4)$$

where, *ISS* is the inverse subthreshold slope and $\Delta$ is an arbitrary constant. The subthreshold operation of OTFT is closely related to the mobility enhancement for carrier hopping [14]. The parameter $\alpha$ is a result of an exponential trap distribution that corresponds to the subthreshold slope. Quantitatively, these trap states ($N_{it}$) can be evaluated as [16]

$$N_{it} = \left[ \frac{SS\,(\log e)q}{k_B T} - 1 \right] \frac{C_{ox}}{q} \qquad (7.5)$$

Trap states are calculated as $8.1 \times 10^{12}$ and $3.7 \times 10^{12}$ cm$^{-2}$V$^{-1}$ for single and dual gate devices, respectively. Higher trap density results in lower mobility and corresponding decrease in performance. The gate field enhancement of the mobility can be explained with a charge carrier density dependent mobility, assuming trap filling and operation of the OTFT in the ohmic region. To extract the value of $\alpha$, a function $f_r$ is used. It is the ratio of the integral of the drain current over the gate bias and the drain current as [14]

$$f_r = \frac{\displaystyle\int_{V_t}^{V_{gs}} I_{ds}^{acc} V_{gs}\, dV_{gs}}{I_{ds}^{acc}} = \frac{V_{gs} - V_t}{\alpha + 2}; \quad V_{ds} < V_{gs} - V_t \qquad (7.6)$$

The ratio $f_r$ holds a linear relationship with gate overdrive voltage, $V_{gs} - V_t$; therefore, the plot of $f_r$ versus $V_{gs}$ is obtained as a straight line. Parameter $\alpha$ can be extracted from its slope, while $V_t$ is the $x$-intersection of its abscissa.

The parameter $\alpha$ is obtained as 0.053 and 0.021 for SG and DG transistors as depicted in Figure 7.5. A higher value of $\alpha$ for an SG device corresponds to the higher trap density [17]. The researchers [16,18] have demonstrated deviation from the ideal behavior of TFT, due to variable grain size and discontinuity in the active layer. It leads to accumulation of defects and increase in trap states

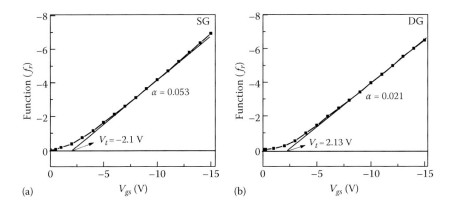

**FIGURE 7.5**   Characteristic plots of function $f_r$ versus $V_{gs}$ for (a) SG and (b) DG transistors, where $f_r$ is obtained by numerical integration of $I_{ds}$ over $V_{gs}$, divided by $I_{ds}$.

that affect not only contact resistance but also mobility, and results in reduction of TFT reliability.

## 7.3  DIODE LOAD LOGIC (DLL) AND ZERO-$V_{gs}$ LOAD LOGIC (ZVLL) CONFIGURATIONS

The performance of a logic circuit is characterized mainly in terms of voltage swing, noise margin, gain, and propagation delay. An all-$p$ type organic circuit does not provide a full swing at the output due to large threshold voltage of the organic transistors. This in turn reduces the noise margin and gain, thereby making it less robust. In order to obtain a sufficiently better performance, a zero-$V_{gs}$ drive load is proposed for the inverter and NAND/NOR gates, wherein the gate and source terminals of the load transistor ($T_L$) are connected. In this logic, driver transistor ($T_D$) strongly pulls up the output voltage toward $V_{DD}$ due to a significantly lower channel resistance in comparison to the load. However, the pull-down transistor acts as a constant current source due to zero-$V_{gs}$, therefore, a large $W/L$ ratio is required for the load to obtain an output close to 0 V. The width of load (2000 μm) is therefore taken 4 times higher than the driver (500 μm) in both SG and DG inverters and NAND/NOR ZVLL configurations. A ZVLL configuration exhibits a high output resistance, thereby providing a higher gain [17]. However, the switching speed is inherently low, since the pull-down current provided by the load is quite low, which in turn produces a large pull-down delay.

One of the ways to increase the switching speed is by connecting the load in diode logic, wherein the gate and drain of the load transistor are shorted. This type of logic, therefore, is more robust and yields a large pull-down current leading to a less stage delay, thereby providing faster circuits [2]. In this load logic, both the driver and load operate in the enhancement mode. Therefore, to obtain a logic high output, the $W/L$ ratio of the driver should be large in comparison to the load transistor. Taking this into account, the width ratio of driver to load is chosen as 4:1 (2000 μm:500 μm) in DLL configuration.

## 7.4  ORGANIC INVERTER CIRCUITS

This section investigates the performance of organic $p$-type inverter circuits in DLL and ZVLL configurations. The schematics of SG and DG OTFTs-based DLL inverters are shown in Figure 7.6. Primarily, front and back gates of driver and load transistors are tied to analyze the inverter circuits. A biasing voltage, $V_{DD}$ of 10 V is applied and output voltage, $V_{OUT}$ is obtained at the drain terminal of the driver for an input supply, $V_{IN}$ swept from 0 to 10 V.

The schematics of ZVLL inverters using SG and DG OTFTs are shown in Figure 7.7, whereas, corresponding voltage transfer characteristics (VTCs) of single and dual gate inverters are illustrated in Figure 7.8 for DLL and ZVLL.

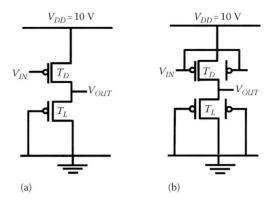

**FIGURE 7.6**  Schematics of all-*p* DLL inverters with (a) SG and (b) DG OTFTs.

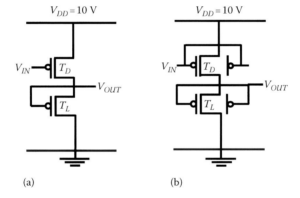

**FIGURE 7.7**  Schematics of all-*p* ZVLL inverters with (a) SG and (b) DG OTFTs.

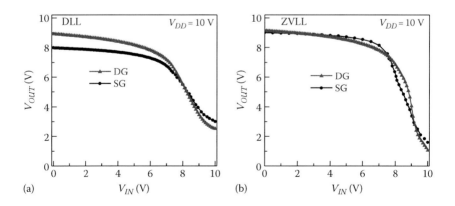

**FIGURE 7.8**  VTCs of SG and DG based all-*p* inverters in (a) DLL and (b) ZVLL modes.

The magnitude of output high voltage, $V_{OH}$, should be high enough to achieve the higher noise margins. However, it attains lower magnitude due to threshold voltage drop in the load transistor [19]. The value of $V_{OH}$ is limited to $V_{DD}-V_{tL}$, where $V_{tL}$ is the threshold voltage of load transistor.

In order to analyze the dynamic behavior, a pulse of 0–10 V at a frequency of 1 KHz is applied at the input terminal, as shown in Figure 7.9 for DLL and ZVLL configurations. Extracted parameters including $\tau_{PLH}$, $\tau_{PHL}$, and $\tau_p$ are summarized in Table 7.3 for both the logics. The results demonstrate, higher gain, and larger output swing for ZVLL inverters, since, in DLL, the pull-down network always appears in the *on* state, and therefore, it resists lowering the output toward zero.

Compared to DLL, the voltage gains in ZVLL are more than 37% and 45% for SG and DG transistors, respectively. A DG OTFT-based inverter demonstrates faster switching behavior. The propagation delay reduces by 63% and 38% for DLL and ZVLL, respectively, as compared to SG OTFT-based configurations.

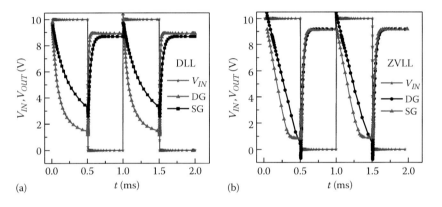

**FIGURE 7.9** Transient response of SG- and DG-based organic inverters using (a) DLL and (b) ZVLL configurations at 1 KHz input.

**TABLE 7.3 Extracted Parameters (Static and Dynamic) of Inverter Circuits in DLL and ZVLL Configurations with SG and DG OTFTs at $V_{DD} = 10$ V**

| | Performance Parameters | | | | | | | |
|---|---|---|---|---|---|---|---|---|
| Inverter's Configuration | Voltage Swing (V) | Noise Margin (V) | Gain | $\tau_{PLH}$ (μs) | $\tau_{PHL}$ (μs) | $\tau_p$ (μs) | $P_S$ (μW) | $P_L$ (μW) |
| DLL (SG) | 5.0 | 3.4 | −1.9 | 32 | 154 | 93 | 34 | 3.3 |
| DLL (DG) | 6.1 | 3.8 | −2.2 | 14 | 54 | 34 | 139 | 14 |
| ZVLL (SG) | 6.9 | 3.9 | −2.6 | 40 | 277 | 159 | 11 | 3.2 |
| ZVLL (DG) | 7.4 | 4.3 | −3.2 | 42 | 153 | 98 | 22 | 4.3 |

Figure 7.10 illustrates the VTCs of DG inverters as a function of $V_{DD}$ for DLL and ZVLL configurations, whereas similar plots for voltage gain are presented in Figure 7.11.

It is observed that the gain increases for higher values of $V_{DD}$. In addition to this, the current consumption, $I_{con}$ measured at the drain terminal of the driver, is high at zero $V_{IN}$ and reduces significantly with increasing $V_{IN}$, as demonstrated in Figure 7.12 for DLL and ZVLL configurations. It is contrary to the current consumption reported for complementary configuration [20], wherein the maximum current is consumed at the trip point, the reason being the load transistor is always *on* in all-*p* inverter circuits. Therefore, maximum current consumption is achieved at $V_{IN} = 0$ V, while in the case of complementary logic, one of the load and driver transistors is always *off* except at the trip point.

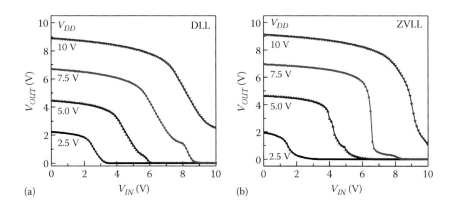

**FIGURE 7.10**    VTCs of DG OTFT-based inverter at different $V_{DD}$ in (a) DLL and (b) ZVLL configurations.

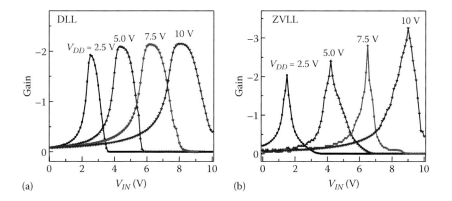

**FIGURE 7.11**    Voltage gain of DG inverter at different $V_{DD}$ in (a) DLL and (b) ZVLL configurations.

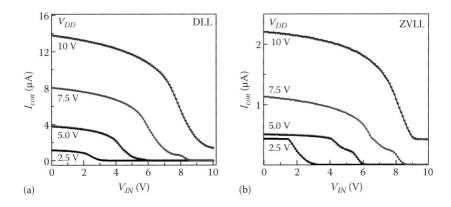

**FIGURE 7.12** $I_{con}$–$V_{IN}$ characteristics of DG inverter using (a) DLL and (b) ZVLL configurations at different $V_{DD}$.

The power consumption of an inverter circuit corresponds to the current consumption. The $I_{con}$ accounts quantitatively for the static power consumption, $P_S$, while both transistors are in the "*on*" state. However, it leads to leakage power, $P_L$, while the driver is in the "*off*" condition. $P_S$ and $P_L$ are estimated through conventional expressions of power consumption, at $V_{IN}$ equal to 0 and 10 V, respectively,

$$P_S = I_{con} \times V_{DD}, \ V_{IN} = 0 \ V \tag{7.7}$$

$$P_L = I_{con} \times V_{DD}, \ V_{IN} = 10 \ V \tag{7.8}$$

It can be inferred from Table 7.3 that by using the DG transistor, the overall $P_S$ and $P_L$ are reduced by 84% and 69%, respectively, for ZVLL as compared to DLL.

## 7.5 IMPROVEMENT IN PERFORMANCE OF ORGANIC DUAL GATE INVERTERS

Noise margin is a figure of merit that accounts for the robustness and can be defined as the maximum tolerable noisy signal by a gate while showing correct operation [1]. Threshold voltage strongly affects the noise margin of an organic inverter [8]. Because organic transistors are not intentionally doped on a single substrate, generally, the transistors exhibit the same $V_t$. Also from the fabrication point of view, it is difficult to create two different doping levels on a substrate through electrical doping. Besides this, a difference in threshold voltage can be created by employing metal gates of different work function, but $V_t$ difference obtained through this technique is very small as compared to operational biasing voltage.

This section analyzes two practical ways for achieving higher noise margin and gain by introducing different $V_t$: (1) applying fixed bias at the back gate and (2) employing the bootstrap technique. These techniques can be considered as realistic options to increase the robustness of digital circuits.

### 7.5.1  ORGANIC INVERTER CIRCUITS WITH BACK-GATE BIASING

Reliable and robust operation of an inverter circuit is directly associated with the noise margin that needs to be sufficiently high to ensure stable characteristics [19]. Organic all-$p$ type logic demonstrates a low noise margin. This technology has constraints regarding integration of larger circuits as a consequence of the parameter's variability [2]. The robustness of these circuits can be enhanced by controlling threshold voltage through the back gate, leading the transistor operation from depletion mode to enhancement mode. Moreover, the back gate $V_t$ control technique can be used to compensate for the shift of $V_t$ due to chemical degradation of the OTFTs [21]. Furthermore, the trip point can be shifted toward $V_{DD}/2$ by appropriate control of $V_t$ [2]. Therefore, back gate control is a viable method to increase the gain and noise margins of the inverter circuit consisting of $p$-type dual gate devices.

The schematics of the DG inverter in the DLL and ZVLL configurations are drawn in Figure 7.13 that incorporate fixed back-gate biasing in the driver transistor, $V_{BG,D}$, while the back gate of the load is coupled to its front gate. It leads to an increase in noise margin, gain, and dynamic response. Ideally, the trip point should be located at $V_{DD}/2$. Its asymmetric position severely limits the noise margin that is due to the same pinch-off voltage of both transistors [1].

The input at which the output toggles from high to low is determined by the driver's pinch-off voltage that needs to be shifted toward more negative values. Altering the geometry ratio, $(W/L)_D/(W/L)_L$, only moves the position of logic 0. VTCs as a function of $V_{BG,D}$ are demonstrated in Figure 7.14 for DLL and ZVLL

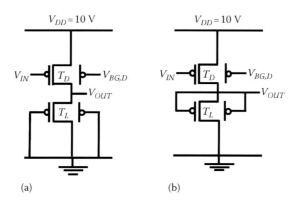

(a)                              (b)

**FIGURE 7.13**    Schematics of DG inverter in (a) DLL and (b) ZVLL with $V_{BG,D}$ at driver.

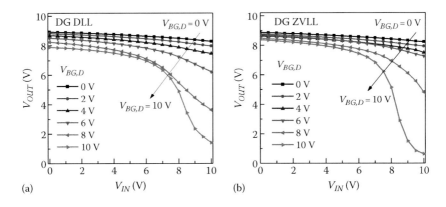

**FIGURE 7.14**   VTCs of DG inverter at different $V_{BG,D}$ in (a) DLL and (b) ZVLL modes.

configurations when the front and back gates of the load are coupled. The characteristics demonstrate the shifting of trip voltage to the left with increasing positive back-gate bias at the driver, resulting in an increase in noise margin of the DLL and ZVLL configurations. Figure 7.15 shows the plots of gain and transient response for both dual gate load logics.

Compared to the results provided in Table 7.3 (for DG inverter circuits, wherein the back gate is connected to the respective front gate in loads and drivers), an increase of 18% and 22% in gains is achieved at $V_{BG,D}$ of 10 V for DLL and ZVLL, respectively. Moreover, an increase of 0.4 and 0.2 V in output swing is obtained for diode and zero-$V_{gs}$ load logic, respectively. Typical values of gain, output voltage swing, and propagation delays for both logics are summarized in Table 7.4.

In addition, the dual gate inverter circuits are analyzed with different biasing at the back gate of load and driver transistors, as represented in Figure 7.16 for DLL and ZVLL [2]. Figure 7.17 shows the corresponding VTC plots with

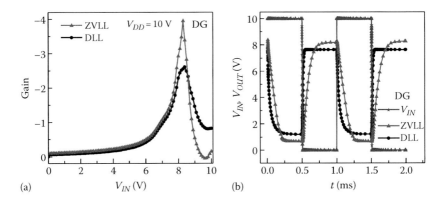

**FIGURE 7.15**   Characteristics curves of (a) gain and (b) transient response for DG inverter in DLL and ZVLL configurations at $V_{BG,D} = 10$ V.

**TABLE 7.4     Comparison between the Parameters of DG Inverter Circuits in DLL and ZVLL Configurations with Back-Gate Biasing**

| DG Inverter | Configuration | Performance Parameters | | | | | |
| | | Voltage Swing (V) | Noise Margin (V) | Gain | $\tau_{PLH}$ (µs) | $\tau_{PHL}$ (µs) | $\tau_p$ (µs) |
|---|---|---|---|---|---|---|---|
| $V_{BG,D}$ = 10 V (back gate tied to front gate for load) | DLL | 6.5 | 4.1 | −2.6 | 14 | 30 | 22 |
| | ZVLL | 7.6 | 5.1 | −3.9 | 38 | 85 | 62 |

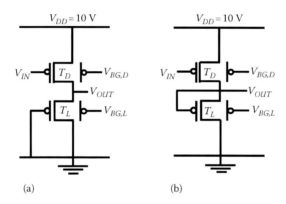

**FIGURE 7.16**   Schematics of organic DG inverter circuits with back gate voltages at both transistors in (a) DLL and (b) ZVLL configurations.

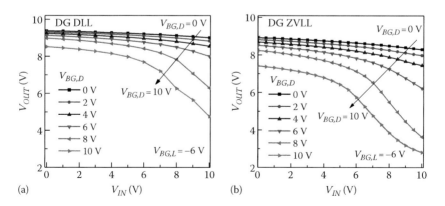

**FIGURE 7.17**   VTCs of DG inverters with different positive $V_{BG,D}$ magnitudes, while $V_{BG,L}$ is kept at −6 V in (a) DLL and (b) ZVLL configurations.

fixed back-gate voltage at load $V_{BG,L}$ of −6 V while sweeping $V_{BG,D}$ from 0 to 10 V, changing the drive transistor mode from depletion to enhancement. For a negative value of $V_{BG,L}$, when $V_{BG,D}$ is increased positively, the trip point shifts toward the ideal value ($V_{DD}/2$) that improves the noise margin.

Figure 7.18 demonstrates the noise margin for different back-gate biasing ($V_{BG,D}$ and $V_{BG,L}$) conditions. As compared to tied back gate (driver and load) conditions, the increase in noise margin is 8% and 19% with $V_{BG,D}$ = 10 V for DLL and ZVLL configurations, respectively. Furthermore, VTCs are analyzed at fixed $V_{BG,D}$ while $V_{BG,L}$ is varied from 0 to 10 V as shown in Figure 7.19 for DLL and ZVLL dual gate inverters. Decreasing positive bias at the load shifts the trip point toward the ideal position. In this mode, the $V_t$ of load transistor changes dynamically with the output node due to coupling of this node to the source. The load is made weaker with increasing positive back-gate bias of the load transistor. It limits the pulling down of the output toward zero resulting in

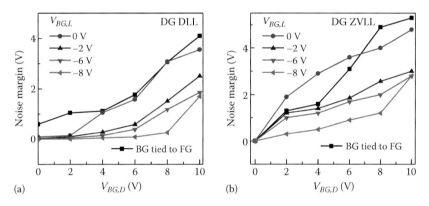

**FIGURE 7.18**    Noise margin analysis of DG inverter circuits in (a) DLL and (b) ZVLL configurations at different $V_{BG,D}$ and $V_{BG,L}$.

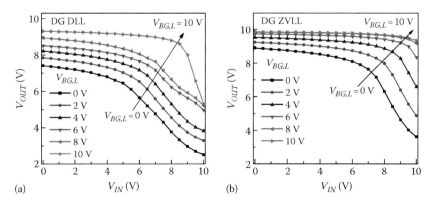

**FIGURE 7.19**    VTCs of organic DG inverters as a function of $V_{BG,L}$, while $V_{BG,D}$ is kept at 10 V in (a) DLL and (b) ZVLL configurations.

a lower noise margin. Therefore, both configurations are preferred with positive $V_{BG,D}$ without applying any bias at the back gate of the load transistor.

## 7.5.2  ORGANIC BOOTSTRAP INVERTER CIRCUITS

Bootstrapping is an effective technique to overcome the threshold voltage drop in digital circuits. It pulls down the output voltage to the minimum level and thus increases the output voltage swing and noise margin [22,23]. This technique includes an additional bootstrapping transistor, $T_B$, and its output is dynamically boosted by providing capacitive coupling through the capacitor, $C_{BC}$ [24]. The bootstrapping circuit using capacitive coupling that reduces the $V_t$ sensitivity is shown in Figure 7.20 for DLL and ZVLL.

Compared to back-gate biasing with $V_{BG,D}$ = 10 V and tied front gate and back gate of the load, this technique demonstrates an improved voltage swing of 26% and 11% for DLL and ZVLL configurations, respectively, as shown in Figure 7.21a. Using bootstrapping, the gain as a function of $V_{IN}$ increases for both DLL and ZVLL inverters as shown in Figure 7.21b. As observed, the gain and noise margin increases by 12% and 20% and 5% and 10%, respectively, for DLL and ZVLL configurations, as compared to the configurations with $V_{BG,D}$ of 10 V. Figure 7.22 shows the dynamic response of DG bootstrap inverters, analyzed at a pulse rate of 1 KHz.

Using the bootstrapping technique, a significant improvement in the $\tau_{PHL}$ is observed due to strong pull-down action by bootstrap and load transistors; however, $\tau_{PLH}$ is almost similar to the former case. Subsequently, this results in a reduction of 18% and 26% in average propagation delay for DLL and ZVLL configurations, respectively, as mentioned in Table 7.5. This technique substantially improves inverter performance.

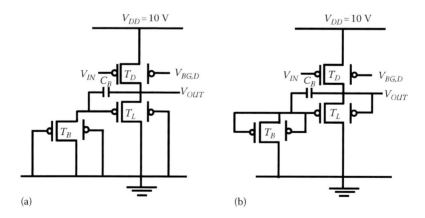

(a)  (b)

**FIGURE 7.20**    Schematics of bootstrap inverter circuits using DG OTFTs in (a) DLL and (b) ZVLL configurations.

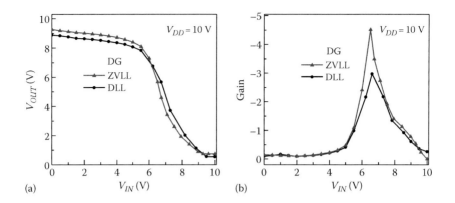

**FIGURE 7.21**    (a) VTCs and (b) gain of organic DG inverters in DLL and ZVLL configurations with bootstrapping technique.

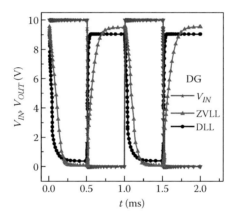

**FIGURE 7.22**    Transient response of DG bootstrap inverter in DLL and ZVLL modes.

**TABLE 7.5    Comparison between the Parameters of DG Inverter Circuits in DLL and ZVLL Configurations with Bootstrapping Technique**

| | | Performance Parameters | | | | | |
|---|---|---|---|---|---|---|---|
| **DG Inverter** | **Configuration** | **Voltage Swing (V)** | **Noise Margin (V)** | **Gain** | $\tau_{PLH}$ **(μs)** | $\tau_{PHL}$ **(μs)** | $\tau_p$ **(μs)** |
| Bootstrapping technique | DLL | 8.2 | 4.3 | −2.9 | 13 | 22 | 18 |
| | ZVLL | 8.4 | 5.6 | −4.7 | 30 | 62 | 46 |

## 7.6  ORGANIC NAND LOGIC GATE

This section presents a detailed analysis of NAND and NOR logic gates based on DG OTFTs. The static and dynamic behaviors of all-$p$ NAND and NOR gates are analyzed under both DLL and ZVLL configurations. The schematics of the NAND gate realized in DLL and ZVLL configurations are shown in Figure 7.23. In an all $p$-type NAND logic gate, the pull-up network consists of two driver transistors, $T_{D1}$ and $T_{D2}$, connected in parallel [9]. This network pulls up the output voltage (logic 1) when at least one of the inputs, $V_1$ or $V_2$, is 0 that forms a conducting path between $V_{DD}$ and $V_{OUT}$. The pull-down action is performed by load TFT, $T_L$, when both driver transistors are turned *off*.

The static and dynamic characteristics of logic gates depend on the data input patterns, therefore, the analysis of the gain and noise margin is more complex than the inverter. In NAND gate, a logic 0 at the $V_1$ or $V_2$ terminal enables the corresponding transistor to be turned *on*. Thus, the circuit reduces to a simple inverter that enables logic 1 in the output. The $k_R$ (driver to load) of the corresponding inverter is expressed as

$$k_R = \frac{\left(\mu\, C_{ox} W/L\right)_{T_{D1}/T_{D2}}}{\left(\mu\, C_{ox} W/L\right)_{T_L}} \tag{7.9}$$

Combined plots of VTCs of NAND gate in DLL and ZVLL configurations are shown in Figure 7.24, while sweeping $V_1$ from 0 to 10 V and keeping $V_2 = 10$ V. Similar characteristics are observed on swapping the two inputs. The NAND gate demonstrates an increment of 16% in voltage swing and 14% in noise margin for ZVLL configuration in comparison to the DLL. Besides this, the gain also improves by 1.5 times in ZVLL mode. Furthermore, when both $T_{D1}$ and

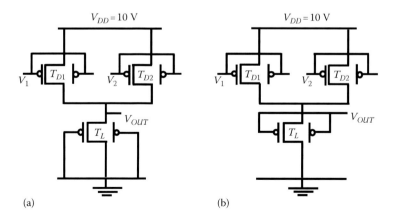

(a)                                              (b)

**FIGURE 7.23**  Schematics of NAND gate in (a) DLL and (b) ZVLL configurations.

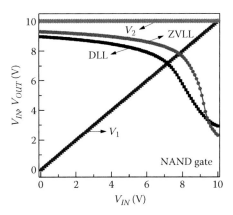

**FIGURE 7.24**    VTCs of NAND gate in DLL and ZVLL with $V_2 = 10$ V and $V_1$; $0{\rightarrow}10$ V.

$T_{D2}$ are turned *on* ($V_1 = V_2 = 0$ V), the load current is expressed as the sum of two driver currents

$$I_{ds,T_L} = I_{ds,T_{D1}} + I_{ds,T_{D2}} \tag{7.10}$$

The NAND logic circuitry with both of its inputs tied to a logic 0 can be replaced by an inverter circuit with the driver-to-load ratio given as

$$k_R = \frac{\left(\mu\, C_{ox}W/L\right)_{T_{D1}} + \left(\mu\, C_{ox}W/L\right)_{T_{D2}}}{\left(\mu\, C_{ox}W/L\right)_{T_L}} \tag{7.11}$$

The $k_R$ is 2 times higher (if both $T_{D1}$ and $T_{D2}$ are of same strength) than the case, wherein only one driver is turned *on* [25]. Hence, the magnitude of $V_{OH}$ is higher. The plots of VTC and gain for the NAND gate (DLL and ZVLL) are presented in Figure 7.25 for different input combinations. Resulting DC characteristics while only one driver TFT is *on* demonstrate a significant shift in the trip point toward left in comparison to the case where both driver TFTs are *on*. This is due to a weaker pull-up action in the former case compared to the latter one. Subsequently, the NAND logic with tied inputs demonstrates an increment of 6% and 4% in the $V_{OH}$ for DLL and ZVLL, respectively, in comparison to the case where only one of the inputs is swept and other remains in a "high" state. The ZVLL configuration outperforms the DLL ones for all the combinations of inputs. Compared to DLL, it shows an improvement of 20%, 68%, and 25% in the voltage swing, gain, and noise margin, respectively, while sweeping both inputs $V_1$ and $V_2$.

In order to analyze the dynamic behavior, a square pulse of 10 V magnitude is applied at a frequency of 1 KHz. Similar to the static response, the propagation

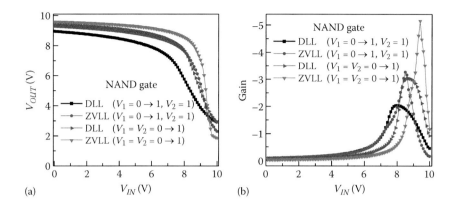

**FIGURE 7.25**  Characteristics plots of (a) VTCs and (b) gain for NAND gate in DLL and ZVLL configurations with different combinations of $V_1$ and $V_2$.

delay is also input dependent. The transient response of NAND gate in DLL and ZVLL configurations is shown in Figure 7.26a for the pulse applied at $V_1$, while $V_2$ is consistent at a logic 1. Similar plots for pulse applied at both $V_1$ and $V_2$ are presented in Figure 7.26b. Evaluated static and dynamic parameters, output swing, noise margin, gain, and delay times for different input combinations are summarized in Table 7.6.

Compared to DLL, the propagation delay is higher in ZVLL mode. For DLL NAND configuration, it is one-third of that of the ZVLL counterpart (for $V_1 = 0 \rightarrow 1/1 \rightarrow 0$ and $V_2 = 1$) due to significantly higher strength of the pull-up network and large pull-down current in DLL. When both inputs are swept simultaneously, a significant reduction of 26% and 16% in propagation delay is observed for DLL and ZVLL, respectively, since the strength of the pull-up network is doubled. Therefore, the output voltage rises quickly as shown in Figure 7.26b,

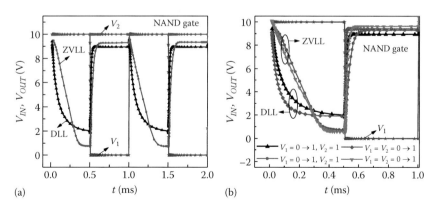

**FIGURE 7.26**  Transient response of NAND gate for (a) $V_2 = 10$ V and $V_1$ sweeping from $0 \rightarrow 10$ V and $10 \rightarrow 0$ V and (b) different combinations of $V_1$ and $V_2$.

**TABLE 7.6    Static and Dynamic Parameters of DG OTFT-Based NAND Gates in DLL and ZVLL Configurations**

| Input Condition | Load Configuration | Performance Parameters | | | | | |
|---|---|---|---|---|---|---|---|
| | | Voltage Swing (V) | Noise Margin (V) | Gain | $\tau_{PLH}$ ($\mu s$) | $\tau_{PHL}$ ($\mu s$) | $\tau_p$ ($\mu s$) |
| $V_2 = 1, V_1 = 0 \rightarrow 1/$ | DLL | 6.1 | 3.7 | −2.1 | 12 | 50 | 31 |
| $1 \rightarrow 0$ | ZVLL | 7.1 | 4.2 | −3.3 | 41 | 152 | 97 |
| $V_1 = V_2 = 0 \rightarrow 1/1 \rightarrow 0$ | DLL | 6.6 | 4.8 | −3.1 | 6 | 40 | 23 |
| | ZVLL | 7.9 | 6.0 | −5.2 | 14 | 147 | 81 |

since both the drivers are turned *on*. The transient behavior of NAND gate for all four input patterns is shown in Figure 7.27.

The NAND gate, when used as an inverter, demonstrates improved performance for the inputs $V_1 = V_2 = 0 \rightarrow 1$ in comparison to the case where $V_1 = 0 \rightarrow 1$ and $V_2 = 1$. This improvement is due to the doubling of pull-up network strength. The NAND gate in DLL (ZVLL) shows an improvement of 8% (11%), 48% (58%), 30% (43%), and 26% (16%) in voltage swing, gain, noise margin, and propagation delay, respectively, for the former case in comparison to the latter one. During low-to-high output transition, the worst-case delay occurs when only one driver is *on*. As a result, the worst-case $\tau_{PLH}$ is 2 and 3 times higher for DLL and ZVLL, respectively, in comparison to the condition when both drivers perform the pull-up action together.

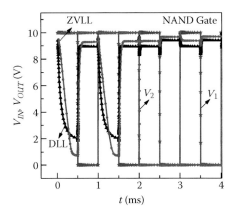

**FIGURE 7.27**  Transient response of NAND gate for all four combinations of $V_1$ and $V_2$.

## 7.7  ORGANIC NOR LOGIC GATE

The schematics of a two-input NOR gate realized in DLL and ZVLL configurations are shown in Figure 7.28, where two dual gate organic drivers, $T_{D1}$ and $T_{D2}$, are connected in series. This series network pulls up the output voltage only when both $T_{D1}$ and $T_{D2}$ are turned *on*. However, if any one of them or both drivers are *off*, the transistor $T_L$ will pull down the $V_{OUT}$.

All three transistors are connected in series, therefore, the current in the NOR circuit with both inputs set to 0 is expressed as

$$I_{ds,T_L} = I_{ds,T_{D1}} = I_{ds,T_{D2}} \tag{7.12}$$

$$\frac{\left(\mu\, C_{ox} W/L\right)_{T_L}}{2}[V_{gs} - V_{t,L}]^2 = \left(\mu\, C_{ox} W/L\right)_{T_{D1}}\left[\left(V_{gs,T_{D1}} - V_{t,T_{D1}}\right)V_{ds,T_{D1}} - \frac{V_{ds,T_{D1}}^2}{2}\right]$$

$$= \left(\mu\, C_{ox} W/L\right)_{T_{D2}}\left[\left(V_{gs,T_{D2}} - V_{t,T_{D2}}\right)V_{ds,T_{D2}} - \frac{V_{ds,T_{D2}}^2}{2}\right] \tag{7.13}$$

Considering both $T_{D1}$ and $T_{D2}$ of the same strength, an equivalent $k_R$ is obtained as

$$k_R = \frac{0.5\left(\mu\, C_{ox} W/L\right)_{T_{D1}/T_{D2}}}{\left(\mu\, C_{ox} W/L\right)_{T_L}} \tag{7.14}$$

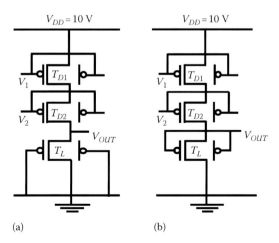

(a)    (b)

**FIGURE 7.28**    Schematics of NOR gate in (a) DLL and (b) ZVLL configurations.

Thus, the NOR logic circuitry with both inputs tied to a logic 0 is similar to an inverter circuit with $k_R$ reduced to half. The DC characteristics of NOR gate, while sweeping $V_1$ (0→1) and keeping $V_2$ at 0, is shown in Figure 7.29. The NOR gate attains a lower magnitude of $V_{OH}$ in comparison to the NAND gate due to voltage drop across two driver transistors connected in series. However, in NAND logic this drop is less due to only one transistor, thereby resulting in a higher $V_{OH}$ comparatively.

Compared to the NAND characteristics ($V_2 = 1$, $V_1 = 0→1$), the NOR gate ($V_2 = 0$, $V_1 = 0→1$) shows a reduction of 9% and 6% in $V_{OH}$ for DLL and ZVLL configurations, respectively. The plots of DC characteristics and gain for the DG OTFT-based NOR gate (DLL and ZVLL) are shown in Figure 7.30 with different combinations of two inputs. Compared to DLL, the gain is 2 times higher in the ZVLL configuration due to a larger output swing and sharper high-to-low

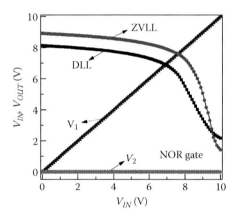

**FIGURE 7.29**    VTC of NOR gate in DLL and ZVLL with $V_2 = 0$ V and $V_1 = 0→10$ V.

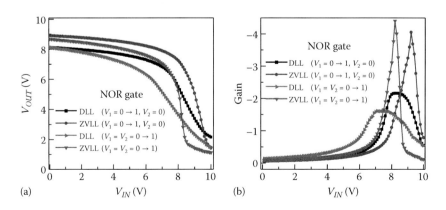

**FIGURE 7.30**    Characteristics plots of (a) VTCs and (b) gain for NOR gate in DLL and ZVLL configurations for different combinations of inputs, $V_1$ and $V_2$.

output transition. Similar to the NAND gate, the NOR characteristics are also dependent on the input combinations. It is observed that the load transistor performs a strong pull-down action if both driver transistors are *off*. Therefore, the magnitude of $V_{OL}$ is slightly higher when only one transistor remains *off*, thereby resulting in a reduction of 9% and 4% in the voltage swing for DLL and ZVLL, respectively. This can be understood on the basis of the leakage currents in organic transistors.

For the case, when both $T_{D1}$ and $T_{D2}$ are *off*, the leakage current is lower due to a significant increase in the equivalent *off*-resistance of two transistors connected in series. However, a higher leakage current flows if only one of the transistors is turned *off* resulting in a comparatively higher $V_{OL}$, thereby shifting the trip point toward the left side as shown in Figure 7.30a. Similar to the prior outcomes, the propagation delay of the NOR gate is lower for the DLL configuration in comparison to ZVLL, as shown in Figure 7.31.

The performance of NOR gate is better in the ZVLL configuration in comparison to DLL mainly in terms of voltage swing, noise margin, and gain, as summarized in Table 7.7. Compared to DLL, it shows an increment of 24% and

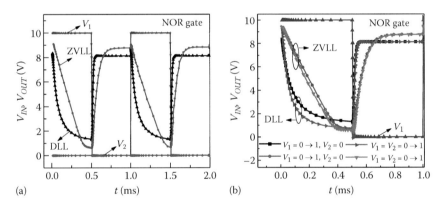

**FIGURE 7.31**   Transient response of NOR gate for (a) $V_2 = 0$ V and $V_1$ sweeping from 0→10 V and 10→0 V and (b) different combinations of $V_1$ and $V_2$.

**TABLE 7.7    Parameters of DG OTFT-Based NOR Gates in DLL and ZVLL Modes**

| | | Performance Parameters | | | | | |
|---|---|---|---|---|---|---|---|
| Input Condition | Load Configuration | Voltage Swing (V) | Noise Margin (V) | Gain | $\tau_{PLH}$ (μs) | $\tau_{PHL}$ (μs) | $\tau_p$ (μs) |
| $V_1 = 0 \rightarrow 1/1 \rightarrow 0,$ | DLL | 5.9 | 4.6 | −2.1 | 17 | 63 | 32 |
| $V_2 = 0$ | ZVLL | 7.3 | 6.0 | −4.1 | 66 | 190 | 128 |
| $V_1 = V_2 = 0 \rightarrow 1/$ | DLL | 6.5 | 4.4 | −1.7 | 18 | 57 | 36 |
| $1 \rightarrow 0$ | ZVLL | 7.6 | 5.7 | −4.4 | 68 | 165 | 117 |

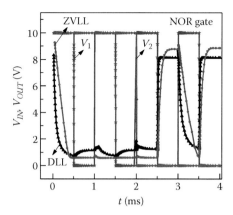

**FIGURE 7.32**  Transient response of NOR gate for all four combinations of $V_1$ and $V_2$.

30% in voltage swing and noise margin, respectively, in ZVLL (at $V_2 = 0$ and $V_1 = 0 \rightarrow 1$). On the contrary, the $\tau_{PLH}$ in DLL is one-quarter of that of the ZVLL, due to significantly larger $W/L$ ratios of the driver transistors in DLL configuration. Besides this, a significant reduction of 67% in $\tau_{PHL}$ is noticed for the DLL in comparison to the ZVLL configuration due to a larger pull-down current. The transient response of NOR gate for all four combinations of $V_1$ and $V_2$ is plotted in Figure 7.32 that validates the NOR logic in both DLL and ZVLL configurations.

## 7.8  CONCLUDING REMARKS

This chapter analyzed the static and dynamic behavior of all-$p$ organic inverter and universal NAND/NOR logic gate circuits in DLL and ZVLL configurations. A comparison presented between inverters based on SG and DG organic transistors shows an improved performance for DG inverters. Besides this, fixed back-gate bias and bootstrap techniques further enhance the performance of DG OTFT-based inverter circuits.

The performance of single and dual gate OTFTs is verified with experimentally reported results. The $I_{ds}$ of SG and DG devices are 0.7 and 1.3 μA, respectively, at $V_{ds} = V_{gs} = -10$ V, which reasonably match to the experimental current with an error of 2.8% and 3.5%, respectively. Similarly, other parameters such as $V_t$, $\mu$, SS, $I_{on}/I_{off}$, and $g_m$ are a close match to the experimental results with an average error of 1%, 9%, 8%, 5%, and 1% respectively.

The performance of a DG OTFT is substantially higher compared to its SG counterpart due to increased conductivity of the semiconducting layer under the effect of double-gate biasing. As a result, the $\mu$ and $g_m$ are 5 and 3 times higher, respectively, for a DG device. Besides this, an improvement of 42% is observed in *on/off* current ratio. Encouragingly, a reduction of 41%

is noticed for *SS* that reveals a lower density of trap states in a DG device than the SG. Consequently, these trap states are calculated as $8.1 \times 10^{12}$ and $3.7 \times 10^{12}$ cm$^{-2}$V$^{-1}$ for SG and DG devices, respectively. Besides this, the analytical value of parameter, $\alpha$ is 0.053 for SG and 0.021 for DG transistors. A higher value of alpha corresponds to a higher trap density and thus inferior TFT reliability.

The performance of DG OTFT-based inverters is substantially higher compared to SG OTFT-based inverters for both DLL and ZVLL. An appropriate tuning of threshold voltage in a dual gate driver transistor leads to a shift in the trip voltage toward $V_{DD}/2$, thereby resulting in an improved noise margin. Compared to SG inverters, an improvement of 22% (7%), 16% (23%), and 12% (10%) in voltage swing, gain, and noise margin, respectively, are observed in DLL (ZVLL) DG inverters. Besides this, the propagation delay of a DG inverter also reduces by 63% and 38% for DLL and ZVLL, respectively. On the contrary, the power consumption is 4 and 2 times higher for DLL and ZVLL DG inverters in comparison to their SG counterparts.

The inverter circuit in ZVLL configuration outperformed the DLL ones mainly in terms of noise margin, gain, and voltage swing. The driver transistor in ZVLL performs a strong pull-up action due to significantly lower channel resistance, leading to a high voltage swing and gain. As a result, the voltage gain in ZVLL is 37% and 45% higher for SG and DG inverters, respectively. In addition to this, the voltage swing and noise margin also exhibit an improvement of 21% and 13% for the DG OTFT-based ZVLL inverter as compared to DLL. In contrast, there is a trade-off in terms of speed, as the pull-down current is quite low due to zero-$V_{gs}$ at the load that in turn produces a large pull-down delay. On the other hand, DLL circuits are faster due to the operation of load at higher $V_{gs}$ resulting in a smaller stage delay. Therefore, compared to ZVLL, the propagation delay in DLL configuration is reduced by 42% and 65% for SG and DG inverters, respectively.

Noise margin of all-*p* organic circuits is severely limited due to asymmetric position of trip point. To enhance the performance of DG OTFT-based inverters in DLL and ZVLL, the back-gate biasing technique is applied. Compared to tied back gate (driver and load) conditions, an improvement of 8% (19%) in noise margin, 7% (3%) in voltage swing, and 18% (22%) in gain are achieved at $V_{BG,D}$ of 10 V for the DG inverter in DLL (ZVLL) configurations.

The bootstrapping technique is also applied to the DG inverter along with back-gate biasing that further improves the voltage swing by 26% and 11%, noise margin by 5% and 10%, and gain by 12% and 20% for DLL and ZVLL, respectively. In addition to this, propagation delay also reduces by 18% and 26% for DLL and ZVLL configurations, respectively, in comparison to the inverter circuits with only back-gate biasing ($V_{BG,D}$ = 10 V).

The static performance of two-input organic universal logic gates is observed to be higher in ZVLL mode, whereas a better dynamic response in DLL mode. The performance of these logic circuits is substantially affected by the two-input patterns due to a significant difference in the driver-to-load transconductance ratio. On realizing the NAND gate as an inverter, the performance is higher for the inputs, $V_1 = V_2 = 0\rightarrow1$ in comparison to the case, wherein $V_1 = 0\rightarrow1$ and $V_2 = 1$. This improvement is due to the doubling of pull-up network strength. The NAND gate in DLL (ZVLL) configuration shows an improvement in voltage swing, gain, noise margin, and propagation delay by 8% (11%), 48% (58%), 30% (43%), and 26% (16%), respectively, for the former case in comparison to the latter one. In addition to this, the worst-case $\tau_{PLH}$ is 2 and 3 times higher for DLL and ZVLL, respectively, when only one of the drivers is *on* in comparison to the condition when both drivers pull up together.

The NOR gate in ZVLL shows an increment of 24% and 30% in voltage swing and noise margin, respectively, in comparison to DLL configuration. However, the propagation delay for DLL NOR configuration is one-quarter of that of its ZVLL counterpart. Realization of this inverter and universal gate circuits using organic TFTs demonstrates the possibility of producing logic circuits on the flexible substrate and thus proves to be a stepping-stone toward low-cost organic electronics.

## PROBLEMS

### MULTIPLE CHOICE

1. What $n$-type organic material has inferior performance to $p$-type material?
   a. Large barrier height in $n$-type
   b. Small barrier height in $n$-type
   c. Low carrier mobility
   d. None of the above
2. Why does a dual gate OTFT have better performance than a single gate OTFT device?
   a. Better control on threshold voltage
   b. Charge carrier modulation
   c. Steep subthreshold
   d. All of the above
3. ZVLL configuration lacks in performance in terms of
   a. Voltage swing
   b. Speed
   c. Leakage current
   d. Noise margin
4. Mobility of a dual gate OTFT device is
   a. 10 times higher than single gate
   b. 5 times higher than single gate
   c. 3 times higher than single gate
   d. Same as single gate
5. The mobility enhancement factor for OSC depends on
   a. Doping density of OSC
   b. Dielectric permittivity of OSC material

c.  Both a and b
d.  None of the above
6.  What is the realistic option to increase the robustness of a digital circuit?
    a.  Increase channel length
    b.  Employing bootstrap technique
    c.  Increasing OSC thickness
    d.  None of the above
7.  Bootstrap technique is effective in
    a.  Overcoming threshold voltage drop
    b.  Increasing the speed
    c.  Reducing cost
    d.  All of the above
8.  A load transistor of a NAND gate performs a strong pull-down action when
    a   Both transistors are *on*
    b.  One transistor is *on*
    c.  Both transistors are *off*
    d.  All of the above

## ANSWER KEY

1. a; 2. d; 3. b; 4. b; 5. c; 6. b; 7. a; 8. c

## SHORT ANSWER

1.  Differentiate between single gate and dual gate structure.
2.  Explain $I_{ds}$–$V_{ds}$ characteristics of single gate and dual gate OTFTs.
3.  Derive an expression for ratio of integral of drain current over gate bias.
4.  Describe ZVLL and DLL configurations in detail.
5.  Explain an inverter circuit using single gate and dual gate for DLL and ZVLL configurations.
6.  Explain how improvements can be made in an organic dual gate inverter using
    a.  Back-gate bias technique
    b.  Bootstrap technique
7.  Describe NAND gate realization using DLL and ZVLL configurations in detail.
8.  Describe NOR gate realization using DLL and ZVLL configurations in detail.

## EXERCISES

1.  Design and simulate a *p*-type dual gate device, using an Atlas TCAD tool, with pentacene as an OSC material and the following parameters: semiconductor thickness ($t_{osc}$) = 200 nm, channel length ($L$) = 25 μm, channel width ($W$) = 800 μm, bottom insulator thickness

$(t_{oxb})$ = 100 nm (SiO$_2$), top insulator thickness $(t_{oxt})$ = 300 nm (SiO$_2$), thickness of source/drain $(t_s/t_d)$ = 80 nm (Au), bottom gate thickness $(t_{gb})$ = 150 nm (Si), and top gate thickness (tgt) = 150 nm (Al). Analyze the performance of the device based on the Poole and Frenkel mobility model. Plot the transfer characteristics of the device for the supply voltage; $V_{ds}$ = –15 V and $V_{gs}$ = 0 to –15 V. Also draw the output characteristic when $V_{gs}$ and $V_{ds}$ varies from 0 to –15 V. Extract the performance parameter in terms of threshold voltage, subthreshold slope, transconductance, mobility, drive current ($I_{on}$), and current *on/off* ratio. Hint:

Transconductance:

$$g_m = \frac{\partial I_{ds}}{\partial V_{gs}} = C_i \mu \frac{W}{L}(V_{gs})$$

Mobility:

$$\mu = \mu_0 (V_{gs} - V_t)^\alpha$$

Current *on/off* ratio:

$$I_{on}\Big/ I_{off} = \frac{C_i \mu (V_{gs} - V_t)^2}{(t_{osc} V_{ds} \sigma)}$$

*Off* current:

$$I_{off} = \left(\frac{W}{L}\right) t_{osc} V_{ds} \sigma$$

Subthreshold slope:

$$SS = \left(\frac{\partial V_{gs}}{\partial \log_{10}(I_{ds})}\right)$$

where, *W/L* is the aspect ratio, $V_t$ is the threshold voltage, $V_{gs}$ is the gate to source voltage, $t_{osc}$ is the thickness of OSC layer, $\sigma$ is the conductivity, $C_i$ is the intrinsic capacitance, $\mu_0$ is the null field mobility, $I_{ds}$ is the drain to source current, and $\alpha$ is the mobility enhancement factor.

2. Using the device parameters as discussed in question 1, analyze and compare the performance of a DG OTFT device in top gate, bottom gate, and dual gate modes of operation. Comment on the difference in device performance.

   Hint:
   - Dual gate mode: Use the gate electrodes in common mode configuration.
   - Top gate mode: Ground the bottom gate and connect the supply to the top gate.
   - Bottom gate mode: Ground the top gate and connect the supply to the bottom gate.

3. Design and simulate an all $p$-type organic dual gate inverter (DG inverter) circuit using device simulated and discussed in question 1. Analyze the static and dynamic behavior of the DG inverter in diode load logic (DLL) configuration and extract the static and dynamic performance parameters. Consider the supply voltage; $V_{DD}$ = 10 V; the $W_D/W_L$ ratio as 1:5, with the width of the driver and load transistors; $W_D$ = 400 μm and $W_L$ = 2000 μm.

   Hint:

$$NM_L = V_{IL} - V_{OL}$$

$$NM_H = V_{OH} - V_{IH}$$

   where, $V_{OH}$ is the maximum value of output voltage taken as logic 1, $V_{OL}$ is the minimum value of the output voltage taken as logic 0, $V_{IH}$ is the minimum value of input voltage to be interpreted as logic 1, $V_{IL}$ is the maximum value of input voltage to be interpreted as logic 0, $NM_H$ is the high noise margin, and $NM_L$ is the low noise margin.

4. Analyze the low and high noise margins ($NM_L$ and $NM_H$) for an all $p$-type transistor based organic DG inverter with the device parameters as discussed in question 1. The inverter should be simulated in zero-$V_{gs}$ load logic (ZVLL) configuration. Consider the supply voltage; $V_{DD}$ = 10 V; the $W_D/W_L$ ratio as 5:1, with the width of the driver and load transistors; $W_D$ = 2000 μm and $W_L$ = 400 μm.

## REFERENCES

1. Spijkman, M.; Smits, E. C.; Blom, P. W.; De Leeuw, D. M.; Come, Y. B. S.; Setayash, S.; Cantatore, E. J. "Increasing the noise margin in organic circuits using dual gate field effect transistors," *Appl. Phys. Lett.* **2008**, 92(14), 143304-1–143304-3.

2. Myny, K.; Beenhakkers, M. J.; Van Aerle, N. A. J. M.; Gelinck, G. H.; Genoe, J.; Dehaene W.; Heremans, P. "Unipolar organic transistor circuits made robust by dual-gate technology," *IEEE J. Solid-State Circuits* **2011**, 46(5), 1223–1230.

3. Guerin, M.; Bergeret, E.; Benevent, E.; Daami, A.; Pannier, P.; Coppard, R. "Organic complementary logic circuits and volatile memories integrated on plastic foils," *IEEE Trans. Electron Devices* **2013**, 60(6), 2045–2051.

4. Wang, W.; Ma, D.; Pan, S.; Yang, Y. "Hysteresis mechanism in low-voltage and high mobility pentacene thin film transistors with polyvinyl alcohol dielectric," *Appl. Phys. Lett.* **2012**, 101(3), 033303-1–033303-5.

5. Oh, M. S.; Hwang, D. K.; Lee, K.; Choi, W. J.; Kim, J. H.; Im, S.; Lee, S. "Pentacene and ZnO hybrid channels for complementary thin film transistor inverters operating at 2V," *J. Appl. Phys.* **2007**, 102(7), 076104–076104-3.

6. Kim, J. B.; Hernandez, C. F.; Hwang, D. K.; Tiwari, S. P.; Potscavage Jr., W. J.; Kippelen, B. "Vertically stacked complementary inverters with solution processed organic semiconductors," *Org. Electron.* **2011**, 12(7), 1132–1136.

7. Narasimhamurthy, K. C.; Paily, R. "Performance comparison of single gate and dual gate carbon nanotube thin film field effect transistors," *IEEE Trans. Electron. Devices* **2011**, 58(7), 1922–1927.

8. Pavlovic, Z.; Manic, I.; Prijic, Z.; Davidovic, V.; Stojadinovic, N. D. "Influence of gate oxide charge density on VDMOS transistor *on*-resistance," *22nd Int. Conf. Microelectronics* (ICMEL-2000), 2, 663–666, **2000**, doi: 10.1109 /ICMEL.2000.838777.

9. Gelinck, G. H.; Veenendaal, E. V.; Coehoorn, R. "Dual gate organic thin film transistors," *Appl. Phys. Lett.* **2005**, 87(7), 073508-1–073508-3.

10. Cui, T.; Liang, G. "Dual gate pentacene organic field-effect transistors based on a nanoassembled SiO$_2$ nanoparticle thin film as the gate dielectric layer," *Appl. Phys. Lett.* **2005**, 86(6), 064102-1–064102-3.

11. Maddalena, F.; Spijkman, M.; Brondijk, J. J.; Fonteijn, P.; Brouwer, F.; Hummelen, J. C.; De Leeuw, D. M.; Blom, P. W. M.; De Boer, B. "Device characteristics of polymer dual gate field effect transistors," *Org. Electron.* **2008**, 9(5) 839–846.

12. Spijkman, M. J.; Myny, K.; Smits, E. C. P.; Heremans, P.; Blom, P. W. M.; De Leeuw, D. M. "Dual gate thin film transistors, integrated circuits and sensors," *Adv. Mater.* **2011**, 23(29), 3231–3242.

13. Brondijk, J. J.; Spijkman, M.; Torricelli, F.; Blom, P. W. M.; De-Leeuw, D. M. "Charge transport in dual gate organic field effect transistors," *Appl. Phys. Lett.* **2012**, 100(2), 023308-1–023308-4.

14. Ha, T. J.; Sonar, P.; Dodabalapur, A. "High mobility top gate and dual-gate polymer thin film transistors based on diketopyrrolopyrrole-naphthalene copolymer," *Appl. Phys. Lett.* **2011**, 98(25), 253305-1–253305-3.

15. Brown, A. R.; Jarrett, C. P.; de Leeuw, D. M.; Matters, M. "Field effect transistors made from solution-processed organic semiconductors," *Sythetic Metals* **1997**, 88(1), 37–55.

16. Carranza, A. C.; Nolasco, J.; Estrada, M.; Gwoziecki, R.; Benwadih, M.; Xu, Y.; Cerdeira, A. *et al.* "Effect of density of states on mobility in small-molecule *n*-type organic thin-film transistors based on a perylene diimide," *IEEE Electron Device Lett.* **2012**, 33(8), 1201–1203.

17. Marinov, O.; Deen, M. J.; Datars, R. "Compact modeling of charge mobility in organic thin film transistors," *J. Appl. Phys.* **2009**, 106(6), 064501-1–064501-13.

18. Marinov, O.; Deen, M. J.; Iniguez, B. "Charge transport in organic and polymer thin-film transistors: Recent issues," *IEE Proc. Circ. Dev. Syst.* **2005**, 152(3), 189–209.

19. Natali, D.; Fumagalli, L.; Sampietro, M. "Modeling of organic thin film transistors: Effect of contact resistances," *J. Appl. Phys.* **2007**, 101(1), 014501-1–014501-12.
20. Marinov, O.; Deen, M. J.; Zschieschang, U.; Klauk, H. "Organic thin film transistors: Part I. Compact DC modeling," *IEEE Trans. Electron Devices* **2009**, 56(12), 2952–2961.
21. Deen, M. J.; Marinov, O.; Zschieschang, U.; Klauk, H. "Organic thin-film transistors: Part II. Parameter extraction," *IEEE Trans. Electron Devices* **2009**, 56(12), 2962–2968.
22. Mittal, P.; Kumar, B.; Kaushik, B. K.; Negi Y. S.; Singh R. K., "Channel length variation effect on performance parameters of organic field effect transistors," *Microelectr. J.*, **2012**, 43(12), 985–994.
23. Resendiz, L.; Estrada, M.; Cerdeira, A.; Cabrera, V. "Analysis of the performance of an inverter circuit: Varying the thickness of the active layer in polymer thin film transistors with circuit simulation," *Jap. J. Appl. Phy.* **2012**, 51(4S), 04DK04-1–04DK04-6.
24. Gupta, D.; Katiyar, M.; Gupta, D. "An analysis of the difference in behavior of top and bottom contact organic thin film transistors using device simulation," *Org. Electron.* **2009**, 10(5), 775–784.
25. Vusser, S. D.; Genoe, J.; Heremans, P. "Influence of transistor parameters on the noise margin of organic digital circuits," *IEEE Trans. Electron Devices* **2006**, 53(4), 601–610.

# Digital Circuit Designs Based on Single and Dual Gate Organic Thin-Film Transistors Using Diode Load Logic and Zero-$V_{gs}$ Load Logic Configurations

**8**

## 8.1 INTRODUCTION

During the past decade, with the advent of novel organic material compounds and inexpensive process requirements, significant interest and tremendous progress in the field of organic electronics have been reported. Organic thin-film transistors (OTFTs) have become the backbone of recent developments in the field of large-area organic electronics for producing promising digital and analog applications [1]. Despite consistent efforts, the use of OTFT devices is restricted to the basic digital and analog benchmark circuits. The key reason behind this status is the lower mobility, instability, and repeatability of the organic devices against their inorganic counterparts [2]. Most organic circuits still make use of $p$-type devices, as $n$-type OTFT devices exhibit poor performance and high susceptibility to air and moisture. Through relentless efforts to improve the performance of analog and digital circuits, designers applied several new techniques such as dual threshold voltage [3], dual gate structure, and different load configurations [4,5]. Among the available $p$-type organic materials, pentacene is the most preferred with a mobility higher than $1.0 \text{ cm}^2/\text{Vs}$.

This chapter emphasizes OTFT-based combinational and sequential circuit designs using single gate (SG) and dual gate (DG) architecture in diode load logic (DLL) and zero-$V_{gs}$ load logic (ZVLL) configurations. A conventional block diagram of combinational as given in Figure 8.1a comprises inputs and outputs without any direct interactions. A sequential circuit consists of a combinational circuit with memory element in the feedback loop, as shown in Figure 8.1b. Different benchmark sequential circuits conventionally called latches (flip-flops) have become core elements of integrated circuits built using inorganic semiconducting materials. The progress with inorganic devices is based on the fact that the responses of these sequential circuits are proven, robust, reliable, and stable in contrast with organic semiconductor (OSCs). Most recent research efforts using OSCs are still in the hunt for innovative material, process technologies, and circuit design techniques to develop stable, robust, and proven organic sequential circuits [6–10].

This chapter comprises six sections including the current introductory section. The next four sections present implementation of various basic sequential building blocks, including SR, D, JK, and T latches using single and dual gate OTFT devices in DLL and ZVLL configurations. All the latches described in these sections are based on NAND gates whose static and dynamic characteristics in DLL and ZVLL configurations were presented in the previous chapter. Also, all the designs discuss synchronous operation using a clock signal applied in combination with latch inputs. The final section describes the comparison of all the designs, which can serve as a platform for selecting the circuit techniques to build larger digital systems using these benchmark organic sequential circuits.

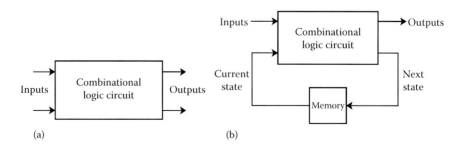

(a)

(b)

**FIGURE 8.1**    Block diagram of (a) combinational and (b) sequential circuits.

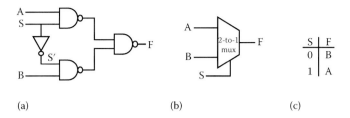

    (a)                         (b)                      (c)

**FIGURE 8.2**   (a) Gate level schematic, (b) logic symbol, and (c) characteristic table of 2-to-1 mux.

## 8.2 COMBINATIONAL CIRCUIT DESIGNS

A multiplexer (or mux), also referred to as a data selector, is a device that selects one signal from several input signals and forwards it to the output line. Conventionally, a mux with $2^n$ inputs has $n$ select lines and one output line. A gate level schematic, logic symbol, and characteristic table for a 2-to-1 mux is given in Figure 8.2. As shown in the gate level schematic, a 2-to-1 mux has two inputs (A and B), one select line (S), and one output line (F). Depending on the logic level (1 or 0) at the select line, the input signal A or B is forwarded to the output line F. The subsequent section describes the SG-based 2-to-1 mux design using DLL and ZVLL configurations.

### 8.2.1 SG-BASED 2-TO-1 MULTIPLEXER DESIGN IN DLL AND ZVLL CONFIGURATIONS

A transistor level schematic of a single gate 2-to-1 mux in DLL configuration is given in Figure 8.3a and the dynamic response of the circuit is plotted in Figure 8.3b.

As shown in the dynamic response of the circuit, the select line (S) is driven using a signal with 6 ms time period, and signals A and B are driven using the signals of 2 ms and 4 ms periodic pulses. The output of the mux circuit (F) responds exactly in accordance with the functionality table shown in Figure 8.3c. However, it is evident that the DLL configuration does not provide a full swing output; in fact, logic 1 is maximally reached at 9.1 V and logic 0 attains a minimum value of 5 V for the given power supply voltage of 10 V. Henceforth, the noise margins of this circuit are lowered.

The ZVLL configuration of the mux is given in Figure 8.4. The select line (S) is driven using 12 ms signal, input signal B is driven by 6 ms signal with 50% duty cycle, and signal A is chosen to have 5 ms time period with 40% duty cycle. The dynamic response plotted in Figure 8.4b, reveals the higher propagation delay in ZVLL configuration, specifically, $\tau_{PHL}$ exhibits a longer duration. However, the voltage swing of the circuit is very near to rail-to-rail, with a

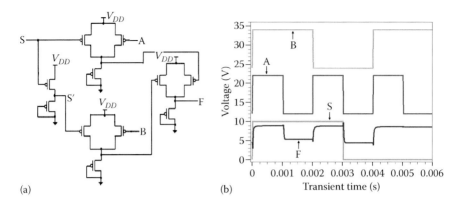

**FIGURE 8.3**    (a) A transistor level schematic and (b) dynamic response of SG-based 2-to-1 mux in DLL configuration.

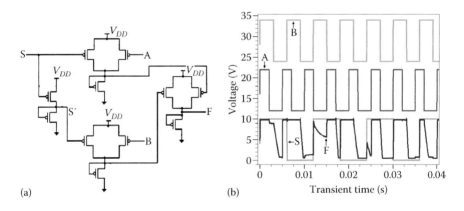

**FIGURE 8.4**    (a) A transistor level schematic and (b) dynamic response of SG-based 2-to-1 mux in ZVLL configuration.

maximum of 9.9 V and minimum of 0.7 V, resulting in the higher noise margins of the circuit.

## 8.3  CLOCKED SEQUENTIAL CIRCUIT DESIGNS BASED ON SG AND DG OTFTs

In the following sections, four benchmark sequential circuit implementations based on SG and DG OTFT devices are presented. All the circuits are primarily NAND-based clocked latch designs (SR, D, JK, and T), implemented using DLL and ZVLL configurations.

## 8.3.1  SR LATCH IMPLEMENTATIONS

A clocked SR latch comprises an additional gating clock signal to facilitate synchronous operation in the basic SR latch, so that the outputs will respond to the input levels during the active clock period only. This section describes an all NAND-gate-based SR latch design. The gate level schematic, logic symbol, and the characteristic table of a clocked NAND-based SR latch are shown in Figure 8.5. The response of the latch circuit can be verified in accordance with the characteristic table with an active clock signal application. If the clock (CK) is at logic 0, the input signals have no influence upon the circuit response; the outputs Q and Qbar preserve the current status. During the inactive period of the clock, the outputs of the front two NAND gates, where the clock signal is coupled directly with the applied input signals, will remain at logic 1, which helps the next two NAND gates in preserving their current states regardless of the S and R input signals. When the clock input goes to logic 1, the logic levels applied to the S and R inputs are permitted to reach the next stage of gates in inverted manner, and the SR latch outputs possibly change their states depending on their current states.

With the input combination S = R = 0, the latch outputs Q and Qbar hold their current states. With both inputs S and R at logic 1, the occurrence of a clock pulse causes both outputs to go at the same logic 1; hence it is not valid and forms the forbidden input combination. When the input combination is given in complementary form with CK = 1, that is, if S = 1 and R = 0, the latch Q

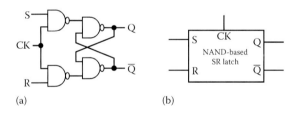

(a)                                                         (b)

| CK | S | R | Q | $\overline{Q}$ | Operation |
|----|---|---|---|---|-----------|
| 0 | X | X | Q | $\overline{Q}$ | Hold |
| 1 | 0 | 0 | Q | $\overline{Q}$ | Hold |
| 1 | 1 | 0 | 1 | 0 | Set |
| 1 | 0 | 1 | 0 | 1 | Reset |
| 1 | 1 | 1 | 1 | 1 | Invalid |

(c)                         Forbidden state

**FIGURE 8.5**    (a) Gate level schematic, (b) logic symbol, and (c) characteristic table of NAND-based SR latch.

output sets to logic 1; or if S = 0 and R = 1 then the Q reset to logic 0. Therefore, the S is termed the set input and R is termed the reset input.

As mentioned in the previous section, an SR latch is built using an all-NAND gate implementation. All the NAND gates are elementarily built using either SG or DG OTFT devices. As explored in the previous chapters the performance of SR latch designs are also analyzed in DLL and ZVLL configurations.

### 8.3.1.1 SG-BASED SR LATCH DESIGNS IN DLL AND ZVLL CONFIGURATIONS

The transistor level schematic of a NAND-based SR latch using $p$-type organic TFT as an elementary device is shown in Figure 8.6a. As depicted in the figure, all the load transistors are configured in DLL mode and the applied power supply voltage ($V_{DD}$) is 10 V. The driver transistor width is 2000 μm and the load transistor width is 400 μm, which is 4 times higher for the better performance in DLL mode. The clock signal (CK) is applied at the frequency of 100 Hz. The S and R inputs are given in the clocked form with the frequency of 250 Hz and 500 Hz, respectively, for analyzing the dynamic response of the design under all possible input combinations. Figure 8.6b shows the dynamic response of the design. It is evident from the response that output swing is lower, which is near to 40% of the supply voltage and swinging between maximum 9 V to minimum 5 V. The poor swing of the DLL configuration is the major reason for the incorrect outputs and false logical results produced by the multistage circuit operations. As the dynamic response shows, the design works well with complementary input combinations during the active period of the clock, i.e., S = 1 and R = 0 or vice versa.

For the forbidden input combination, i.e., for S = R = 1, both outputs stuck to logic 1. For S = R = 0, both outputs should hold the present state. However, in DLL configuration both outputs try to attain the same value. During the inactive period of the clock the outputs should hold the present states, regardless of

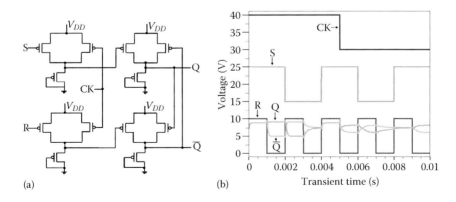

(a)    (b)

**FIGURE 8.6**    (a) A transistor level schematic and (b) dynamic response of SG-based clocked SR latch in DLL configuration.

the changes in the inputs. However, dynamic responses signify that the complementary outputs react to the changes in the inputs, and their levels are varying.

The rise and fall times of the outputs are small, hence, faster switching is feasible for the design. The propagation delay times $\tau_{PHL}$ and $\tau_{PLH}$ are calculated to 80 μs and 50 μs, respectively, as shown in Figure 8.7. Therefore, at the supply voltage of 10 V, overall propagation delay time $(\tau_p)$ is 65 μs. These results of the SR latch signify the faster switching for the DLL configuration; however, there is poor output swing.

For comparatively better voltage swing and correct output upon application of different input combinations, the load transistors are configured in ZVLL mode in the SR latch design. A transistor level schematic is given in Figure 8.8a with the supply voltage of 10 V, driver transistor width of 400 μm, and load

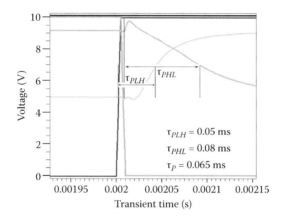

**FIGURE 8.7**    Calculation of propagation delay of SG-OTFT-based SR latch in DLL configuration.

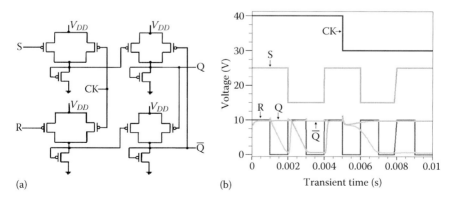

**FIGURE 8.8**    (a) A transistor level schematic and (b) dynamic response of SG-OTFT-based SR latch in ZVLL configuration.

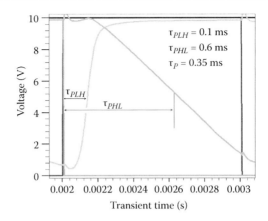

**FIGURE 8.9** Propagation delay calculation of SG OTFT-based SR latch in ZVLL configuration.

transistor width of 2000 μm, which ensures the pull-down transistor output will reach close to 0 V.

A dynamic response of the latch is given in Figure 8.8b. For fair comparison of the dynamic responses of SG OTFT-based SR latches in DLL and ZVLL configurations, both designs were given the same inputs with the clock frequency of 100 Hz, S at 250 Hz, and R at 500 Hz. The dynamic response showed a better output voltage swing greater than 90%, much higher than the 40% of that available in DLL configuration. The higher output voltage swings allow the design to serve further stages with higher noise margins. The input stimuli generate all possible combinations so as to verify the correctness of the output response. During the active period of the clock, with S = 1 and R = 0, the output signal Q set to logic 1; whereas for S = 0 and R = 1, the output Q reset to logic 0 and approaches close to 0 V. For the forbidden set of inputs, the both output signals attain the logic 1 showing nonvalidity. With CK = 1 and S = R = 0, Q and Qbar hold their current states. With CK = 0, both outputs stay at their current complementary levels regardless of the changes in the inputs. These observations verify the correctness of the SR latch operation for all possible input combinations.

The propagation delay ($\tau_p$) of the latch in ZVLL mode is approximately 350 μs, as calculated from Figure 8.9, which is higher compared to the DLL configuration. The $\tau_{PLH}$ is comparable to the DLL configuration. The $\tau_{PHL}$ is very large in particular due to very low pull-down current capability of the load transistor, hence, taking more time to reach to logic 0. Therefore, overall switching is slower in ZVLL configuration.

### 8.3.1.2 DG-BASED SR LATCH DESIGNS IN DLL AND ZVLL CONFIGURATIONS

This section describes the use of dual gate OTFT devices for the designing of SR latch in DLL configuration. For each transistor, both gate terminals are tied together. The transistor level schematic of the design is shown in Figure 8.10a. The widths of all the transistors are kept the same as described in the SG-based latch design with DLL configuration. The design is analyzed under the same input stimuli depicted in the SG counterpart for better comparison.

The dynamic response of the DG-based latch is presented in Figure 8.10b. It can be seen that the DG-based latch design in DLL configuration shows very little improvement in the voltage swing and no improvement in the correctness of the output response compared to the SG-based design.

Similar to the SG-based design, during the active clock period the outputs work well with complementary input stimulus. However, if both the inputs are applied with equal level then both outputs try to attain the same higher level. Also, during the inactive period of the clock signal, output Q and Qbar show changes with the inputs. Propagation delay is observed to be approximately the same for the design. Therefore, the DG implementation of the latch in DLL mode shows negligible improvement as compared to the SG counterpart.

This following analyzes the DG-based SR latch response in ZVLL configuration. The transistor level schematic of the design is shown in Figure 8.11a. The widths of all the devices and the input stimulus conditions are kept the same as the SG counterpart for better comparison.

The dynamic response of the DG-based SR latch in ZVLL configuration is shown in Figure 8.11b. Output swing is seen to be improved and the pull-down transistor attains low logic very close to 0. Regarding the correctness of the output signals, Q and Qbar show resemblance to the SG counterpart and correctly demonstrate all four operations (i.e., hold, set, reset, and invalid) mentioned in the characteristics table of the latch. However, the switching becomes slower than the SG counterpart. The propagation delay ($\tau_p$) of the DG-based latch in ZVLL configuration is approximately 550 μs as calculated from Figure 8.11b, which is almost double as compared to the ZVLL configuration in SG-based design. The $\tau_{PLH}$ is 120 μs and $\tau_{PHL}$ is approximately 1000 μs, which is very large.

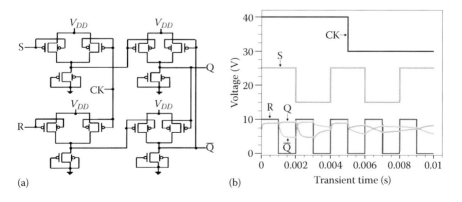

(a)  (b)

**FIGURE 8.10**  (a) A transistor level schematic and (b) dynamic response of DG OTFT-based SR latch in DLL configuration.

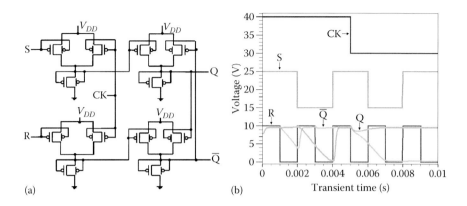

(a)                                                                 (b)

**FIGURE 8.11**    (a) A transistor level schematic and (b) dynamic response of DG OTFT-based SR latch in ZVLL configuration.

### 8.3.2  D-LATCH IMPLEMENTATIONS

A gate level representation, logic symbol, and characteristic table of a NAND-based clocked D-type latch are given by Figure 8.12. D-latch is the special case of SR latch, where S and R are always applied in complementary form, and the D-stimulus is directly applied to the S-input and the inverted signal of it to the R-input of the SR latch. The D-type latch is also referred to as the transparent latch for the reason that the logic level of the D-input is propagated to output Q as long as the clock input remains at logic 1; otherwise, Q preserves it previous state. The D-latch finds many applications in digital circuit design, such as storage of data and delay element.

#### 8.3.2.1  SG-BASED D-LATCH DESIGNS IN DLL AND ZVLL CONFIGURATIONS

A transistor level schematic of an SG-OTFT-based clocked D-latch in DLL configuration is shown in Figure 8.13a. All the physical dimensions of the driver (W = 2000 μm) and load (W = 400 μm) transistors are the same as in the SG-based SR latch implementation described earlier and the supply voltage is 10 V.

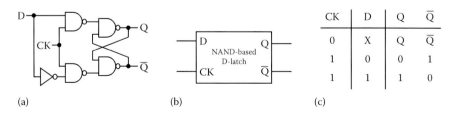

(a)                              (b)                              (c)

**FIGURE 8.12**    (a) Gate level schematic, (b) logic symbol, and (c) characteristic table of a NAND-based clocked D-latch.

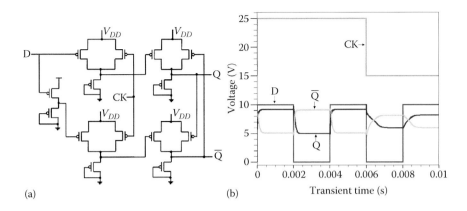

(a)                                                                                          (b)

**FIGURE 8.13**   (a) A transistor level schematic and (b) dynamic response of SG OTFT-based D-latch in DLL configuration.

The dynamic response of the D-latch is plotted in Figure 8.13b. The clock input signal (CK) is applied at the frequency of 100 Hz with 60% duty cycle and D-input at the frequency of 250 Hz. As the response plot of the latch reveals, Q output follows the changes in the D-input as long as the clock period is active, i.e., CK = 1. The output voltage swing is around 40%, wherein the Q (Qbar) swings between a maximum of 9 V and minimum of 5 V. The propagation delay time is approximately the same as calculated for the SG-based SR latch implementation. The correctness of the output shows close resemblance with the previous DLL design, where Q varies with input changes even during the inactive period of the clock.

A transistor level schematic of SG-based clocked D-latch in ZVLL configuration is given in Figure 8.14a with the supply voltage of 10 V, driver transistor

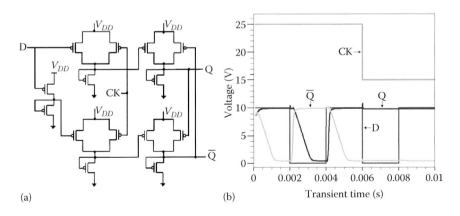

(a)                                                                                          (b)

**FIGURE 8.14**   (a) A transistor level schematic and (b) dynamic response of SG OTFT-based D-latch in ZVLL configuration.

width of 400 μm, and load transistor width of 2000 μm, which are same as the SG-based SR latch design.

A dynamic repsonse of the latch is given in Figure 8.14b. For fair comparison of the dynamic responses of the latches in DLL and ZVLL configurations, both designs were given the same stimuli of inputs with the clock frequency of 100 Hz and D-input of 250 Hz. The dynamic response plots given in Figure 8.14b reveals the same nature of the outputs as achieved in SR latch with a better output voltage swing greater than 90%. During the active period of the clock, the output signal Q follows the changes in the D-input signal creating a transparent "window" for propagating the input signal to the output terminal. During the inactive period of the clock signal, both outputs hold their levels to the current status irrespective of the changes in the input signal. These observations verify the correctness of the latch operation for all possible input conditions. The propagation delay time is calculated to be approximately 400 μs, which is more than DLL-based latch designs.

### 8.3.2.2 DG-BASED D-LATCH DESIGNS IN DLL AND ZVLL CONFIGURATIONS

A transistor level schematic of DG-based clocked D-latch in DLL configuration is shown in Figure 8.15a. Both gate terminals are tied together for all the OTFT devices. The widths of all the transistors are kept the same as described in the SG-based D-latch design with DLL configuration.

The design is analyzed under the same input stimuli depicted in the SG counterpart for better comparison. Similar to the SG-based design, during the active clock period, the Q and Qbar output signals confirm the correct operation of the D-latch. However, during the inactive period of the clock signal, output Q and Qbar show changes with the inputs. The dynamic response of the latch is given in Figure 8.15b. Propagation delay is calculated to be approximately

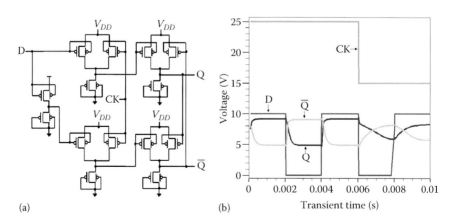

(a)    (b)

**FIGURE 8.15**    (a) A transistor level schematic and (b) dynamic response of DG OTFT-based D-latch in DLL configuration.

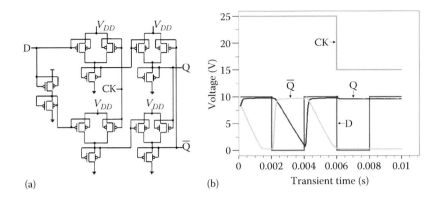

(a)                                                                    (b)

**FIGURE 8.16**   (a) A transistor level schematic and (b) dynamic response of DG OTFT-based D-latch in ZVLL configuration.

100 μs for the design. The voltage swing is negligibly more than the SG counterpart latch design. Therefore, the DG implementation of the latch in DLL mode shows negligible improvement as compared to the SG counterpart.

This following analyzes the DG-based clocked D-latch response in ZVLL configuration. The transistor level schematic of the design is shown in Figure 8.16a. The widths of all the devices and the input stimulus conditions are kept the same as that of the SG counterpart for better comparison.

The dynamic response of the DG-based clocked D-latch in ZVLL configuration is shown in Figure 8.16b. Output swing is improved and the pull-down transistor attains a low logic very close to 0. The output signals Q and Qbar confirm the correct operation of the D-latch resemblance to the SG counterpart, and during the inactive clock period the outputs preserve their states. However, the switching becomes slower than the SG counterpart. The propagation delay ($\tau_p$) of the DG-based latch in ZVLL configuration is approximately 550 μs as calculated from Figure 8.16b, which is more than the ZVLL configuration in SG-based design. The $\tau_{PLH}$ is 120 μs and $\tau_{PHL}$ is approximately 1000 μs, which is very large.

### 8.3.3  JK LATCH IMPLEMENTATIONS

A gate level schematic, logic symbol, and characteristic table of a NAND-based clocked JK latch are given in Figure 8.17.

A JK latch is a modified version of the SR latch to validate the restricted input condition S = 1 and R = 1 in the SR latch. The extension of Q output is fed back as the third input with the K and CK inputs. Similarly, the extension of Qbar is fed back with the J and CK inputs. With J = 1 and K = 1 inputs, the JK latch enters into "toggle" mode, where output Q toggles at every active clock period. If the clock period is made longer than the highest propagation delay time of the

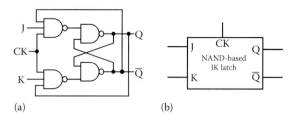

(a)                              (b)

| CK | J | K | Q | $\overline{Q}$ | Operation |
|----|---|---|---|----------------|-----------|
| 0  | X | X | Q | $\overline{Q}$ | Hold |
| 1  | 0 | 0 | Q | $\overline{Q}$ | Hold |
| 1  | 1 | 0 | 1 | 0              | Set |
| 1  | 0 | 1 | 0 | 1              | Reset |
| 1  | 1 | 1 | $\overline{Q}$ | Q          | Toggle |

(c)

**FIGURE 8.17**    (a) Gate level schematic, (b) logic symbol, and (c) characteristic table of NAND-based clocked JK latch.

latch, then it enters into oscillation mode, where the output switches to 1 and 0 and vice versa. To avoid this "race around" condition, the clock period should be made less than the propagation delay time of the latch. For other input combinations, the Q and Qbar outputs present the same response as the SR latch.

### 8.3.3.1 SG-BASED JK LATCH DESIGNS IN DLL AND ZVLL CONFIGURATIONS

A transistor level schematic of SG-OTFT-based clocked JK latch in DLL configuration is shown in Figure 8.18a. All the physical dimensions of the driver (W = 2000 μm) and load (W = 400 μm) transistors are the same as in the SG-based SR latch implementation described in the previous section, and the supply voltage is 10 V.

The dynamic response of the JK latch is plotted in Figure 8.18b. To avoid the oscillation condition while applying J = K = 1 input combination, the clock signal (CK) frequency of 200 Hz with 20% duty cycle is applied.

The J-input is stimulated with a 10 ms time period signal and 60% duty cycle. Similarly, the K-input signal is applied with a time period of 8 ms and 43% duty cycle. The dynamic response plot of the latch confirms the various operations mentioned in the characteristic table. However, as can be seen from the response plots, the output voltage swing reduces to 30% making the design ineffective for multiple stages of implemetation. Also, during the inactive period of the clock both outputs show the changes with the input signal changes, which reflect the incorrect results for various input combinations. The propagation

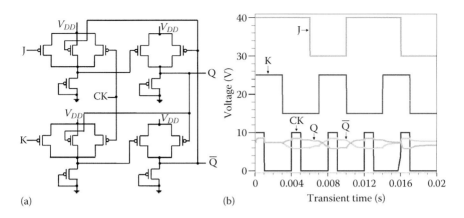

**FIGURE 8.18**    (a) A transistor level schematic and (b) dynamic response of SG OTFT-based JK latch in DLL configuration.

delay is longer and differ in a larger manner compared to the previous SG or DG OTFT-based clocked latch designs in DLL configuration due to incorrect signal levels propagated to the subsequent stages.

A transistor level schematic of SG OTFT-based clocked JK latch in ZVLL configuration is shown in Figure 8.19a. All the physical dimensions of the driver (W = 400 μm) and load (W = 2000 μm) transistors are the same as in the SG-based SR latch implementation in ZVLL configuration described earlier with the supply voltage of 10 V.

The dynamic response of the JK latch is plotted in Figure 8.19b. All the input stimuli are kept the same as the SG-based JK latch in DLL configuration for fair comparison of the two designs. The response plots confirm the various operations mentioned in the characteristic table of the latch. Unlike the response of

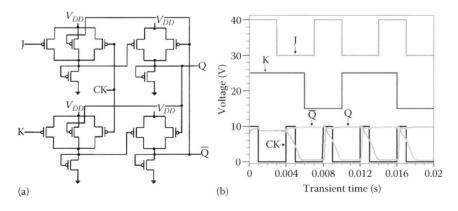

**FIGURE 8.19**    (a) A transistor level schematic and (b) dynamic response of SG OTFT-based JK latch in ZVLL configuration.

the JK latch in DLL configuration discussed in the previous section, response for the ZVLL configuration shows proper JK latch operations for various input combinations mentioned in the characteristic table of the latch and states of the outputs change only at the active duration of the clock signal.

Propagation delay of the latch is approximately 750 μs, which is more in comparison to the previous designs. The reason for the longer propagation delay is the addition of the one input to the front NAND gate, which is an extension of output signal fed back to the input.

### 8.3.3.2 DG-BASED JK LATCH DESIGNS IN DLL AND ZVLL CONFIGURATIONS

A transistor level schematic of DG-OTFT-based clocked JK latch in DLL configuration is shown in Figure 8.20a. The dimensions of all the transistors are kept the same as described in the SG-based latch design with DLL configuration.

The response of the latch is given in Figure 8.20b, which is identical to the response of the SG-based JK latch in DLL configuration. The present DLL configuration shows incorrect operation for the active and inactive period of the clock with very low output swing, making the design nonapplicable for the integral part of any digital system or as an individual element.

A transistor level schematic of DG OTFT-based clocked JK latch in ZVLL configuration is shown in Figure 8.21a. The widths of all the transistors are kept the same as that described in the SG-based latch design with ZVLL configuration. The dynamic response of the latch is presented in Figure 8.21b. The clock input signal with the period of 5 ms and duty cycle of 20% is applied for the avoidance of the oscillations when J = K = 1 are stimulated at the input terminals.

The dynamic response shows correctness of the output signals for various excitations tabulated in the characteristic table of the latch. The output signal

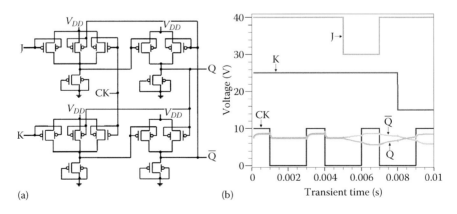

(a)    (b)

**FIGURE 8.20**    (a) A transistor level schematic and (b) dynamic response of DG OTFT-based JK latch in DLL configuration.

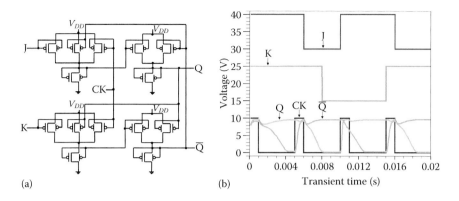

(a)                                                                      (b)

**FIGURE 8.21**   (a) A transistor level schematic and (b) dynamic response of DG OTFT-based JK latch in ZVLL configuration.

changes its state only during the active period of the clock signal, and during the inactive period of the clock signal it preserves the current state. Voltage swing of the design is slightly improved as compared to its SG design counterpart, and the pull-down transistor output reaches very close to logic 0. The propagation delay is higher and calculated to be 800 μs.

### 8.3.4  T-LATCH IMPLEMENTATIONS

A gate level schematic, logic symbol, and characteristic table of the T-latch is shown in Figure 8.22. T-latch is the special case of JK latch, where the J and K inputs are tied together. During the active period of the clock signal, if T = 0 then both outputs preserve their states, and if T = 1 then both outputs set to the complementary states of the previous state, i.e., as long as T = 1, outputs toggle. Hence, the name "toggle" (T) latch. During the inactive period of the clock, both outputs hold their current states.

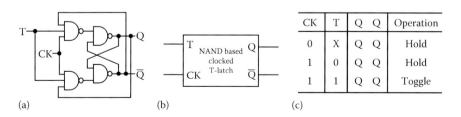

| CK | T | Q | Q | Operation |
|----|---|---|---|-----------|
| 0  | X | Q | Q | Hold      |
| 1  | 0 | Q | Q | Hold      |
| 1  | 1 | Q | Q | Toggle    |

(a)                                  (b)                                  (c)

**FIGURE 8.22**   (a) Gate level schematic, (b) logic symbol, and (c) characteristic table of NAND-based clocked T-latch.

### 8.3.4.1 SG-BASED T-LATCH DESIGNS IN DLL AND ZVLL CONFIGURATIONS

A transistor level schematic of the SG-OTFT-based T-latch in DLL configuration is presented in Figure 8.23a. All the dimensions are the same as mentioned for the SG-based JK latch. The dynamic response of the T-latch in DLL configuration is plotted in Figure 8.23b. The design shows incorrect operation for T = 0 as well as T = 1. The output swing is reduced to 20%, which may not be sufficient to discriminate between logic 0 and 1. Hence, the DLL configuration consistently shows poor and incorrect performance.

A transistor level schematic of SG-based clocked T-latch in ZVLL configuration is shown in Figure 8.24a. With all equal dimensions of the transistors mentioned in the SG-based JK latch counterpart, the T-latch gives better and correct performance. The dynamic performance of Figure 8.24b reveals this fact. The

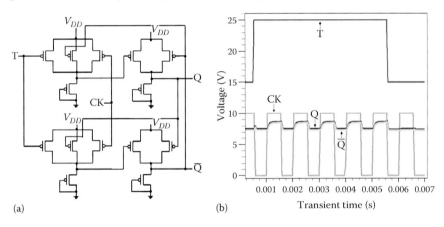

**FIGURE 8.23**   (a) A transistor level schematic and (b) dynamic response of SG OTFT-based T-latch in DLL configuration.

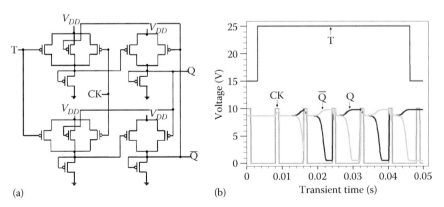

**FIGURE 8.24**   (a) A transistor level schematic and (b) dynamic response of SG OTFT-based T-latch in ZVLL configuration.

output swing is greater than 90%. The propagation delay $\tau_{PLH}$ is approximately 200 µs and $\tau_{PHL}$ is 5000 µs. The overall propagation delay is 2600 µs.

### 8.3.4.2  DG-BASED T-LATCH DESIGNS IN DLL AND ZVLL CONFIGURATIONS

A transistor level schematic of the DG-based T-latch is given in Figure 8.25a. The dynamic response of the latch is given in Figure 8.25b, which is similar to the SG-based T-latch response.

A transistor level schematic of the DG-based T-latch in ZVLL configuration is given in Figure 8.26a. With the same dimensions of the transistors mentioned for SG-based T-latch design, the dynamic performance of Figure 8.26b shows the correct operation verifying the characteristic table for the latch. As predicted, it gives a propagation delay of approximately 2100 µs.

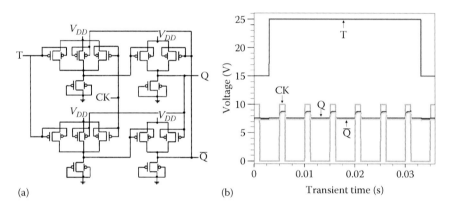

(a)    (b)

**FIGURE 8.25**    (a) A transistor level schematic and (b) dynamic response of DG OTFT-based T-latch in DLL configuration.

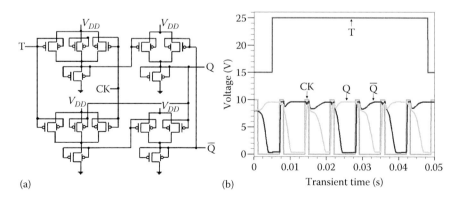

(a)    (b)

**FIGURE 8.26**    (a) A transistor level schematic and (b) dynamic response of DG OTFT-based JK latch in ZVLL configuration.

## 8.4 CONCLUDING REMARKS

This chapter summarizes the dynamic responses of benchmark sequential circuits such as RS, D, JK, and T latches designed using single and dual gate OTFT devices. All the circuits are built using universal NAND gate incorporating synchronous clock operation in conjunction with latch inputs.

A comparison of output swing and propagation delay of all the latch designs based on SG and DG structures in DLL and ZVLL configurations is given in Table 8.1. As shown in the dynamic responses of the designed latches presented in the previous sections, DLL configuration fails to reproduce defined outputs for several input combinations and synchronous operations with the clock. On the contrary, all the designs in ZVLL configuration produce the expected results for the respective applied inputs verifying the characteristics presented in the form of tables for respective latches.

The following are the concluding remarks based on the simulation results obtained for the different latch designs:

- Responses are presented for DLL and ZVLL configuration separately for SG and DG OTFT-based latch designs. Clearly, the latch designs in ZVLL configuration outperform the DLL configuration. The output swings for the designs based on ZVLL configuration are observed in

**TABLE 8.1  Comparison of NAND-Based Latch Responses in Terms of Their Output Swings and Propagation Delays**

| Type of Latch | Load Configuration | Voltage Swing (V) | $\tau_{PLH}$ (µs) | $\tau_{PHL}$ (µs) | $\tau_P$ (µs) |
|---|---|---|---|---|---|
| SR latch | DLL (SG) | 4.10 | 50 | 80 | 65 |
|  | ZVLL (SG) | 9.20 | 140 | 650 | 395 |
|  | DLL (DG) | 4.40 | 70 | 90 | 80 |
|  | ZVLL (DG) | 9.40 | 200 | 900 | 550 |
| D-latch | DLL (SG) | 4.15 | 50 | 100 | 75 |
|  | ZVLL (SG) | 9.40 | 175 | 625 | 400 |
|  | DLL (DG) | 4.20 | 90 | 160 | 125 |
|  | ZVLL (DG) | 9.40 | 200 | 1100 | 650 |
| JK latch | DLL (SG) | 3.00 | – | – | – |
|  | ZVLL (SG) | 9.35 | 200 | 1200 | 700 |
|  | DLL (DG) | 3.00 | – | – | – |
|  | ZVLL (DG) | 9.35 | 200 | 1200 | 700 |
| T-latch | DLL (SG) | 1.50 | – | – | – |
|  | ZVLL (SG) | 9.30 | 200 | 5000 | 2600 |
|  | DLL (DG) | 1.50 | – | – | – |
|  | ZVLL (DG) | 9.30 | 200 | 4000 | 2100 |

the range of 92% to 94% of the supply voltage, which is very high in comparison to the poor output swings of 15% to 40% available in DLL configuration.

- The accuracy of the responses of the designs built using OTFT devices for all possible combinations of input signals is the major concern for researchers who are in the process of innovating stable, robust, and reliable devices and circuit design techniques.

- A comparison of output swing and propagation delay of all the latch designs based on SG and DG structures in DLL and ZVLL configurations is summarized in Table 8.1. As shown by the dynamic responses of the designed latches presented in the previous sections, DLL configuration fails to reproduce defined outputs for several input combinations and synchronous operations with the clock. On the contrary, all the designs in ZVLL configuration produce the expected results for the respective applied inputs verifying the characteristic tables of respective latches.

- Regarding the propagation delay of the designs calculated from the responses, DLL configuration shows better performance as compared to ZVLL configuration. However, for JK and T-latch designs, propagation delay calculations are not measurable due to incorrect responses. Hence, DLL configuration requires some modifications to improve the output voltage swings and correct logic result production. On the other hand, ZVLL produces correct logical results with higher voltage swings, although with longer propagation delay. For T-latch, it reaches up to 2.6 ms.

- All ZVLL-configuration-based designs show very long $\tau_{PHL}$ as compared to DLL configuration. The main reason for the slower response of the ZVLL configuration is the very weak pull-down current available while the output terminal swings to logic 0. Furthermore, DG implementation of all the latches shows negligible improvement in voltage swing compared to the SG counterpart. However, propagation delay increases for all the designs.

- Overall, from the designer's selection point of view the ZVLL configuration offers higher swing and correct output as per the characteristic table of the latches; however, DLL configuration still needs modifications in the design parameters or enhancement in the configuration to improve the voltage swing for correct output production in multistage circuit operations.

## PROBLEMS
## MULTIPLE CHOICE

1. The use of OTFT devices is restricted to the basic digital and analog benchmark circuits due to
   a. Lower mobility
   b. Instability

   c.  Repeatability
   d.  All of these
2. To improve the performance of analog and digital circuits, designers apply several new techniques such as
   a.  Dual threshold voltage
   b.  Dual gate structure
   c.  Different load configurations
   d.  All of the these
3. The major reason for the incorrect outputs and false logical results of the DLL configuration is
   a.  Increase in width of load transistor
   b.  Faster switching
   c.  Poor swing
   d.  None of the above
4. The forbidden input condition of SR latch is
   a.  $S = 0, R = 0$
   b.  $S = 0, R = 1$
   c.  $S = 1, R = 0$
   d.  $S = 1, R = 1$
5. The propagation delay of ZVLL is
   a.  Lower than DLL
   b.  Higher than DLL
   c.  Equal to DLL
   d.  None of these
6. In DLL configuration, the propagation delay of the dual gate SR latch is
   a.  Higher than single gate
   b.  Lower than single gate
   c.  Comparable to single gate
   d.  None of these
7. The switching speed of the dual gate SR latch in ZVLL configuration is
   a.  Lower than single gate
   b.  Equal to single gate
   c.  Higher than single gate
   d.  None of these
8. The oscillation condition in JK flip-flop occurs when
   a.  $J = 1, K = 1$
   b.  $J = 0, K = 1$
   c.  $J = 1, K = 0$
   d.  $J = 0, K = 0$
9. Which of the following configurations has the largest voltage swing?
   a.  Dual gate D latch in DLL
   b.  Dual gate SR latch in ZVLL
   c.  Single gate JK latch in DLL
   d.  Single gate T latch in DLL

10. Which of the following configurations has highest propagation delay?
    a. Single gate SR latch in DLL
    b. Single gate D latch in ZVLL
    c. Dual gate JK latch in ZVLL
    d. Dual gate T latch in ZVLL

**ANSWER KEY**

1. d; 2. d; 3. c; 4. d; 5. b; 6. c; 7. a; 8. a; 9. b; 10. d

**SHORT ANSWER**

1. With the help of diagrams, explain single gate-based 2:1 mux in DLL and ZVLL configurations.
2. Discuss the single gate OTFT-based clocked T latch in DLL configuration.
3. Explain the single gate SR latch in ZVLL configuration.
4. Discuss dual gate OTFT-based clocked JK latch in ZVLL configuration.
5. Differentiate dual gate OTFT-based D latch in DLL and ZVLL configuration.

## REFERENCES

1. Jacob, S.; Abdinia, S.; Benwadih, M.; Bablet, J. "High performance printed N and P-type OTFTs enabling digital and analog complementary circuits on flexible plastic substrate," *Solid-State Electron.* **2013**, 84, 167–178.
2. Koo, J. B.; Ku, C. H.; Lim, J. W.; Kim, S. H. "Novel organic inverters with dual gate pentacene thin film transistor," *Org. Electron.* **2007**, 8, 552–558.
3. Nausieda, I.; Ryu, K. K.; He, D. D.; Akinwande, A. I.; Bulovic, S.; Sodini, C. G. "Mixed signal organic integrated circuits in a fully photolithographic dual threshold voltage technology," *IEEE Tran. Electron. Dev.* **2011**, 58, 865–873.
4. Kumar, B.; Kaushik, B. K.; Negi, Y. S.; Goswami, V. "Single and dual gate OTFT based robust organic digital design," *Microelectronics Reliability* **2014**, 54, 100–109.
5. Myny, K.; Beenhakkers, M. J.; Van Aerle, N. A. J. M.; Gelinck, G. H.; Genoe, J.; Ehaene W. *et al.* "Unipolar organic transistor circuits made robust by dual-gate technology," *IEEE J. Solid-State Circuit* **2011**, 46, 1223–1230.
6. Martin, D.; Patchett; E. R.; Williams, A.; Neto, N. J.; Ding, Z.; Assender, H. E.; Morrison, J. J.; Yeates, S. G. "Organic digital logic and analog circuits fabricated in a roll-to-roll compatible vacuum-evaporation process," *IEEE Trans. Electron. Dev.* **2014**, 61(8), 2950–2956.
7. Takeda, Y.; Yoshimura, Y.; Adib F. "Flip-flop logic circuit based on fully solution-processed organic thin film transistor devices with reduced variations in electrical performance," *Japn. J. Appl. Phys.* **2015**, 54, 04DK03.
8. Mittal, P.; Kumar, B.; Kaushik, B. K.; Negi, Y. S.; Singh, R. K. "Channel length variation effect on performance parameters of organic field effect transistors," *Microelectronic J.* **2012**, 43(12), 985–994.

9. Kumar, B.; Kaushik, B. K.; Negi, Y. S. "Design and analysis of noise margin, write ability and read stability of organic and hybrid 6-T SRAM," *Microelectronics Reliability* **2014**, 54(12), 2801–2812.

10. Kumar, B.; Kaushik, B. K.; Negi, Y. S. "Static and dynamic characteristics of dual gate organic TFT based NAND and NOR circuits," *J. Computat. Electronics* **2014**, 13(3), 627–638.

# Static Random Access Memory Cell Design Based on All-$p$ Organic, Hybrid, and Complementary Organic Thin-Film Transistors

9

## 9.1 INTRODUCTION

Over the last decade, memory devices based on organic materials have attracted lots of interest due to their better structural flexibility and low-cost solution processability. Inexpensive and fast memory devices characterized by longer data retention time and higher density are in huge demand. Organic memory demonstrates unique characteristics, including being nonvolatile, flexible, fast, inexpensive, lightweight, and capable of printing components directly onto the flexible substrates. This combines the best features of three memory types: (1) nonvolatility of flash memory; (2) rapidity of dynamic random access memory (DRAM); and (3) the cycling endurance, high data density, and lower cost of hard disk drives. These memory devices are potentially useful in organic radio-frequency identification tags (transponder chips), smart cards, and disposable circuitry.

Semiconductor memory arrays capable of storing large digital information are essentially required in all digital systems. Random access memory (RAM) allows users to retrieve as well as modify data bits already stored in the memory array. Operating strategy classifies RAM into dynamic RAM (DRAM) and static RAM (SRAM). A DRAM cell consists of a capacitor for storage that requires periodic read and rewrite operations due to charge leakage through the capacitor. On the other hand, SRAM is realized as a bistable latch circuitry to store the data bits. SRAM is mainly used as the cache memory in microprocessors, workstations, routers, and mainframe computers due to its high speed and low power consumption. Besides this, they are frequently used as high speed registers, hard disk buffers, and microcontrollers (capacity of 32B–128KB).

Storage capacity and operational speed are the two major factors in determining the efficiency of memory. The performance of silicon-based memory is quite high; however, the production cost is apparently a big limitation. Organic SRAM provides better designs in terms of mechanical flexibility and is an economically viable solution capable enough to attain a reasonable performance. Although memory is a key element in many organic thin-film transistor (OTFT)-based applications, few efforts have been devoted to creating the organic SRAM cell [1–3]. To the best of my knowledge, SRAM cells fabricated with different material combinations for driver, load, and access transistors have not been reported. Such a SRAM cell based on optimized heterogeneous material could demonstrate better performance. It is therefore imperative to look into such SRAMs that address the effect of different materials and configurations.

Organic circuits are often created in all-$p$ configuration due to comparatively inferior performance of organic $n$-type transistors. However, in recent years fullerene, $C_{60}$-based, high-performance $n$-type organic transistors have been reported [4,5]. Besides this, inorganic transistors based on zinc oxide, ZnO $n$-type semiconductors also reportedly [6] can be used to create flexible complementary memory circuits due to their fabrication at lower temperature. The memory circuits and fabrication methodology need to be driven toward more compact design rules with high data storage density. In Chapter 6, the performance of inverter circuits based on different organic and inorganic thin-film transistors (TFTs) was analyzed, wherein pentacene–$C_{60}$, pentacene–ZnO, and pentacene–pentacene combinations exhibited reasonably good static and dynamic performances. Motivated by the outcomes, the pentacene–$C_{60}$, pentacene–ZnO, and pentacene–pentacene combinations are used for designing organic all-$p$, organic complementary, and hybrid complementary SRAM cell configurations, respectively. The performance of a SRAM cell made up of organic and inorganic material combinations is analyzed and compared through 2-D numerical device simulation. Static noise margin (SNM), read stability, and write access time are obtained for different SRAM cell designs at different cell and pull-up ratios. Additionally, a comparison of the performance of organic complementary SRAM cells based on *n*-type and *p*-type access transistors is discussed in this chapter.

This chapter presents design and comparison of all-*p* organic (pentacene), organic complementary (pentacene–$C_{60}$), and hybrid (pentacene–ZnO) SRAM cells. Performance analysis of different SRAM cell designs mainly in terms of static noise margin, read stability, and write ability at different cell ratios and pull-up ratios is discussed. Furthermore, verification of analytically obtained SNM of all-*p* organic SRAM cells with respect to the simulation results is demonstrated. Additionally, optimization of cell ratios and pull-up ratios for read and write operations are carried out. An analysis of organic complementary SRAM cells using pentacene-based *p*-type access transistors in place of *n*-type to improve the overall performance is also discussed.

This chapter is arranged in five sections along with the current introductory Section 9.1. The basic operation of all-*p* and hybrid/organic complementary SRAM cells is discussed in Section 9.2, and the performance of SRAM cells with different thin-film transistor combinations is analyzed for different cells and pull-up ratios in Section 9.3. A comparison between organic SRAM cells with *p*- and *n*-type access transistors is presented in Section 9.4. Finally, important outcomes of the chapter are summarized in Section 9.5.

## 9.2  STATIC RANDOM ACCESS MEMORY (SRAM) CELL CONFIGURATION

A static RAM cell consists of a latch unit (two inverters connected back to back) with two stable states 0 and 1 (1 and 0) at node 1 (Data) and node 2 (Datab) (complement of data), respectively, that effectively keeps the storage as long as the power remains *on*. Figure 9.1 shows the block diagram of one bit SRAM cell, wherein two access transistors are connected to the bistable latch circuitry. These access transistors are turned *on* when a word line (WL; rows) is activated for read and write operation, connecting the cell to the complementary bit-lines

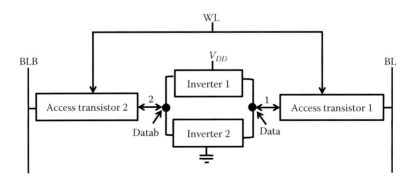

**FIGURE 9.1**  Block diagram representation of SRAM cell architecture.

(BL and BLB; columns). When the cell is not addressed, the data is kept to a stable state latched within the flip-flop [7].

The conventional 6-TFT SRAM cell shown in Figure 9.2 consists of two complementary inverters (*P1-N1* and *P2-N2*) connected back to back, wherein *n*-type and *p*-type TFTs are used as the driver and load, respectively. The inverters are connected to the BL and BLB through *n*-type access transistors (*N3* and *N4*) operated by the word line voltage (10 V). In the standby mode, biasing voltage at WL is kept low (0 V), thus turning *off* the access transistors, which in turn isolate the bit lines from the cross-coupled inverter pair.

An all-*p* organic SRAM cell (Figure 9.3) consists of *p*-type transistors as the driver (*P1* and *P2*), load (*P3* and *P4*), and access (*P5* and *P6*) transistors, wherein load transistors are connected in zero-$V_{gs}$ configuration that produces high noise margin. The access transistors enable the stored data to read or write the desired data when these transistors are turned *on* by keeping *WL* at 0 V. Otherwise, the cell remains in standby state, which maintains its storage until

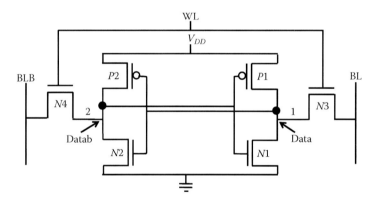

**FIGURE 9.2**   Schematic of hybrid/organic complementary SRAM cell.

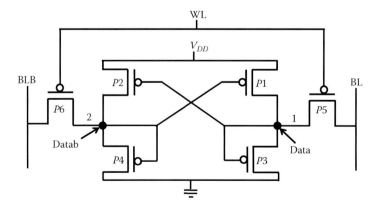

**FIGURE 9.3**   Schematic of all-*p* organic SRAM cell.

$V_{DD}$ is kept at 10 V. The performance of the all-$p$ 6-TFT SRAM configuration is analyzed using pentacene-based OTFTs. In addition, the conventional SRAM architecture is created with two different material combinations for $p$- and $n$-type TFTs: (1) pentacene (organic $p$-type) and ZnO (inorganic $n$-type); and (2) pentacene (organic $p$-type) and $C_{60}$ (organic $n$-type). Independent designs and performance of different SRAM configurations are discussed in the following section.

## 9.3 PERFORMANCE OF SRAM CELL WITH DIFFERENT THIN-FILM TRANSISTOR COMBINATIONS

The performance of an SRAM cell is characterized mainly in terms of SNM during standby and read operations. Besides this, the time required to read the stored data correctly without flipping it and the speed of modifying the stored data during write operation are other important performance-measuring parameters. It is becoming increasingly challenging to maintain the performance of cells considering downscaling of the feature size. While selecting the $W/L$ ratios for the three TFTs (pull-up, pull-down, and access), two basic requirements—nondestruction of the stored information during read operation and modification of the stored data during write operation—must be met. The transconductance ratio of the pull-down transistor to the access transistor is known as the cell ratio ($\beta_c$) that determines the cell stability during a read operation. However, the transconductance ratio of the pull-up to the access transistor is termed as the pull-up ratio ($\beta_p$), which determines the writing ability of the cell [8]. The performance of an SRAM cell in terms of SNM and ability to read and write the data is discussed next for different combinations of TFTs used to design the cell.

### 9.3.1 STATIC NOISE MARGIN

The stability of an SRAM cell is a crucial functional parameter that determines the ability of stored data retention both in standby and read access modes. The stability, usually defined by the SNM, is a measure of maximum static noise voltage that can be tolerated by the cell without altering the stored data [9]. This can be represented graphically by plotting superimposed voltage transfer characteristics (VTCs) of two symmetrical inverters resulting in a two-lobed curve. The area inside the two lobes is a measure of sensitivity of the cell to the external noise, whereas the SNM is represented by the side of the maximum possible nested square between the two superimposed curves. In an all-$p$ SRAM cell, it is necessary that the drain current of load TFT be kept constant, which can be achieved by connecting the gate and source terminals of the load leading to a ZVLL (zero-$V_{gs}$ load logic) configuration. In this configuration, the load TFT always has a conducting channel regardless of input and output voltage levels, since the condition $V_{gs} > V_t$ is satisfied.

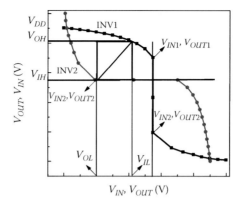

**FIGURE 9.4**  VTCs of two symmetrical inverters of *p*-type SRAM in ZVLL configuration.

Figure 9.4 shows the VTCs of two symmetrical inverters of an all-*p* SRAM cell. In the VTC, $V_{IN1}$ represents the input voltage required to transit the conduction of driving TFT from linear to the saturation regime while the load TFT is operating in the saturation region. Similarly, voltage $V_{IN2}$ is the input voltage when the load TFT switches from the saturation to linear regime, while the driving TFT remains in the saturation. When $V_{IN}$ lies between 0 and $V_{IN1}$, the driver and load TFTs operate in the linear and saturation regime, respectively. Considering the same current flowing in the two TFTs, the output voltage [10] can be expressed as

$$V_{OUT} = f_{lin,sat}(V_{IN}) = V_{IN} - V_{t,D} + \sqrt{(V_{DD} + V_{t,D} - V_{IN})^2 - \left(\frac{1}{k_R}\right)V_{t,L}^2} \; ; 0 < V_{IN} < V_{IN1}$$

(9.1)

where $V_{t,D}$ and $V_{t,L}$ are the threshold voltage of driver and load TFTs, respectively, and $k_R$ is the transconductance ratio of the driver to the load. Furthermore, the driver and load TFTs operate in the saturation and linear regime when the input voltage lies between $V_{IN2}$ and $V_{DD}$ leading to the output voltage expression as

$$V_{OUT} = f_{sat,lin}(V_{IN}) = V_{t,L} - k_R \sqrt{\frac{1}{k_R^2}V_{t,L}^2 - \frac{1}{k_R}(V_{DD} + V_{t,D} - V_{IN})^2} \; ; V_{IN2} < V_{IN} < V_{DD}$$

(9.2)

For an input voltage between $V_{IN1}$ and $V_{IN2}$, both TFTs operate in the saturation region. Since most organic TFTs exhibit high output resistance, $V_{IN1}$ and $V_{IN2}$ can be assumed to have approximately the same potential, i.e., $V_{IN}$ [11].

The current relation for driver and load TFTs operating in the saturation region can be expressed as

$$\frac{k_L}{2}(0-V_{t,L})^2 = \frac{k_D}{2}(V_{IN}-V_{DD}-V_{t,D})^2 \tag{9.3}$$

where $k_L$ and $k_D$ are the respective transconductance of load and driver transistors. $V_{IN1}$ and $V_{IN2}$ can be expressed as

$$V_{IN} = V_{IN1} = V_{IN2} = V_{DD} + V_{t,D} - \sqrt{\left(\frac{1}{k_R}\right)}V_{t,L} \tag{9.4}$$

Additionally, the corresponding output voltage, $V_{OUT1}$ and $V_{OUT2}$, can be derived by substituting $V_{IN1}$ and $V_{IN2}$ in Equations 9.1 and 9.2, respectively,

$$V_{OUT1} = V_{DD} - \sqrt{\left(\frac{1}{k_R}\right)}V_{t,l} \tag{9.5}$$

$$V_{OUT2} = V_{t,l} \tag{9.6}$$

Furthermore, to calculate the noise margin, a point ($V_{IN2}$ and $V_{OUT2}$) is marked on the transfer characteristics of inverter 2. From this point, a straight line ($y = x$) is drawn that forms the right-top corner (at the intersection point of the straight line with the VTC of inverter 1) of the largest possible nested square in the lobe. With the known expression [10,11] of the transfer curve, the SNM in terms of side length of the fitted square can be obtained as

$$SNM = V_{DD} - \sqrt{\left(V_{DD} + V_{t,D} - \sqrt{\left(\frac{1}{k_R}\right)}V_{t,L}\right)^2 + \left(\frac{1}{k_R}\right)V_{t,L}^2} \tag{9.7}$$

Noise margin strongly depends on the $V_{DD}$, $k_R$, and threshold voltages of driver and load TFTs. As previously discussed, the width of the load transistor (connected in zero-$V_{gs}$ mode) should be much higher than the driver so as to achieve an adequate pull-down response. Therefore, the width ratio of the load to the driver is 15. The channel length of TFTs is 90 μm throughout SRAM cell designs.

The resulting butterfly curves for the inverters of a *p*-SRAM cell during standby and read (access) modes are shown in Figure 9.5a. The SNM during hold state is 1.46 V, which closely matches to the analytically obtained value of 1.5 V with an error of 2.6%. To achieve the high noise margins, the output high

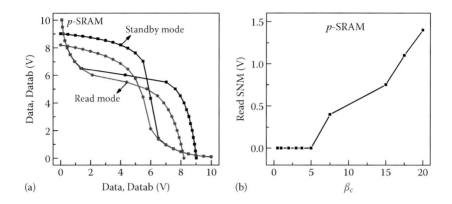

**FIGURE 9.5**    (a) Butterfly curves during standby and read mode and (b) read SNM (RSNM) as a function of the cell ratio for an all-$p$ SRAM cell.

voltage ($V_{OH}$) should be high enough. However, in only $p$-type designs, the value of $V_{OH}$ is limited to $V_{DD}-V_{t,L}$ [10]. It is due to a threshold voltage drop in the load TFT, since it always remains in the *on* condition, thereby attaining a lower magnitude even in the standby mode. During a read operation, the high-to-low transition of the inverter is not as steep as that obtained during standby mode. It is due to a reduction in the strength of the pull-down transistor. A slow pull-down action reduces the area of the lobe and thus SNM too.

The plot of SNM during a read operation at different cell ratios is shown in Figure 9.5b for the $p$-SRAM cell. The noise margin is near to the desired value (SNM in standby mode) at $\beta_c$ of 20 and reduces significantly with scaling down the cell ratio. This cell is found unstable at $\beta_c \leq 5$, since the potential difference of the two complementary storage points is zero, which is due to stronger access transistors. This in turn produces ambiguity in reading the storage at two nodes.

The combined butterfly curves obtained during the hold and read access modes for the hybrid and organic complementary SRAM cells are plotted in Figure 9.6. The width of pull-up transistors, $P1$ and $P2$, in all the proposed SRAM cells is 500 μm, whereas the width ratio of the pull-down to the pull-up transistor is 10 and 18 for the hybrid and organic complementary SRAM cell, respectively.

The hold SNM of SRAM cells based on pentacene–ZnO TFTs is more than 2 times higher than the all-$p$ SRAM. High noise immunity due to larger noise margins is one of the major benefits of using complementary SRAM designs. Compared to the hybrid cell, the organic complementary SRAM cell shows a reduction of 18% in SNM, since the high-to-low transition is not that steep. It is due to a reduction of 15% and 72% in the mobility and capacitance of $C_{60}$-based TFTs as compared to ZnO. Combine plots of read SNM for hybrid and organic SRAM with respect to the cell ratio are illustrated in Figure 9.7. Compared to organic complementary cells, the read SNM of hybrid cells is higher, and a

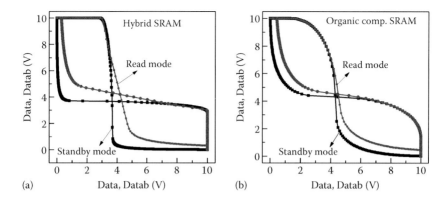

**FIGURE 9.6**    Butterfly curves during standby and read mode for (a) hybrid and (b) organic complementary SRAM cells.

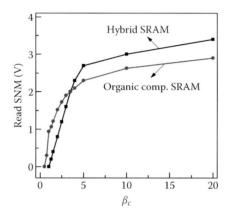

**FIGURE 9.7**    Read SNM with respect to the cell ratio, $\beta_c$, for hybrid and organic complementary SRAM cells.

proportionate reduction is observed for both cells with scaling down the cell ratio up to 5. In addition, the hybrid cell exhibits a sharp fall in the read SNM for the cell ratio below 5. It is due to a high leakage current of ZnO-based transistors as well as a significant change in the potential of stored data with the effect of increasing the strength of access transistors.

### 9.3.2  READ OPERATION

During a read operation, the cell is coupled to the bit lines through the access transistors. The word line voltage turns *on* the access transistors that allow the logical state of the cell to be sensed through the bit lines. Prior to turning *on* the access transistors, both bit lines are precharged to the $V_{DD}/2$ level. When the access TFTs are tuned *on*, it allows the flow of the charge from *BL* to node 1 as well as from node 2 to BLB (if 0/1 is stored at node 1/2). Therefore, a voltage divider is formed through access and pull-down transistors between the precharged bit line potential and ground. The potential difference of bit lines is accessed by the sensing amplifier [7], as shown in Figure 9.8.

The sensing amplifier sends an alert to the word line to be reverted from the current state after sensing the potential difference in the bit lines. This in turn isolates the bit lines from the cell, thereby leading the node potentials to return back to their previous values. During a read access, the node voltage deviates from its standby state, which strongly depends on the cell ratio. A low cell ratio translates to a reduction of potential difference of the complementary storage, thereby increasing the possibility of flipping the state [8]. The operational circuit of a complementary SRAM cell with storage 0 at node 1 (Data) and 1 at node 2 (Datab) is illustrated in Figure 9.9. A low cell ratio can force an unintended change in the stored information by increasing the node potential (storage 0) above the threshold voltage of pull-down TFT. Therefore, to avoid the violation of stored data during a read access, the condition that must be satisfied can be given as

$$V_{max,node-1} < V_{t,N2} \tag{9.8}$$

where $V_{max,node-1}$ and $V_{t,N2}$ represent the maximum voltage at node 1 and the threshold voltage of the *N2* transistor, respectively. To access the stored data during a read operation, it is required to build a voltage difference between the

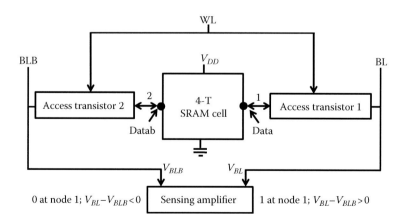

**FIGURE 9.8**   Block diagram of SRAM cell along with sensing amplifier in read mode.

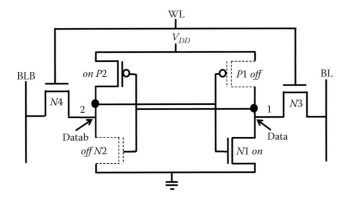

**FIGURE 9.9**    Operational circuit schematic of hybrid/organic SRAM cell for storage 0 at Data and 1 at Datab nodes.

bit lines without altering the node voltages too much. To meet this requirement, the following condition should be satisfied

$$R_{on}(N_1) < R_{on}(N_3) \tag{9.9}$$

where $R_{on}(N_1)$ and $R_{on}(N_3)$ are the respective resistance of the pull-down TFT ($N1$) and access TFT ($N3$). This implies that the $N1$ transistor should be stronger than the $N3$, and that can be achieved by setting the width of the pull-down transistor to be high enough than the access transistor. During a read access, the transistor $N3$ operates in saturation, whereas $N1$ operates in the linear region. Considering the same current in both TFTs

$$\frac{k_{N3}}{2}(V_{DD} - V_{node-1} - V_{t,n})^2 = \frac{k_{N1}}{2}\left\{2(V_{DD} - V_{t,n})V_{node-1} - V_{node-1}^2\right\} \tag{9.10}$$

$$\frac{k_{N3}}{k_{N1}} = \frac{(W/L)_{N3}}{(W/L)_{N1}} < \frac{2(V_{DD} - 1.5V_{t,n})V_{t,n}}{(V_{DD} - 2V_{t,n})^2} \tag{9.11}$$

By symmetry, the same relation is considered for TFTs $N2$ and $N4$, as specified by Equation 9.11. Transient response of a pentacene–ZnO-based cell during reading of 0 (1) and 1 (0) at the Data (Datab) node is shown in Figure 9.10. The respective waveforms for a pentacene–$C_{60}$ SRAM cell are presented in Figure 9.11. After turning *on* the access transistors, the bit line voltage ($V_{BL}$) is decreased (Figures 9.10a and 9.11a) in both cells, but the potential of the Data node is maintained close to the 0 level, since TFT $N1$ is operating in the linear mode. On

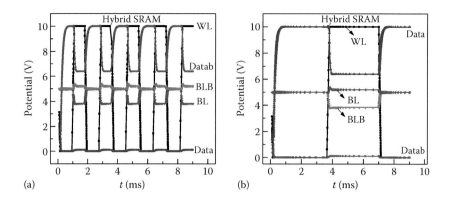

**FIGURE 9.10**   Transient response of pentacene–ZnO SRAM cell during read access ($\beta_c = 20$) for (a) storage 0 at Data and 1 at Datab and (b) storage 1 at Data and 0 at Datab nodes.

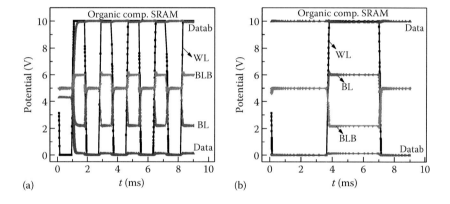

**FIGURE 9.11**   Transient response of pentacene–$C_{60}$ SRAM cell during read access ($\beta_c = 20$) for (a) storage 0 at Data and 1 at Datab and (b) storage 1 at Data and 0 at Datab nodes.

the other hand, potential at the Datab node declines quickly in the hybrid cell, whereas a minute drop is observed in the organic complementary cell.

This is due to a large leakage current in the ZnO-based TFT. This current is 7 times higher in comparison to the $C_{60}$-based TFT, as summarized in Table 9.1. This enables the conduction in transistor $N2$ even at zero $V_{gs}$ that in turn reduces the node (Datab) potential. Therefore, the potential of the complementary bit line ($V_{BLB}$) raises minutely. The $p$-SRAM cell operates in a slightly different mode to the conventional cell, wherein the load TFTs conduct at the constant current source. Considering the same storage as discussed earlier (0 at Data and 1 at Datab nodes), the potential at the Data node always lies near the threshold voltage of load transistor; therefore, the output voltage does not pull

**TABLE 9.1**    *On/Off* **Current Ratio and Leakage Current of ZnO-, C$_{60}$-, and Pentacene-Based TFTs**

| TFT Device | $I_{on}/I_{off}$ | | Leakage Current (A) | |
| --- | --- | --- | --- | --- |
| (Semiconductor) | Simulated | Experimental[a] | Simulated | Experimental[a] |
| Pentacene | $1.34 \times 10^4$ | $2.1 \times 10^4$ | $2.6 \times 10^{-10}$ | $1.8 \times 10^{-10}$ |
| ZnO | $0.7 \times 10^4$ | $1.2 \times 10^4$ | $4.2 \times 10^{-10}$ | $2.1 \times 10^{-10}$ |
| C$_{60}$ | $0.84 \times 10^5$ | $1 \times 10^5$ | $6.2 \times 10^{-11}$ | $5 \times 10^{-11}$ |

[a]   Data from Na, J. H.; Kitamura, M.; Arakawa, Y., "High performance *n*-channel thin film transistors with an amorphous phase C$_{60}$ film on plastic substrate," *Appl. Phys. Lett.* 2007, 91(19), 193501-1–193501-3; Oh, M. S.; Hwang, D. K.; Lee, K.; Choi, W. J.; Kim, J. H.; Im, S.; Lee, S., "Pentacene and ZnO hybrid channels for complementary thin film transistor inverters operating at 2V," *J. Appl. Phys.* 2007, 102(7), 076104–076104-3.

down to the ground level. Additionally, the voltage pulls up below the $V_{DD}$ level even in standby mode. During read operation, the potential of the Data node increases from $V_{t,L}$ due to a flow of charge from BL to node 1, thereby reducing the potential $V_{BL}$. Similarly, a fall is observed in the potential of the Datab node that results in a corresponding rise in the magnitude of $V_{BLB}$ as shown in Figure 9.12a and b correspondingly for storage 0 and 1 at the Data node.

The read access time of a cell is characterized by the time taken for $|V_{BL} - V_{BLB}|$ potential to attain a threshold level after turning *on* the access transistors. Here, the threshold level of 1.5 V is considered, since a slight increment in this difference voltage can reduce the cell stability of hybrid cell substantially, as shown in Figure 9.10a. The hybrid SRAM cell shows an increment of 46% in the read access time in comparison to the organic complementary SRAM cell, as summarized in Table 9.2.

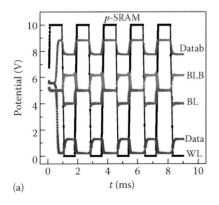

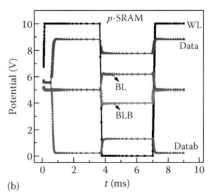

**FIGURE 9.12**    Transient response of *p*-SRAM cell during read access ($\beta_c = 20$) for (a) storage 0 at Data and 1 at Datab and (b) storage 1 at Data and 0 at Datab nodes.

**TABLE 9.2    Reading Time and Magnitude of Stored Data during Read Operation ($\beta_c = 20$) for Different SRAM Cells**

| SRAM Design | Reading Time ($\mu$s) | Magnitude of Stored Data during Read Access (V) | | Difference in Magnitude of Storage (V) |
| --- | --- | --- | --- | --- |
| | | Storage 1 | Storage 0 | |
| Organic all-*p* | 95 | 8.1 | 0.5 | 7.6 |
| Hybrid (pentacene–ZnO) | 105 | 8.1 | 0.2 | 7.9 |
| Organic (pentacene–C$_{60}$) | 72 | 9.9 | 0.3 | 9.6 |

This is due to a change in the potential of one bit line only during read operation. However, in the organic SRAM cell, the potential of both bit lines changes simultaneously, thereby requiring less time to attain the required difference voltage. Additionally, an increment of 26% and 22% in the cell stability is observed for the organic SRAM cell in comparison to the all-*p* and hybrid SRAM cells.

### 9.3.3  WRITE OPERATION

To perform a write operation, appropriate biasing voltages need to be applied at the bit lines that force the cell to be turned out in the required logical state. For writing 1 at node 1 and 0 at node 2, a biasing of magnitude $V_{DD}$ and 0 is applied at the BL and BLB, respectively. Before the write operation (0/1 at node 1/2), $P2$ and $N1$ operate in the linear region, whereas $P1$ and $N2$ remain *off*, as shown in Figure 9.9. When the access transistors are turned *on* by applying the appropriate biasing at the word line, the potential of node 2 must be reduced below the threshold voltage of the $N1$ transistor to flip the state. At $V_{node-2} = V_{t,n}$, TFTs $P2$ and $N4$ operate in the linear and saturation regions, respectively. The current in the potential dividing network formed by $P2$ and $N4$ can be expressed as

$$\frac{k_{P2}}{2}(0 - V_{DD} - V_{t,p})^2 = \frac{k_{N4}}{2}\left\{2(V_{DD} - V_{t,n})V_{t,n} - V_{t,n}^2\right\} \tag{9.12}$$

Upon solving, the transconductance ratio can be expressed as

$$\frac{k_{P2}}{k_{N4}} < \frac{2(V_{DD} - 1.5V_{t,n})V_{t,n}}{(V_{DD} + V_{t,p})^2} \tag{9.13}$$

$$\frac{(W/L)_{P2}}{(W/L)_{N4}} < \frac{\mu_n C_{ox,n}}{\mu_p C_{ox,p}} \times \frac{2(V_{DD} - 1.5V_{t,n})V_{t,n}}{(V_{DD} + V_{t,p})^2} \tag{9.14}$$

Similarly,

$$\frac{(W/L)_{P1}}{(W/L)_{N3}} < \frac{\mu_n C_{ox,n}}{\mu_p C_{ox,p}} \times \frac{2(V_{DD} - 1.5V_{t,n})\, V_{t,n}}{(V_{DD} + V_{t,p})^2} \tag{9.15}$$

where $\mu_p$ ($\mu_n$) and $C_{ox,p}$ ($C_{ox,n}$) are the mobility and capacitance of $p$-type ($n$-type) transistors, respectively.

Figures 9.13 through 9.15 illustrate the transient response during write operation of hybrid, organic complementary, and all-$p$ SRAM cells, respectively. As discussed earlier, the access transistor should be stronger than the pull-up transistor to attain the potential close to the BL and BLB terminals. The write-1

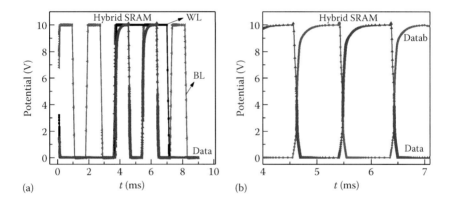

**FIGURE 9.13**    Transient response of hybrid SRAM cell during write operation (a) 1 and 0 at the Data node and (b) Data and Datab potential at $\beta_p = 0.2$.

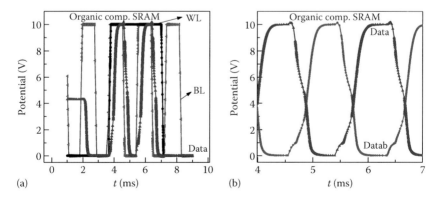

**FIGURE 9.14**    Transient responses of organic complementary SRAM cell during write operation (a) 1 and 0 at the Data node and (b) Data and Datab potential at $\beta_p = 0.2$.

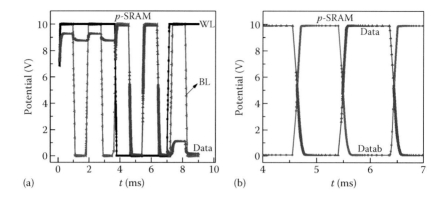

**FIGURE 9.15** Transient responses of an all-$p$ SRAM during write operation (a) 1 and 0 at the Data node and (b) Data and Datab potential at $\beta_p = 0.2$.

and write-0 access times are defined as the time taken to attain 90% and 10% levels of $V_{DD}$, respectively, after applying appropriate word pulse ($V_{DD}$ and 0 in case of hybrid/organic and all-$p$ SRAM). Write-0 time is lesser in comparison to write-1 for all three SRAM cells, since the potential of the node that contains a high storage (node 2) discharges first. It turns *off* the $N1$ transistor as the potential reaches below its threshold voltage. Thereafter, node 1 attains the potential close to $V_{DD}$. The pentacene–ZnO-based SRAM cell shows a reduction of 72% and 86% in the write-1 and write-0 time, respectively, as compared to the pentacene–$C_{60}$-based cell due to higher transconductance of the ZnO-based transistor.

For the SRAM cell, two major sources of defining cell instability are the read and the write failure. The read failure is defined as flipping the state during read access, whereas the write failure is characterized by failure to flip the cell during a write cycle. The cell stability during read access is quantified by the minimum voltage difference between the complementary storage over the time. However, during write operation, it is characterized by the time required to flip the state of the cell.

The major design issue in the SRAM cell is the conflict between the read and write stability. As previously discussed, the cell ratio should be large enough to enable the read access without altering the stored data. However, it degrades the writing ability due to a reduction in the strength of the access transistor that in turn raises the write access time. Similarly, the pull-up ratio should be small enough to increase the write ability, but it deteriorates the performance of cell during read access. Figure 9.16 illustrates the combined plots of write-1 access time and potential difference of the complementary storage (during read operation) with respect to the cell ratio for hybrid and organic SRAM cells. A similar plot for an all-$p$ SRAM cell is placed in Figure 9.17.

Table 9.3 summarizes the write access time (1 and 0), potential difference of complementary storage during read operation, and the optimized cell and

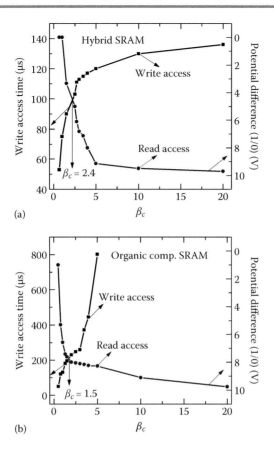

**FIGURE 9.16**    Combined plots of write-1 access time and potential difference of complementary storage during read access for (a) hybrid and (b) organic complementary SRAM cells with respect to the cell ratio.

pull-up ratios to perform both read and write operations for all three proposed TFT combinations. Compared to pentacene–ZnO TFT-based cells, the pentacene–$C_{60}$ combination shows inferior writing ability with an approximately 2 times higher write-1 time. On the contrary, an improvement of 70% in its read stability is observed in comparison to the hybrid cell, which makes it more robust against the external noise. Additionally, an all-$p$ SRAM cell shows a reduction of 32% in the write-1 time and an increment of 37% in the read stability in comparison to the hybrid cell.

The read stability is highest for the pentacene–$C_{60}$ combination due to the lowest leakage current of the $C_{60}$-based TFT in comparison to the ZnO- and pentacene-based TFTs [5,6]. The optimized cell ratio and pull-up ratio for hybrid SRAM cell is 2.4 and 0.24, respectively. On the other hand, the optimized cell (pull-up) ratio is obtained as 1.5 (0.083) and 17.5 (1.2) for pentacene–$C_{60}$ and all-p organic SRAM cells, respectively.

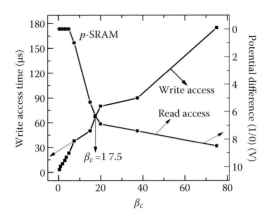

**FIGURE 9.17** Plot of write-1 access time and potential difference of complementary storage during read access for all-$p$ SRAM with respect to the cell ratio.

**TABLE 9.3    Write Access Time (1 and 0), Potential Difference of Complementary Storage during Read, Optimized Cell Ratio, and Pull-Up Ratio for SRAM Cells**

| SRAM Design | Write-1 Access Time ($\mu$s) | Write-0 Access Time ($\mu$s) | Potential Difference in 1 and 0 during Read (V) | Optimized $\beta_c$ | Optimized $\beta_p$ |
|---|---|---|---|---|---|
| Hybrid comp. (pentacene–ZnO) | 99 | 33 | 4.6 | 2.4 | 0.24 |
| Organic comp. (pentacene–$C_{60}$) | 195 | 90 | 7.8 | 1.5 | 0.083 |
| Organic all-$p$ | 67 | 10 | 6.3 | 17.5 | 1.2 |

## 9.4  COMPLEMENTARY SRAM CELL WITH $p$-TYPE ACCESS TRANSISTORS

SRAM cells based on three combinations of TFTs were discussed in the previous section. The pentacene–ZnO-based SRAM cell exhibited the highest SNM and good writing ability, whereas the organic complementary cell showed good reading ability with the lowest reading time. On the other hand, the all-$p$ SRAM cell showed a balance between the reading and writing ability but exhibited very low SNM, which increases the possibility of disturbing the cell stability with the effect of small external noise voltage. To achieve a fairly good performance in all three aspects, the organic complementary SRAM cell is designed with $p$-type access transistors in place of $n$-type, since in pentacene–$C_{60}$-based cells major difficulty arises due to weaker strength of the access transistors.

The schematic of an organic complementary SRAM cell with $p$-type access TFTs ($P5$ and $P6$) is shown in Figure 9.18; however, other TFTs are similar to the organic SRAM cell discussed earlier.

The butterfly curve of the cross-coupled inverters during read access ($\beta_c = 20$) is illustrated in Figure 9.19a. Compared to the organic complementary cell with $n$-type access transistors, this cell shows a reduction of 27% in the SNM during read access due to an increment in the strength of access transistors, but the magnitude of storage 1 and 0 remains almost similar to the previous cell, as shown in Figure 9.19b. It is due to the very low leakage current of the $C_{60}$-based pull-down TFT that helps to retain the potential of nodes containing a high storage as high as in the standby mode.

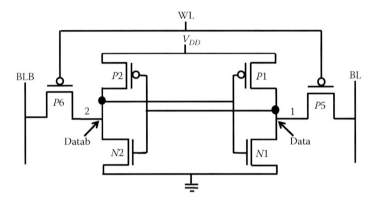

**FIGURE 9.18**   Schematic of organic complementary SRAM cell with $p$-type access TFTs.

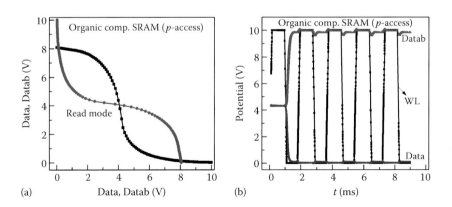

**FIGURE 9.19**   (a) Butterfly curve during read access ($\beta_c = 20$) and (b) transient response during read operation for pentacene–$C_{60}$ organic SRAM cell with $p$-type access TFTs.

During write operation, the write-1 access time of a pentacene-$C_{60}$ SRAM cell with $n$-type access TFTs is 371 μs at $\beta_p = 0.2$ (Figure 9.14a). However, a significant reduction of 79% in the write time is observed for the pentacene–$C_{60}$ SRAM cell with $p$-type access TFTs at the same $\beta_p$. Moreover, the cell with $p$-type access TFTs demonstrates reasonably good behavior at $\beta_p$ of 0.28, as shown in Figure 9.20. On the other hand, at $\beta_p = 0.28$, the cell with $n$-type access TFTs shows a poor write response due to a reduction in the strength of access transistors with an increase in the pull-up ratio. Figure 9.21 illustrates the transient response of pentacene–$C_{60}$-based organic SRAM cell consisting of $n$-type access TFTs and the transition of states at Data and Datab nodes with respect to time. It is observed that the organic cell with $n$-type access TFTs shows an

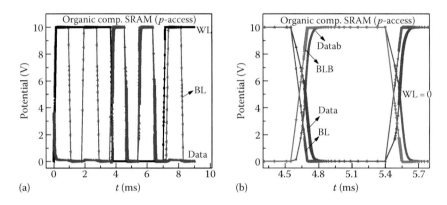

**FIGURE 9.20**     Transient response of organic SRAM cell with $p$-access TFTs during write operation ($\beta_p = 0.28$) (a) 1 and 0 at the Data node and (b) Data and Datab potential.

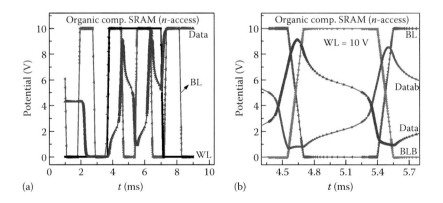

**FIGURE 9.21**     Transient response of organic complementary SRAM cell with $n$-access TFTs during write operation ($\beta_p = 0.28$) (a) 1 and 0 at the Data node and (b) Data and Datab potential.

**TABLE 9.4    Comparison between the Performance of Organic SRAM Cell with *p*- and *n*-Type Access OTFTs at $\beta_c = 10$ and $\beta_p = 0.55$**

| Access OTFTs (Organic Complementary SRAM) | Write Access Time | | Read Stability | |
|---|---|---|---|---|
| | Write-1 | Write-0 | Potential Difference of Storage 1 and 0 (V) | RSNM (V) |
| $C_{60}$-based *n*-type | Infinite | Infinite | 9.5 | 2.5 |
| Pentacene-based *p*-type | 131 µs | 115 µs | 9.2 | 1.7 |

inability to follow the bit line signal properly and attains the voltage levels of 9 V and 1 V against 10 V and 0, respectively, that too have quite large write access times of 802 and 730 µs, respectively. On the other hand, the cell with *p*-type access TFTs shows good writing capability with adequate follow-up to the bit line potential based on time. This cell shows a significant improvement of 85% and 88% in the write-1 (115 µs) and write-0 (86 µs) access times, respectively.

The optimized cell and pull-up ratios for an organic SRAM cell with *p*-type access transistors are obtained as 10 and 0.55, respectively. The width of the *p*-type access TFT is one-sixth of the width of the *n*-type access TFT, whereas the dimensions of pull-up and pull-down TFTs are similar in both the cases. Table 9.4 summarizes the performance of the organic cell with both *p*- and *n*-type access transistors at $\beta_c = 10$ and $\beta_p = 0.55$. The organic complementary cell with *p*-type access TFTs demonstrates a reduction of 32% in the read SNM in comparison to the cell with *n*-type TFTs; however, it is still more than 2 times higher in comparison to the all-*p* SRAM cell. Furthermore, in comparison to the cell with *n*-type access TFTs, a slight reduction of 3% in the read stability is observed using *p*-type access TFTs. Nevertheless, it is 2 times higher in comparison to the hybrid cell. Besides this, a major benefit of using *p*-type access TFTs is observed in terms of writing ability with reasonably lower write-1 and write-0 times, whereas the cell with *n*-type access TFTs is found unable to flip the state of the cell at the same cell ratio (10) and pull-up ratio (0.55). As a result, the performance of the pentacene–$C_{60}$-based organic SRAM cell is reasonably good in terms of all three aspects (read SNM, read stability, and writing capability) and the *p*-type access transistors have a substantially lower width in comparison to their counterparts.

## 9.5  CONCLUDING REMARKS

This chapter analyzed the performance of organic all-*p*, hybrid, and organic complementary SRAM cells using pentacene-, ZnO-, and $C_{60}$-based TFTs. The performance of different SRAM configurations is assessed mainly in terms of static noise margin, read stability, and write ability at different cell and pull-up ratios. Besides this, the organic complementary cell is designed using *p*-type access TFTs that showed an overall improved performance in comparison to the other proposed SRAM cells.

The organic complementary SRAM cell shows good reading ability with reasonably high read SNM and the lowest reading time. This cell exhibits a reduction of 31% and 24% in reading time (at $\beta_c = 20$) as compared to pentacene–ZnO (hybrid) and all-$p$ (organic) SRAM cells, respectively. In addition, the cell stability during read access is higher by 26% and 22% than the organic all-$p$ and hybrid SRAM cells. On the contrary, the writing capability is inferior due to significantly lower mobility and capacitance of $C_{60}$-based OTFT in comparison to its counterparts.

The pentacene–ZnO hybrid complementary SRAM cell demonstrates good writing ability with reasonably lower write time. At $\beta_p = 0.2$, this cell exhibits 72% and 86% reduced write-1 and write-0 time, respectively, in comparison to the pentacene–$C_{60}$ SRAM cell. In addition, the SNM of this hybrid cell is 1.2 and 2.5 times higher as compared to the organic complementary and all-$p$ SRAM cells. On the other hand, it shows an inadequate performance during read access due to a high leakage current in the ZnO-based TFT.

The all-$p$ SRAM cell shows a balanced characteristic for the read and write operations but exhibits very low SNM. It is due to a lower magnitude of $V_{OH}$ limited to $V_{DD}-V_{t,L}$. Besides this, the high-to-low transition of the all-$p$ inverter is not as steep as that obtained in hybrid and organic complementary inverters. It is a result of the slower pull-down action due to zero-$V_{gs}$ of the load, thereby reducing the area of the lobe, and thus, SNM too.

To achieve overall improved performance, the organic complementary SRAM cell is designed using pentacene-based $p$-type access transistors in place of $n$-type. Favorably, this organic SRAM design shows 79% lower write access time in comparison to the cell with $n$-type access OTFTs. Moreover, this cell shows adequate SNM and read stability at substantially lower widths of $p$-type access OTFTs.

The optimized cell ratio and pull-up ratio of the organic complementary cell with $p$-type access OTFTs are 10 and 0.55, respectively. However, the cell with $n$-type access TFTs with similar cell and pull-up ratios is unable to even flip the state of the cell. This is due to the substantially lower strength of the $C_{60}$ $n$-type OTFT in comparison to the pentacene $p$-type OTFT, thereby requiring widths 6 times higher to suitably perform during both read and write operations.

## PROBLEMS
### MULTIPLE CHOICE

1. Organic memory is better due to
   a. Structural flexibility
   b. Low-cost solution process ability
   c. Lightweight
   d. All of the above

2. Performance of the SRAM cell made up of organic/inorganic material combinations is analyzed through
   a.  2-D numerical device simulator
   b.  3-D numerical device simulator
   c.  1-D numerical device simulator
   d.  All of the above
3. Static random access memory cell has
   a.  One access transistor connected to the bistable latch circuit
   b.  Two access transistors connected to the bistable latch circuit
   c.  None of the access transistors are connected to the bistable latch circuit
   d.  None of the above
4. Organic memory can be useful in
   a.  RFID tags
   b.  Sensors
   c.  Smart cards
   d.  All of the above
5. Cache memory in microprocessors uses
   a.  SRAM
   b.  DRAM
   c.  Both a and b
   d.  None of the above
6. Writing ability of a cell is defined by the
   a.  Cell ratio
   b.  Transconductance ratio
   c.  Pull-up ratio
   d.  Static noise ratio
7. The transconductance ratio of the pull-down transistor to the access transistor is known as the
   a.  Cell ratio
   b.  Transconductance ratio
   c.  Pull-up ratio
   d.  Static noise ratio
8. For all-$p$-type SRAM cell, the access transistors are enabled for data read or write operation when the
   a.  Transistors are turned *off* by keeping *WL* at 0 V
   b.  Transistors are turned *on* by keeping *WL* at 1 V
   c.  Transistors are turned *on* by keeping *WL* at 0 V
   d.  Transistors are turned *off* by keeping *WL* at 1 V
9. Stability of the SRAM cell is crucial for determining the
   a.  Speed of operation
   b.  Data density
   c.  Ability of stored data retention
   d.  All of the above

**ANSWER KEY**

1. d; 2. a; 3. b; 4. d; 5. a; 6. c; 7. a; 8. c; 9. c

**SHORT ANSWER**

1. What are the advantages and applications of organic memory?
2. Describe in detail the structure and operation of the static random access memory cell.
3. Draw the structure of hybrid/organic complementary and all-*p*-type SRAM.
4. Describe static noise of the SRAM cell.
5. Explain the read and write operation of hybrid/organic SRAM cell with a diagram.
6. Explain the transient response of
    a. Pentacene–ZnO SRAM cell during read access
    b. Hybrid SRAM cell during write operation
7. Explain the performance of the SRAM cell in terms of
    a. Cell ratio
    b. Pull-up ratio

**EXERCISES**

1. Calculate the cell ratio and pull-up ratios of a single SRAM cell with $W/L = 625/100$ for the load TFT and $W/L = 2500/50$ for the access and driver TFTs.
2. Design an organic all-*p* 6T SRAM cell using p-type transistor with pentacene as an OSC material of the following parameters: semiconductor thickness $(t_{osc}) = 50$ nm, channel length $(L) = 90$ µm, channel width $(W) = 500$ µm, insulator thickness $(t_{ox}) = 10 + 2.5$ nm $(Al_2O_3 + SiO_2)$, thickness of source/drain $(t_s/t_d) = 10$ nm (Au), and gate thickness $(t_g) = 10$ nm $(n^+si)$. The ratio of the transistors is driver transistors (P1, P2) $W/L = 500/90$, load transistors (P3, P4) $W/L = 7500/90$, and access transistors (P5, P6) $W/L = 450/90$.
3. Design an organic complementary 6T SRAM cell with *p*-access TFT's using the p-type transistor simulated in question 2. The dimensions of the n-type device with $C_{60}$ as an OSC material are semiconductor thickness $(t_{osc})$ 60 nm, channel length $(L) = 100$ µm, channel width $(W) = 1000$ µm, insulator thickness $(t_{ox}) = 8$ nm $+ 132$ nm $+ 4$ nm $(SiO_2 + Ti\ SiO_2 + SiO_2)$, thickness of source/drain $(t_s/t_d) = 1 + 100$ nm (LiF + Al), and gate thickness $(t_g) = 30$ nm (*p*-si). The ratio of the transistors is driver transistors (P1, P2) $W/L = 500/90$, load transistors (N1, N2) $W/L = 1800/90$, and access transistors (P3, P4) $W/L = 9000/90$.
4. Design a hybrid 6T SRAM cell using the *p*-type device in question 2. The dimensions of the *n*-type device using ZnO as an OSC material are semiconductor thickness $(t_{osc}) = 60$ nm, channel length $(L) = 90$ µm,

channel width ($W$) = 500 μm, insulator thickness ($t_{ox}$) = 12.5 nm ($SiO_2$ + $Al_2O_3$), thickness of source/drain ($t_s/t_d$) = 10 nm (Al), and gate electrode thickness ($t_g$) = 10 nm ($p$-Si). The ratio of the transistors is driver transistors (P1, P2) $W/L$ = 500/90, load transistors (N1, N2) $W/L$ = 5000/90, and access transistors (N3, N4) $W/L$ = 2000/90.

5. Calculate the static noise margin (SNM) for the SRAM cell designed in question 4.

Hint:

$$SNM = V_{DD} \sqrt{\left( V_{DD} + V_{t,D} - \sqrt{\frac{1}{k_R}} V_{t,L} \right)^2 + \left( \frac{1}{k_R} \right) V_{t,L}^2}$$

where $k_R = \dfrac{\left| V_{T,load} \left( V_{OL} \right) \right|^2}{2 \left( V_{OH} - V_{TO} \right) V_{OL} - V_{OL}^2}$, $V_{t,D}$ is the threshold voltage of the

driver transistor, $V_{t,L}$ is the threshold voltage of the load transistor,

and $V_{DD}$ is the supply voltage.

6. Analyze the effect of variations in cell ratio and pull-up ratio on the performance of SRAM cell discussed in the question 4. Also analyze the effect of variation of cell ratio and pull-up ratios.

## REFERENCES

1. Takamiya, M.; Sekitani, T.; Kato, Y.; Kawaguchi, H. "An organic FET SRAM with back gate to increase static noise margin and its application to braille sheet display," *IEEE J. Solid-State Circuits* **2007**, 42(1), 93–100.
2. Guerin, M.; Bergeret, E.; Benevent, E.; Daami, A.; Pannier, P.; Coppard, R. "Organic complementary logic circuits and volatile memories integrated on plastic foils," *IEEE Trans. Electron Devices* **2013**, 60(6), 2045–2051.
3. Fukuda, K.; Sekitani, T.; Zschieschang, U.; Klauk, H.; Kuribara, K.; Yokota, T.; Sugino, T. *et al.* "A 4 V operation, flexible braille display using organic transistors, carbon nanotube actuators, and organic static random-access memory," *Adv. Funct. Mater.* **2011**, 21(21), 4019–4027.
4. Schwabegger, G.; Ullaha, M.; Irimia-Vladu, M.; Baumgartner, M.; Kanbur, Y.; Ahmed, R.; Stadler, P.; Bauer, S.; Sariciftci, N. S.; Sitter, H. "High mobility, low voltage operating $C_{60}$ based *n*-type organic field effect transistors," *Synthetic Metals* **2011**, 161(19–20), 2058–2062.
5. Na, J. H.; Kitamura, M.; Arakawa, Y. "High performance *n*-channel thin film transistors with an amorphous phase $C_{60}$ film on plastic substrate," *Appl. Phys. Lett.* **2007**, 91(19), 193501-1–193501-3.
6. Oh, M. S.; Hwang, D. K.; Lee, K.; Choi, W. J.; Kim, J. H.; Im, S.; Lee, S. "Pentacene and ZnO hybrid channels for complementary thin film transistor inverters operating at 2V," *J. Appl. Phys.* **2007**, 102(7), 076104–076104-3.

7. Taur, Y.; Ning, T. H. *Fundamentals of Modern VLSI Devices*, 2nd ed., Cambridge University Press, Cambridge, UK, 2009.

8. Bhavnagarwala, A. J.; Tang, X.; Meindl, J. D. "The impact of intrinsic device fluctuations on CMOS SRAM cell stability," *IEEE J. Solid-State Circuits* **2001**, 36(4), 658–665.

9. Grossar, E.; Stucchi, M.; Maex, K.; Dehaene, W. "Read stability and write-ability analysis of SRAM cells for nanometer technologies," *IEEE J. Solid-State Circuits* **2006**, 41(11), 2577–2588.

10. Vusser, S. D.; Genoe, J.; Heremans, P. "Influence of transistor parameters on the noise margin of organic digital circuits," *IEEE Trans. Electron Devices* **2006**, 53(4), 601–610.

11. Cui, Q.; Si, M.; Sporea, R. A.; Guo, X. "Simple noise margin model for optimal design of unipolar thin-film transistor logic circuits," *IEEE Trans. Electron Devices* **2013**, 60(5), 1782–1785.

# Applications and Future Perspectives

# 10

## 10.1 ORGANIC DEVICE APPLICATIONS

Organic transistors have found usage in numerous applications, such as organic digital, analog, and mixed signal circuits; memories; solar cells; photovoltaics; and sensors. They are often used as the backplane driver in organic display circuits. Moreover, they are frequently employed as the rectifier-modulator unit in organic radio frequency identification systems. A few important applications of these transistors are described next.

### 10.1.1 DIGITAL LOGIC CIRCUITS

The concept of realizing Boolean operations through multiple input logic gates brought great evolution in the field of high-end electronic applications. An inverter is considered to be the most elemental and benchmark circuit in digital circuit design. Most inverters based on organic thin-film transistors (OTFTs) are designed using only $p$-type transistors [1,2] due to the higher mobility and better intrinsic stability of $p$-type organic materials in comparison to $n$-type. Compared to single gate (SG) inverters, the dual gate (DG) OTFT-based inverters outperform in voltage swing, gain, and propagation delay because of their low threshold voltage and high $on$-current. Spijkman $et$ $al.$ reported noise margins of 0.6 and 5.9 V for single and dual gate inverters, respectively, based on the poly(triarylamine) (PTAA) organic semiconductor (OSC) [3]. Similarly, Myny $et$ $al.$ compared the pentacene-based SG and DG OTFTs, wherein the DG OTFT improved the voltage swing and noise margin by 13% and 143%, respectively [4]. Inverters with different load configurations such as diode load logic (DLL) and zero-$V_{gs}$ load logic (ZVLL) were discussed earlier in this book. Complementary inverters using small molecules of $p$-type pentacene and $n$-type hexadecafluorocopper-phthalocyanine ($F_{16}$CuPc) [5] or inkjet-printed $p$-type poly(3-hexylthiophene) (P3HT) and $n$-type poly([N,N-9-bis(2-octyldodecyl)naphthalene-1,4,5,8-bis (dicarboximide)-2,6-diyl]-alt-5,59-(2,29-bithiophene) (P(NDI2OD-T2)) [6] organic materials have been demonstrated. An inverter using a novel design style—pseudo-CMOS with single $V_t$—proposed by Huang *et al.* gives a performance comparable to the dual-$V_t$ designs [7]. Furthermore, Raval *et al.* observed an increase in gain and voltage swing by 13% and 40%, respectively, for P3HT-based inverters by applying the bootstrapping technique [1].

Universal logic gates, NAND and NOR, are the backbone of integrated circuits for the reason that any Boolean expression and higher-level digital integrated circuits (latch, flip flop, decoder, multiplexer, arithmetic logic unit, etc.) can be realized with the network of either of these two gates. Brown *et al.* demonstrated two-input universal gates based on all-$p$ pentacene OTFTs, operating at −15 V [8]. Similarly, Kim *et al.* in 2009 reported mechanically flexible NAND and NOR logic gates using a self-assembled monolayer of dielectric on the plastic substrate [9]. Recently, Guerin *et al.* in 2013 reported that organic NAND and NOR gates operated at 40 V using poly(triarylamine)-based $p$-type and acene-diimide-based $n$-type OTFTs [10]. The operating frequency of these logic gates was limited to 450 Hz due to 26% reduced mobility of $n$-type transistor in comparison to its $p$-type counterpart.

Different static and dynamic combinational and sequential digital circuits using OTFTs have been demonstrated by different research groups. In the early phase of the development, a divide-by-two frequency divider operating at a supply voltage of 25 V and 1.1 kHz was reported in 2000 [11]. Wu *et al.* demonstrated dynamic inverters; NAND, NOR, and XOR gates; and an 8-bit ripple-carry adder based on PMOS-only pre-discharge (POPD) logic [12]. The most explored and successfully implemented organic circuit during the initial advancements was the ring oscillator. A five-stage ring oscillator with oscillation frequency of around 1.3 kHz (propagation delay of 73 µs per stage) at the supply voltage of −80 V and level shifting at +50 V was reported in 1999 [13]. Several publications demonstrating a ring oscillator with maximum oscillation frequency ranging from a few hertz to kilohertz have been reported with downscaling of the supply voltage. However, for the five-stage ring oscillator with the higher supply voltage of 13 V, a minimum delay of 230 ns per stage, which is equivalent to 0.4 MHz of operating frequency, was observed [14]. On the contrary, Zschieschang *et al.* demonstrated a 11-stage organic ring oscillator using 2,9-didecyl-dinaphtho [2,3-b:2′,3′-f] thieno[3,2-b]thiophene (C10-DNTT) operating at a lower supply voltage of 3 V, which gives a propagation delay of 420 ns per stage [15]. Besides this, the design of various OTFT-based logic gates, multiplexers, decoders, latches, flip-flops, and 4-bit arithmetic logic unit have been reported [16]. Schwartz *et al.* compared static and dynamic inkjet-printed organic shift registers [17].

In the persistent process of developing embedded systems based on flexible electronic technology, Myny *et al.* [18] proposed an 8-bit organic microprocessor on plastic foil. The microprocessor comprising 3381 OTFT devices was made operational at 10 V supply voltage with power consumption of 92 µW. The clock frequency was 6 Hz and used single instruction foil to provide 06 operations per second. A year later in 2012, the same group of researchers proposed an 8-bit microprocessor with 40 instructions per second made operational at the supply voltage of 10 V and with little increase in the power consumption of 100 µW [19]. In 2014, they demonstrated an 8-bit microprocessor on very thin polyimide film with read-once write-many programmable memory using complementary transistor technology. This processor was operating at a maximum of 2.1 kHz clock frequency, i.e., operations per second and minimum power supply voltage of 6.5 V [20].

## 10.1.2 ANALOG AND MIXED SIGNAL CIRCUITS

Despite prevalent and incessant research efforts, analog and mixed signal organic circuit development is still lagging the progress of digital logic organic circuit designs. During the initial phase of the development, organic analog circuit design progress was restricted to differential amplifier circuits. In 2000, Kane *et al.* [11] exhibited a source-coupled pair organic differential amplifier on polyester film substrates with supply voltages of $V_{DD}$ = +20 V and $V_{SS}$ = –10 V. The maximum differential gain of the amplifier was recorded to be very low at 8.5 due to the small product of overall transconductance ($g_m$) and output impedance ($r_o$) of the circuit, wherein the need of different circuit design techniques such as a cascode amplifier, and new materials and processes were depicted to improve the intrinsic gain of the amplifier. In 2006, Nicolas Gay *et al.* [21] implemented a cascode amplifier with gain of approximately 8 dB, unity gain bandwidth of 1.4 kHz, and a differential amplifier with gain-bandwidth product of 3.6 kHz, and reported little improvement in the differential gain owing to the fact that OTFT devices show large process variations with fabrication, hence, greater device mismatch and large parasitic capacitances. In order to make threshold voltage ($V_t$) insensitive circuit design, in 2010 Marien *et al.* [22] presented a single-ended differential amplifier using common-mode feedback (CMFB), bootstrapped gain enhancement, cascoding, and back gate steering circuit techniques. The differential amplifier provided a gain of 18 dB and unity gain bandwidth of a few kilohertz. Furthermore, the amplifier presented good stability after several months with moderate degradation in the gain of 4 dB. In 2011, Guerin *et al.* [23] demonstrated full printed complementary organic differential amplifiers using low-cost screen printing technology on flexible plastic substrate providing a high differential gain of 22 dB and a gain-bandwidth product of 13.36 kHz. Chang *et al.* [24] recently proposed a high-gain organic differential amplifier with positive cum negative feedback using novel fully additive printing technology. The amplifier exhibited a high differential gain of 27 dB and low bandwidth of 70 Hz. Apart from these differential amplifiers, current mirror designs have been explored, which are serving as a circuit element for testing the device match [25] or as a current source injecting the same current as the reference in the given circuit [26].

The majority of the mixed signal circuits comprising three subcircuits: a transducer, which provides an interface with the real world; an analog signal conditioner; and a data converter that can transform the signal from analog to digital or digital to analog. The rate of conversion and resolution are the key parameters that determine the superiority of the data converters. The rate of conversion primarily depends on how fast the circuit provides the stable output on which the conversion result is produced. As OTFT devices inherently exhibit large parasitic capacitance, the increased settling time results in slower conversion speed of the data converter. The prime concern in designing the data converters is the resolution that is decided by the linear transfer function between input and output. The linearity of the circuit mainly depends on the device matching, which determines accuracy, and hence, the resolution. OTFT device fabrication presents large process variations producing devices with greater mismatch, which increases the nonlinearity resulting in the reduced resolution of the data converters.

Organic data converters normally called digital-to-analog converters (DACs) or analog-to-digital converters (ADCs) have been successfully implemented only within the past 7–8 years. Xiong *et al.* [27], in 2009, presented a 6-bit DAC using *p*-type dinaphtho-thieno-thiophene (DNTT) and *n*-type hexadecafluorocopperphthalocyanine ($F_{16}$CuPc) to form complementary OTFT devices and power supply voltage as low as 3 V on glass substrate. In order to reduce the process variation sensitivity, the DAC was implemented using the switched capacitor technique. The switched capacitance output, which is less sensitive to the exactitude of their currents, presents higher precision than the OTFT transistor output. Furthermore, the DAC used C-2C structure to circumvent the effect of large parasitic capacitance of the OTFT structure. After calibration, the DAC exhibited DNL (differential nonlinearity) and INL (integrated nonlinearity) of −0.6 and −0.8 LSB, respectively, with the conversion rate of 100 Hz. In 2010, the same group of researchers demonstrated a 6-bit successive approximation register (SAR) ADC using complementary OTFT devices on glass substrate at the power supply of 3 V [28]. The ADC design was comprised of an organic DAC, organic comparator, and inorganic digital SAR logic. The digital SAR logic was implemented using the external field programmable gate array (FPGA) and interfaces loaded with appropriate value of capacitances to imitate the OTFT parasitic environment. The ADC produced DNL of 0.6 LSB and INL of −0.6 LSB at the sampling frequency of 10 Hz. However, at the sampling frequency of 100 Hz, the DNL and INL were degraded to 1.5 LSB and 3.0 LSB, respectively. At the sampling frequency of 100 Hz, the power consumption of DAC and comparator was measured to be 0.7 and 2.9 μW, respectively. Marien *et al.* [29], in 2011, presented a fully integrated first-order continuous time ΔΣ ADC on plastic substrate using *p*-type OSC. The careful design of subcircuits included a $V_t$-insensitive single-stage differential amplifier, a three-stage operational amplifier, an integrator, a comparator, and a level shifter. The ADC achieved an SNR of 26.5 dB at the clock frequency of 500 Hz; hence, the ENOB (effective number of bits) of the design was around 4 bits. Zaki *et al.* [30], in 2012, demonstrated a 3.3 V, 6-bit binary weighted current steering DAC using $p$-type OTFT devices on glass substrate. As the DAC used the current steering technology, it was prone to device mismatch resulting into higher nonlinearity and reduced precision; however, it achieved a higher sampling rate of 100 kS/s, 1000 times faster than reported by Xiong *et al.* [27]. At the sampling rate of 1 kHz, the measured DNL and INL were 0.69 and −1.16 LSB, respectively. Raiteri *et al.* [31] proposed a 6-bit, voltage-controlled oscillator (VCO)-based ADC using $p$-type OSC. The simplistic design comprised two main subcircuits: a VCO constructed by seven-stage inverters followed by a transconductor, and a 10-bit D-flip-flop-based ripple counter. The linearity of the proposed design depends on electrical properties of the transconductor rather than the matching of the OTFT devices or the capacitances. The design achieved an SNR of 48 dB and the measured DNL and INL of the design were 0.6 and 1.0 LSB, respectively, even before the calibration. Recently, Chang *et al.* [24] in 2014, proposed a fully additive 4-bit DAC, wherein layer printing involves a deposition of layer upon a layer rather than etching off the layer. The design exhibited slower speed of conversion inherently due to the very low carrier mobility of the printed organic semiconductor. The DNL and INL of the design measured at the sampling frequency of 300 Hz were 0.8 and 0.9 LSB, respectively.

OTFT devices are still not mature and established in comparison to their inorganic counterparts, and still need some aggressive and ubiquitous efforts with the inception of novel materials and processing steps to achieve comparable performance.

## 10.1.3 MEMORY CIRCUITS

Organic memory devices are potentially useful in organic radio-frequency identification (RFID) tags, smart cards, and disposable circuitry. Organic memory transponder chips are the backbone of RFID systems, since they contain data that allows the items to be identified [32]. Furthermore, future efforts can be focused to produce low-cost flexible memory chips on paper and plastic. Fukuda *et al.* [33] in 2011 reported a DNTT (dinaptho[2,3-b:2 0, 3 0-f]thieno[3,2-b]thiophene), $p$-type OSC-based 6-T SRAM cell with a carbon-nanotube-based actuator operated at a bias supply of 4 V. This all-$p$ organic SRAM exhibited the write access time and static noise margin of 2 ms and 0.44 V, respectively. Furthermore, Takamiya *et al.* [34] demonstrated a pentacene-based 5-T organic $p$-type SRAM cell that significantly reduced the SRAM cell area by 20% and write time by 69% over the conventional 6-TFT cell. Recently, Guerin *et al.* [10] in 2013 reported an organic SRAM cell operating at 50 Hz using poly(triarylamine)-based $p$-type and acene-diimide-based $n$-type organic TFTs.

Several researchers have reported single organic memory devices under different device structures including organic field-effect transistors (FETs) based on ferroelectric and chargeable gate dielectrics [35–39], embedding metal nanoparticles (NPs) into gate dielectrics [40,41] and organic bistable devices [42,43]. Naber *et al.* [36] in 2005 reported an organic memory device based on ferroelectric polymer dielectric material that exhibited an $I_{on}/I_{off}$ of $10^4$ with a programming time of 0.3 ms along with the storage stability of more than one week. Mabrook *et al.* [44] demonstrated organic thin-film memory devices based on pentacene OSC and PMMA dielectric, wherein gold NPs were used as the charge storage element. Under a gate pulse of 1 Hz, these gold NPs were charged and discharged, which in turn resulted in a significant threshold voltage shift. These devices are simple to fabricate but exhibit inferior performance due to direct deposition of the OSC material onto the metal NPs, thereby resulting in an adverse orientation and inappropriate alignment of the OSC molecules grown on the dielectric/NPs surface. To circumvent this, Wang *et al.* [40] reported a sandwiched structure by placing the silver nanoparticles in between two pentacene semiconducting layers. This structure achieved a high memory window of 90 V, higher by 300% as compared to the conventional structure, wherein silver particles were directly deposited on the $SiO_2$ dielectric surface. Additionally, Ma *et al.* [45], Ouyang *et al.* [46], and Bozano *et al.* [47] demonstrated bistable memory devices using gold NPs in a sandwiched structure formed by the blend of NPs and a polymeric material. Katz *et al.* [37] reported a polarizable gate dielectric based organic memory device that induced a floating gate-like behavior. In these devices, metal NPs act as the floating gate that requires a precise control on the size, species, and the density of metal NPs to ensure good device performance.

### 10.1.4 LIGHT-EMITTING DIODES AND TRANSISTORS

The organic light-emitting diodes (OLEDs) previously commercialized are now making way for low-cost, large-area, flexible displays. They have been successfully employed for small displays, such as mobile phones, programmable digital arrays, MP players, and modern digital cameras. The driving force behind this success is the advantages exhibited by OLEDs such as low cost, lightweight, compact, flexible, ease of fabrication, large color selection, and better efficiency. Besides this, it has also proven to be promising enough for improved color quality, sharp image, intensive background, large sight angle, rapid switching, and lower voltage operation. These properties can be utilized for fabricating better flexible displays of electronic paper, since it requires high $I_{on}/I_{off}$, low $SS$, and low $V_t$.

An OLED consists of thin emissive layers prepared from the organic compounds. It shows high luminous efficiency, since it does not require any backlight function. These OLEDs generate photons through the emissive layer by the principle of generation and recombination of electron-hole pairs. When a current is passed through the multiple layers, it transforms into light.

An OLED structure and its OTFT-based driving circuit as a part of a display system are shown in Figure 10.1. The OTFT1 in Figure 10.1b is used for charging and discharging the capacitor (C), whereas, OTFT2 is the driver transistor. OTFT1 is required to have high *on*-current and low *off*-current to enable the capacitor to charge and discharge efficiently. To compensate the charge leakage from the capacitor, the capacitor size should be large enough to retain the minimum desired data voltage. A quick charging of this large capacitor requires

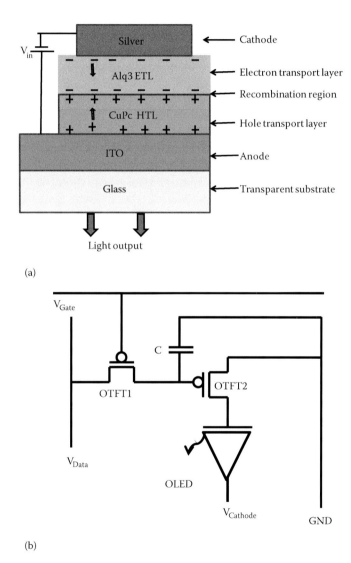

(a)

(b)

**FIGURE 10.1** Organic light-emitting diode (a) schematic structure and (b) OTFT-based driver circuit.

a high *on*-current. Generally, a large *on/off* current ratio is required for operating a large display.

Jackson *et al.* [48] reported a high-performance pentacene-based stacked TFT with a mobility of 1.5 cm$^2$/Vs and a high $I_{on}/I_{off}$ of 10$^8$ comparable to that of an a-Si:H (hydrogenated amorphous silicon) transistor. Such a high mobility enables the display to address at least a thousand lines and the high *on/off* current ratio allows long frame times with minimum charge leakage. In 2006, Zhou *et al.* [49] reported the first fully organic active matrix organic light-emitting diode (AMOLED) display that contained 48 × 48 bottom-emission OLED pixels with two pentacene transistors per pixel.

Li *et al.* [50] reported a fully printed multilayer organic LED fabricated through the polymer inking and stamping technique by transferring a layer of PEDOT material on PES substrate using a stamp of polydimethylsiloxane material. The device emitted yellow light from the bottom of the transparent ITO substrate by applying a voltage of about 7 V. OTFT devices fabricated at low temperature often exhibit low mobility (<0.5 cm$^2$/Vs) that limits the designing of a pixel circuitry [51]. Therefore, a compensatory unit is needed for low mobility OTFT circuits to maintain a constant driving current that can achieve adequate electrical operation of an AMOLED display. Liu *et al.* [52] investigated a novel voltage driving pixel circuit for AMOLED displays, consisting of four switches, one driving transistor, and a capacitor. To narrow the charging time, they developed a complementary voltage-induced coupling driving mechanism by reducing the voltage drop between the source and the drain terminals of an organic transistor, especially suitable for OTFT-based display circuitry with low field effect mobility.

## 10.1.5  RADIO FREQUENCY IDENTIFICATION (RFID) TAGS

Presently, radio frequency identification tags are being aggressively used for applications such as supply chain management, toll bridges, medical science, and in the defense sector. During the last few years, organic RFID tags have received immense interest due to their low cost and flexibility. The cost of organic RFID tags per unit area is almost 3 orders lower than their silicon counterparts. The tags need to be operated in the radio frequency range to identify the item/person and information that can be transmitted without adhering to line of sight.

Baude *et al.* [53] in 2003 reported a pentacene-based RFID circuit, patterned with polymeric shadow mask on a 2 × 2 inch glass plate. This low-cost circuit was directly powered by a radio frequency signal without applying a rectifier stage. This organic RFID circuit responded adequately at 125 KHz and even up to a high frequency of 6.5 MHz. Later, in 2007, Cantatore *et al.* [54] reported an organic transponder based RFID system operating at 13.56 MHz frequency. Myny *et al.* [55] in 2009 demonstrated an improved organic RFID circuit fabricated on the plastic substrate for RF communication at the same carrier frequency of 13.56 MHz. Furthermore, they reported an inductively coupled 64-bit and 128-bit [56] pentacene-based RFID tag that supported a data rate of 787 and 1529 bits/sec. Myny *et al.* [4] recently reported a 64-bit organic RFID transponder chip based on dual gate OTFT that operated at a supply voltage of 10 V. This chip yielded a high data rate of 4300 bits/sec, realized on a small area of 45.38 mm$^2$. Currently, the researchers are focusing toward performance improvement of organic RFID tags for item-level tracking of individual goods at low cost to establish control over check-in/-out of an inventory.

A block diagram of a typical organic RFID tag is shown in Figure 10.2. It consists of three different functional blocks: a transmitter/receiver (antenna and reader unit), rectifier/modulator unit, and an organic integrated circuit (RFID tag). A schematic of a capacitive coupled organic rectifier-modulator circuit is shown in Figure 10.3a that consists of a transmitter/receiver, a rectifier-modulator unit and an RFID tag [57]. In an RFID system, the signal flows in two directions: (1) from a reader to the tag, when transmitter sends the command to read a code from the tag, and (2) from the tag to the reader, when the tag sends the code back to the receiver. For reading data from the tag, the antenna receives the voltage signal of the RF frequency from the reader through coupling capacitors, $C_{COUPLE1}$ and $C_{COUPLE2}$. This high-frequency AC signal is rectified (full-wave) by two half-wave rectifying diodes realized using *p*-type organic transistors, OTFT1 and OTFT2. These transistors are configured in diode load logic, wherein, drain and gate terminals are shorted. The capacitor $C_{DECOUPLE}$ filters out the redundant AC ripples, thereby, producing a smooth DC signal to be fed to the tag.

To send back the code from an RFID tag to the reader, OTFT2 performs a modulation of electric signal generated by the tag. OTFT1 works as a feedback unit providing a return path to the modulated signal. Using an antenna, this signal is transmitted to the reader section, where it is demodulated to regenerate the signal.

A schematic of a typical 64-bit RFID transponder chip (code generator) is shown in Figure 10.3b [4]. It consists of a 19-stage ring oscillator that generates a clock signal for the binary counter, line selector, and output register. A binary 3-bit counter drives an 8:1 multiplexer that selects a particular 8-bit row from the memory and delivers it to the output shift resister. This 3-bit counter is also

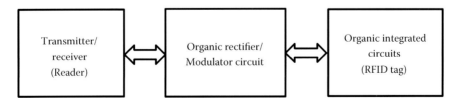

**FIGURE 10.2**    Signal flow setup of radio frequency identification (RFID) tag with a transmitter/receiver unit, rectifier/modulator circuit unit, and organic transponder integrated circuit.

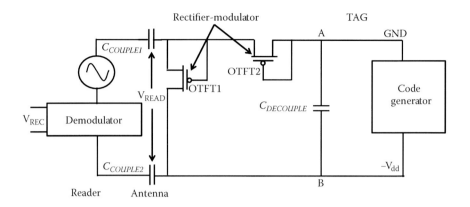

(a)

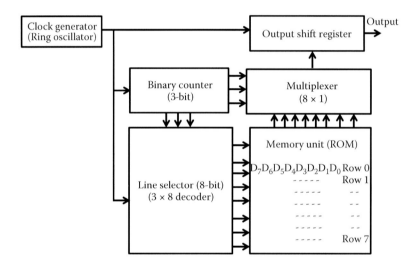

(b)

**FIGURE 10.3** Schematics of (a) a capacitive coupled organic rectifier-modulator and (b) a transponder circuit of an organic RFID.

used to operate the 8-bit line coder that makes a selection of a new row after completing the transmission of earlier 8-bits.

## 10.1.6 SENSORS

Among the exciting developments of organic electronics applications during the last two decades, organic sensor applications in the fields of medical diagnosis, industrial safety, household security, food safety, and environmental observations have drawn tremendous attention [58]. Researchers foresee the tremendous capability of organic devices, especially, OTFTs made up of organic semiconducting materials with virtually endless combinations and compounds to embed system-on-chips using low cost, low temperature, and efficient process techniques. Inorganic/organic sensors can be classified broadly into categories such as biological, chemical, and physical sensors. Depending on the transducing principle, sensors can be further classified into several categories such as resistive, capacitive, field effect, colorimetric, and resonating. Commercially available sensors based on metal-oxide or conducting polymers (chemiresistors) use the resistive transducing principle. The field-effect transducer principle, normally associated with the field-effect transistors, shows the change in work function, threshold voltage, or conductivity of the channel depending upon the exposure and type of analyte [59]. Capacitive transducers normally exhibit variations in dielectric constant. The colorimetric principle demonstrates the modification in the optical spectrum with the presence of analytes. The resonating transducing principle displays change in the resonating frequency of the resonator [60].

The sensors based on organic semiconductors are more suitable inherently for chemical, biological, and gas sensing applications compared to their inorganic counterparts. The reasons behind the fact that inorganic sensing devices have limited selectivity, moderate sensitivity, higher operating temperature, and lack of flexibility for some specific sensing applications such as strain or pressure sensors. The sensors based on organic semiconductors have drawn greater attention of researchers due their low fabrication cost, lower operating temperature, portability, disposability, and multiparametric sensing characteristics. Also their weak Van der Waals force makes them reproducible upon the application of an appropriate cleaning substance [61]. On the other hand, there are several inherent drawbacks of organic semiconductors such as lower carrier mobility and deprived environmental stability make them less suitable as direct sensing elements.

OTFT-based sensors, specifically with an upside-down structure, have drawn immense research interest and emerged as a most promising candidate for sensing applications. OTFT sensors are capable of producing amplification of sensed quantity, hence they are suitable for direct sensing application as the active organic material directly reacts with the analytes depending on the adsorption, diffusion, or absorption properties of the OSC. Furthermore, the reactive properties of OSCs can be modified with the inception of receptor molecules as per the requisition during the synthesis process to enhance the selectivity and sensitivity of the sensor [62], which would be rather difficult to achieve using their inorganic counterpart. The OTFT sensor holds all the essential properties as a sensing element such as selectivity, sensitivity, disposability, and reproducibility at very low cost and low operating temperature. The foremost advantage of the OTFT sensor is the multiparametric sensing operation, wherein various electrical parameters such as the bulk conductivity, two-dimensional field-induced conductivity, threshold voltage, and field-effect mobility of the OTFT device are employed to characterize the sensing output [63–66].

Researchers have demonstrated OTFT-based sensors for detecting various parameters or quantitative levels of substances such as humidity, strain, pressure, inorganic gas substances, DNA, and explosive substances. DNA sensors based on organic materials have turned out to be promising devices for transforming a chemical binding event into the electrical signal that can be easily measured, analyzed, and amplified. In an organic sensor, DNA molecules are made immobilized by the physical absorption through hydrophobic interactions between the OSC and DNA molecules when applied to the organic active layer such as pentacene. The immobilization of DNA molecules alters the doping density of the charge carriers on the basis of charge polarity of the molecules, which in turn modulates the conductivity of the channel [67]. It concludes that these molecules greatly affect the electrical performance of the OTFT. Figure 10.4 shows the immobilization of DNA molecules at the surface of the organic active layer.

Among all the types of sensors, researchers have shown keen interest in the development and performance enhancement of inorganic gas sensors as the low cost, reversible, and reliable sensing technology is the need of the chemical industry today. The most common inorganic toxic gases are nitrogen monoxide (NO), nitrogen dioxide ($NO_2$), carbon monoxide (CO), carbon dioxide ($CO_2$), and hydrogen sulfide ($H_2S$). Ammonia ($NH_3$) is less toxic and less dangerous to human beings. Novel mechanisms, structures, and active materials are being employed to enhance the performance of the OTFT-based sensors. The sensitivity and selectivity of chemical sensors can be enhanced by inserting small receptor molecules in the active layer after the fabrication process [62]. Detection of wide range of analytes is possible with better stability, sensitivity,

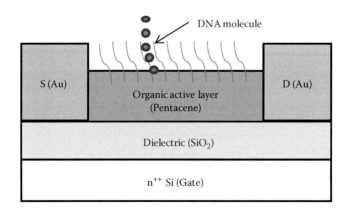

**FIGURE 10.4** Schematic showing the immobilization of DNA molecules in the grain boundaries due to hydrophobic interactions in BGBC OTFT.

and reproducibility by fabricating sensor array with different active layers. An artificial intelligent electronic nose has been demonstrated using this technique [68]. A study of the morphological effect on the performance of the OTFT sensor has been carried out [69].

Modification of the device dimensions and novel structural designs of the OTFT play crucial roles in the enhancement of the performance, sensitivity, and selectivity of the sensors. Scaling of channel length of an OTFT device significantly increase the conductivity of the channel. L. Wang *et al.* [70] in 2006 reported an increase in drain current by a factor >5 with nanoscale channel length compared to the microscaled channel length of OTFT devices. Also, the contact-limited carrier transport effect in very short-channel devices allows the analyte molecules, which are able to diffuse near the source/drain contacts, to change the behavior of the contact. Unlike the inorganic sensors, an upside-down structure is preferred in OTFT sensors, as analytes can directly modulate the behavior of the active organic material.

Despite relentless efforts, innovative applications of organic material compounds, devices, and technologies to design an inexpensive system-on-chip or lab-on-chip in terms of sensing applications is still challenging work for researchers. In addition to sensitivity, selectivity, and robustness, low power consumption and compact size are stringent requirements that sensor devices have yet to be fulfilled.

## 10.1.7 SPINTRONICS

During the last decade, spintronics has gained a lot of research interest because of its prospective advantages for high device integration capability, speed, and smaller device dimensions with low power consumption in comparison to charge-based devices. Depending on the control of the spin orientation of the charge carriers, several new phenomena and interesting applications have been recently introduced. A "spin valve" is a prime spintronic device, formed by a sandwich layer of a nonmagnetic spacer between soft and hard magnetic layers. Conventionally, spintronic devices such as spin valve (SV) and magnetic tunnel junction (MTJ) are constructed using two ferromagnetic electrodes separated by a nonmagnetic spacer. The devices work on the principle of injection and detection of spin-polarized charge carriers.

A new dimension is given to the research field with the inception of organic materials in developing spintronic devices, with the promise to improve functionality and processability. The spin polarized transport mechanism in organic material has gained considerable attention due to its large spin relaxation time, that is, charge carriers can sustain their spin for a longer time in the organic material compared to their inorganic counterparts. However, aggressive research work is still required to understand the spin transport phenomenon in organic materials/semiconductors for successful implementation of organic spintronic devices and their applications.

Two main factors are responsible for the perturbation of spin orientation in materials. The first factor is a spin–orbit interaction, which is basically an interface between the electron spin and the nucleus charge. Spin–orbit interaction is directly proportional to the atomic number ($Z$) of the material used; in particular, the magnitude $Z^4$. The second factor is a hyperfine interaction, which reflects an interface between the electron spin and the nucleus spin. Both factors are dependent on the atomic number of the material used, that is, the heavier the material the more spin–orbit and hyperfine interactions. As the organic materials are a combination of carbon and hydrogen, which are normally lighter elements, they possess very weak spin–orbit and hyperfine interactions resulting in a long spin-relaxation time ($\tau_s$) without spin perturbation. A long spin-relaxation time in the range of milliseconds to 1 s has been measured for some organic semiconducting materials, including rubrene and $Alq_3$. However, the spin diffusion length is a function of mobility. Owing to the very low mobility of the OSC, the spin diffusion length is in the range of few tens of nanometers for OSCs.

The main difference between the organic MTJ (OMTJ) and organic SV (OSV) is the thickness of the organic spacer. Generally, the organic spacers of OSVs are thick, in the range greater than 10 nm. The spin transport mechanism is primarily affected by the thickness and interface properties of the organic layer. The hole or electron transport in an OSV starts with injection through thermionic emission or the tunneling mechanism from the magnetic electrode to the organic material. Thereafter, the charge carriers hop through the thick organic layer and are collected by the other magnetic electrode. On the contrary, an OMTJ possesses a very thin layer of organic spacer, typically less than 5 nm. Due to the very thin layer, the injected charge carriers tunnel through the organic sandwich layer and reach the opposite electrode. The probability of retention of the spin orientation after transport is increased due to the very thin tunneling barrier crossed by the charge carriers.

In inorganic and hybrid spintronic devices, several materials, including Fe, Co, Ni, and half-metals ($CrO_2$, $Fe_3O_4$, and $La_{0.7}Sr_{0.3}MnO_3$ [LSMO]) are utilized as ferromagnetic electrodes. LSMO has become a material of choice due its nearly 100% spin polarization property and operational stability against oxidation. However, the curie temperature of the LSMO is near to room temperature, hence, the magnetization level reduces for a temperature near or above 300 K. For the sandwiched spacer layer, among numerous organic materials, small molecules such as pentacene ($C_{22}H_{14}$), rubrene ($C_{42}H_{28}$), copper phthalocyanine (CuPc), tris(8-hydroxyquinolinato)aluminum (Alq3); and polymers such as regioregular poly(3-hexylthiophene) (RR-P3HT), regiorandom poly(3-octylthiophene) (RRa-P3OT), poly(triarylamine) (PTAA) are the most commonly used organic materials.

A general technique to investigate spin injection and transport in the spintronic devices is to observe the magnetoresistance (MR). Depending on the relative magnetization of the ferromagnetic materials, the device resistance of spintronic devices varies provided that the spin diffusion length is larger than the spacer layer thickness. If the relative magnetization of the ferromagnetic materials is in parallel, then the device exhibits lower resistance ($R_p$), otherwise a higher device resistance ($R_{Ap}$) is observed for anti-parallel orientation. The MR of a spintronic device is defined as

$$MR\ (\%) = \left( \frac{R_{AP} - R_P}{R_P} \right) \times 100 \tag{10.1}$$

Depending on the thickness of the organic spacer layer in the organic spintronic device, the spin transport mechanism changes. Generally, MR is classified into two categories: giant magnetoresistance (GMR) and tunneling magnetoresistance (TMR). For relatively thick spacer thickness, i.e., ≥10 nm, the MR is observed due to the injection and transport of the spin-polarized charge carriers through the organic spacer. This is generally referred to as GMR. In contrast, using a thin spacer with a thickness less than 5 nm, the spin transport in the spintronic device is occurring due to tunneling of the charge carriers through the organic spacer and this is referred to as TMR. TMR is observed at a low bias applied to the device. In the past few years, another type of MR has been observed, particularly in organic spintronic devices, called organic magnetoresistance (OMR). Under the influence on an external magnetic field, the quasi-particles present in organic materials such as polarons and excitons exhibit spin-dependent interactions resulting in a change in current or resistance of the organic spintronic device. This magnetic field-dependent magnetoresistance is referred to as OMR. The origin of OMR mechanisms has yet to be fully understood.

MR of a spintronic device is strongly dependent on the operational temperature and it decreases with the rising temperature. In the initial phase of the development of organic spintronic devices, MR was observed at the near absolute zero temperature (0 K). Xiong *et al.* [71] first reported MR of 40% using LSMO/Alq3/Co spin valve at the temperature of 11 K. However, it reduced to zero at well below room temperature. Wang *et al.* [72] observed an MR of 5% using Fe/Alq3/Co spin valve at 11 K; however, it reduced to zero around a temperature of 100 K. Majumdar *et al.* [73], measured the MR of around 1.5% at room temperature, which was reported 80% at 5 K. Thereafter, Santos *et al.* [74] observed a TMR of 4% at room temperature with improved interface using a Co/Al$_2$O$_3$/Alq3/NiFe layered device. Recently, Li *et al.* [75] observed a higher MR of 90% at 4.2 K and 6.8% at room temperature using a high mobility *n*-type semiconducting polymer—P(NDI2OD-T2). In comparison to these MR ratios achieved using organic materials, very high GMR and TMR are obtained using their inorganic counterparts. A GMR in the range of 50% to 100% and TMR in the very high range of 100% to 600% [76] have already been obtained using inorganic spintronic devices at room temperature. However, Mermer *et al.* [77] reported a high OMR of 10% with an external magnetic field of 10 mT at room temperature. Therefore, it is apparent that more aggressive and intense research efforts are still required to produce organic spintronic devices with high MR ratios making them more comparable to their inorganic counterpart.

A major application of organic spintronic devices is to generate and detect the spin-polarized current using organic ferromagnetic materials. However, organic spintronic applications have shown better outcomes using magneto-resistive devices. The most recent use of organic spintronic devices has been in the field of memory and sensor applications. Depending on the relative orientation of the ferromagnetic materials, the organic spintronic devices exhibit bistable operation. For parallel configuration, it shows low resistance and high resistance for anti-parallel mechanism. This bistability property is most suitable for memory storage applications. A sensitive magnetic field sensor can be implemented using organic spintronic devices. In principle, one electrode with pinned anti-ferromagnetic layer and the other electrode with soft magnetic material and free to move properties can be forced to produce MR under the influence of the external magnetic field whose strength is to be measured.

## 10.1.8 ORGANIC SOLAR CELLS

The need for electrical energy is increasing very rapidly worldwide for all kinds of electronic applications. In fact, in the era of industrialization and globalization its demand has increased exponentially. Fossil fuels, which are the prime elements of electrical energy generation, are on the verge of being exhausted. Also, these fuels are the source of severe environmental pollution. Therefore, it is utmost necessary to harness energy from nonconventional, clean energy resources. Recently, there have been great advancements in the organic semiconductor technology, including organic solar cells/organic photovoltaic cells (OPVs) as an alternative to the conventional energy sources. Due to their low-cost fabrication, simple working principle, synthetic variability, and a capability to incorporate advanced materials, OPVs have gained the attention of researchers. As OLEDs for display and street lighting applications had been commercialized, OPV devices have also improved from less than 1% efficiency to 10% efficiency to fulfill the requirements of the energy-starved global market [78]. Moreover, the charge transport mechanism is simpler in organic materials as compared to inorganic materials. The charge mobility of organic materials is low due to the localization of the charged states, but their low-cost large-scale production may compensate it.

### 10.1.8.1 HISTORICAL BACKGROUND OF ORGANIC SOLAR CELLS

Solar cells based on organic materials are efficient, which implies favorable charge generation as well as light-absorption properties. Organic solar cells based on polymer materials are processable, lightweight, and flexible with the capability to use low-cost processing techniques at low temperature for mass production. The conversion of light energy into electrical energy known as photovoltaic effect was given by Becquerel in 1839 using liquid electrolyte. The transformation of photovoltaic effect into technology to convert sunlight into electricity was reported by Chapin *et al.* on *p-n* junction based on a silicon with the efficiency of 6% [79].

Organic solar cell research area had shown significant growth since 1980 when organic molecules were synthesized at low temperature using physical vapor deposition techniques. Tang explained a two-layer device in which copper phtalocyanine (CuPc) was the donor and perylene tehacarboxylic derivative was the acceptor, and showed a power efficiency of 1% in 1986 [80]. Third-generation technologies include, first, dye-sensitized solar cells traced by Gratzel, which require an electrolyte and work as electrochemical cells. Second are hybrid approaches in which inorganic materials are doped in a semiconductor polymer or by combining organic material with a nanostructured inorganic material such as $TiO_2$. Today, the highest reported power efficiencies are in the range of 6.71%–8.72% for small molecules, 8.40%–10.61% for polymer OPVs, and 7.1%–15.2% for perovskite OPVs [81].

### 10.1.8.2  OPERATING PRINCIPLE OF ORGANIC SOLAR CELLS

The working principle of OPV is shown in Figure 10.5. When the photons present in the sunlight strike the organic semiconductor through a transparent or meshed anode, they generate excitons in the lattice of the semiconductor. These excitons consist of both electrons and holes, and have a lifetime during which, if they are not separated they recombine to release an energy, and no electricity is produced. On the other hand, if they are separated in time by electrostatic force, then they constitute a current. In a single-layer organic solar cell, a strong electric field is achieved due to the presence of depletion region in "Schottky contact." Dissociation of excitons in organic solar cells depends mainly on gradients for the potential across the acceptor–donor interface that results in a photo-induced transfer of charge between donor and acceptor materials [82].

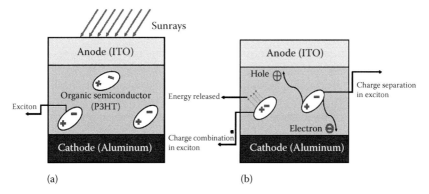

**FIGURE 10.5**   Single-layer organic photovoltaic cell (OPV) with (a) excitons generation and (b) charge transportation phenomenon.

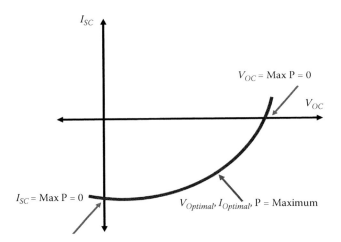

**FIGURE 10.6**    Graphical representation for maximum power curve of OPV.

The performance of the organic solar cell is characterized by the fill factor (*FF*), which is determined from measurement of the *I–V* curve and is defined as the maximum power: $P_{MAX}$ ($P_{MAX} = V_{Optimal} I_{Optimal}$) divided by the product of $V_{OC} I_{SC}$. The parameters; $V_{Optimal}$ and $I_{Optimal}$ are the highest values of the voltage and current simultaneously, whereas, the $V_{OC}$ and $I_{SC}$ are the open circuit voltage and the short circuit current, respectively. Mathematically, the fill factor can be calculated as

$$FF = \frac{V_{Optimal} I_{Optimal}}{V_{OC} I_{SC}} \tag{10.2}$$

Furthermore, the ratio of output power, $P_{MAX}$ (electrical), and the input power, $P_{IN}$ (light), is expressed as the power efficiency, η, which can be expressed as

$$\eta = \frac{P_{MAX}}{P_{IN}} = \frac{V_{Optimal} I_{Optimal}}{P_{IN}} = \frac{FF V_{OC} I_{SC}}{P_{IN}} \tag{10.3}$$

The maximum output power curve is shown in Figure 10.6. The available photocurrent due to transport of charges is electric field dependent. It implies that the short circuit current is negligible at the maximum value of the open circuit voltage. Hence the power is reduced to the minimum level. In corollary, at the maximum $I_{SC}$, the $V_{OC}$ is minimum and thus the minimal power is obtained. Therefore, the maximum power is obtained only when the open circuit voltage and short circuit current are at their optimal level of values as depicted in Figure 10.6 and Equation 10.3.

### 10.1.8.3  BULK HETEROJUNCTION SOLAR CELL

During recent years, researchers have shown tremendous interest in the development of organic solar cells due to their low cost, more versatile fabrication

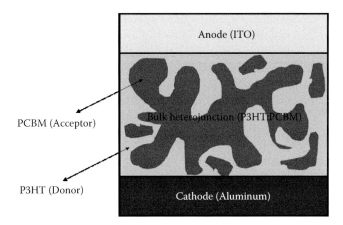

**FIGURE 10.7**    Example of heterojunction OPV cell.

processes, and less complex mechanisms for harnessing solar energy. The production of electricity from sunlight is a result of successive reactions inside the solar cells. Among various types of organic solar cells, the bulk heterojunction solar cell is considered as the most prominent solar device. Compared to the single layer and bilayer solar cells, the heterojunction solar cell is a multilayer device. The bilayer devices, which were better than the single-layer devices, have not achieved the efficiency they were deemed to achieve. Therefore, a new type of device structure was developed to enhance the reliability and efficiency of OPV cells. In heterojunction devices, a blend of donor and acceptor materials is used as a single organic layer. Due to which, an exciton in the active layer is generated, and it quickly reaches the nearby junction and the electrons and holes get separated instantly. Due to this phenomenon, the efficiency of devices increases in comparison to single and bilayer devices. An example of heterojunction OPV cells is presented in Figure 10.7 with P3HT and PCBM as donor and acceptor materials, respectively. Blend molecules exhibit a large interface area and most excitons manage to reaches donor–acceptor interfaces, where the hole moves toward anodes and electrons toward cathodes. The concept of bulk heterojunction can be considered as a breakthrough toward low-cost and highly efficient power production in the future [83,84].

## 10.2  FUTURE PERSPECTIVES

Organic electronics has been a subject of intense commercial and academic interest over the last two decades. This book presents just a small portion of the great potential of organic electronic-based materials, devices, and circuits. OTFTs could find widespread usage in flexible displays with applications to e-newspapers, e-magazines, and e-books. Aggressive efforts are being made to increase the applicability of organic devices in widely used circuits and applications such as the differential amplifier [11], ADC and DAC [27–30], shift register [17], energy-saving organic LED [50], organic memory [34], ring oscillator [11–13], and organic solar cell [82]. The future scope of the organic transistor in compact, portable, and lightweight circuits includes the smart phone, flexible display, and biosensor. The following sections further discuss the scope for enhancing the performance of devices and circuits in the future.

### 10.2.1 NOVEL MATERIALS AND STRUCTURES

The performance of organic semiconductors is governed by how molecules or polymer chains are arranged in the solid state. Regardless of the latest developments in $p$-type and $n$-type materials, there is an ample scope for synthesizing novel materials with stable and high mobility under ambient conditions, ease of synthesis, reliability under processing as well as operational conditions, and the ability to effectively inject and collect charge carriers at the interface. Solubility and mobility are other important concerns for semiconducting organic materials. Although P3HT exhibits outstanding solubility, it is limited by lower mobility. In contrast, pentacene exhibits higher hole mobility but demonstrates instability in the air and lower solubility in solvents that makes it inappropriate for fabricating through low-cost printing methods. Some novel solutions can be helpful to fill this technological gap. Dielectric materials need to be synthesized that would have a high dielectric constant as well as low leakage current. Besides this, the role of border traps needs to be addressed together with oxide and interfacing traps. Furthermore, realization of novel contact materials that can easily inject the charge carriers is urgently required, especially in the lowest unoccupied molecular orbital (LUMO) of $n$-type materials. For flexible devices, novel conducting polymers, such as PSS, PEDOT:PSS, and PANI-CSA, can find their use to make the source, drain, and gate contacts. In addition to this, aggressive research is required to understand the transport phenomenon in organic materials for their successful implementation in spin-based devices. Moreover, the application of organic materials in spintronics can be as organic ferromagnetic materials for generating and detecting spin-polarized current. Development of modified structures for OTFTs, such as dual gate, floating gate, multigate, nanowire FETs, vertical, cylindrical, and organic Fin-FET has become essential for achieving higher speed, drivability, and mobility. On the other hand, power consumption is on the rise in these devices that can be resolved by proper selection of materials to make them suitable for portable systems. Furthermore, most of the complementary circuits are realized using $p$- and $n$-type TFTs connected horizontally. However, performance can be improved through a vertically stacked structure as it is beneficial in reducing the interconnect distance among TFTs that in turn reduces the parasitics. Moreover, vertical stacking may lead to an increase in density of circuits. Realizing novel structures with novel materials and their circuits may open new dimensions in organic electronics with an aim to make them suitable for photovoltaics, wearable medical monitoring smart shirts, and highly sophisticated integrated circuits.

## 10.2.2  OPTIMIZATION OF DEVICE PERFORMANCE

Performance optimization of a device/transistor is a key challenge that includes factors such as series resistance, gate-induced drain leakage, process-induced variation, mobility, and charge injection at the contacts/semiconductor interface. Mobility can be improved by obtaining large grains of active material by optimizing the deposition process. Often, increased order in the molecular packing results in high mobility. A surface treatment and self-assembled monolayer of the dielectric can be helpful in modifying the microstructure of organic thin-films chemically. Moreover, the morphology of the active layer can be organized, especially to the monolayer closest to the insulator interface by tuning the rate of deposition and temperature of the substrate. Performance improvement of the bottom contact structures is an additional challenge. Insertion of an additional high-doping region near the contacts can increase the performance of these structures. In fact, adding an extra semiconductor layer between the contact and semiconductor can surprisingly enhance the performance. Furthermore, the threshold voltage can be reduced substantially by using high-k dielectrics without compromising the high *on-off* current ratio. In organic single gate TFTs, asymmetric S/D contacts can be employed with low and high work function metals for significantly lowering the *off*-current. The performance can be further enhanced in terms of stability, operating bias, cost economy, longevity, temperature dependency, and power dissipation. However, from a theoretical point of view, the accurate prediction of the properties of organic devices, including true current–voltage curves with as few adjustable parameters as possible, still represents a formidable challenge.

## 10.2.3  ANALYTICAL MODELS

The understanding of charge transport in organic devices is critical and important for continuing the miniaturization of electronic devices. Researchers have demonstrated a comparison between the top and the bottom contact structures in terms of drain current and mobility. In the bottom contact structure, a shadow of metal can create a hindrance for accurate deposition of the active layer near the contacts that results in a heterogeneous/assorted morphology of the semiconductor. This results in low effective mobility and high contact resistance for the bottom contact structure. Moreover, this structure exhibits a discontinuity in the active layer between the channel and contacts. Understanding the mechanism behind the microstructure arrangement and its effect on carrier mobility is still one of the formidable challenges. To model these morphological disorders, some calibrated standards need to be developed on the basis of low mobility regions near the contacts and high contact resistance. Efforts can be devoted to creating a model that would represent accurate static and dynamic electrical characteristics in all operating regimes. Mapping of the overlapping region (gate–source/gate–drain) to the resistance would be helpful in developing analytical models based on the thickness of the active layer. Regardless of the few available models, it is required to develop the analytical models for dual gate structures based on the mapping of charge accumulation phenomena through second gate bias. Suitable models are also required to understand the charge transport mechanism in the vertical channel, the floating gate, and the organic FinFET structures. At present, very few models are available that address the charge transport issues in nanowire, multigate, and cylindrical gate organic TFTs. Therefore, the circuit simulation is only limited to the technology computer-aided design (TCAD) mixed-mode, which is extremely time consuming. Therefore, the compact models for these novel structures need to be developed that can be included in the circuit level simulators.

## 10.2.4 ENVIRONMENTAL DEVICE STABILITY

The characteristics of organic devices should not vary during prolonged operation and environmental conditions. Due care must be taken against exposure of devices to moisture and oxygen, which can be achieved by depositing a film of inorganic oxides. It is imperative to explore the physical mechanism that obstructs device stability. Through encapsulation of the devices, a longer functioning lifetime can be achieved. Attaching a metal or glass lid to the substrate with a low-permeation adhesive is one of the encapsulation methods that can be used to protect the organic materials (especially for OLED displays). Additionally, a thin barrier coating at the top and bottom side of the device can offer noteworthy improvement in device performance. Hysteresis behavior is an important challenge associated with the operational stability of organic transistors that is often observed during sweeps of the gate source voltage. This is due to numerous defects associated with OTFTs, such as charge trapping in the OSC, polarization of the dielectric, charge leakage from the OSC to dielectric, and the presence of dipoles in the OSC/insulator interface. A high hysteresis effect results in a significant variation in the threshold voltage, thereby affecting the overall performance of the transistor. Researchers have shown low hysteresis behavior of an OTFT by applying surface treatment to the dielectric layer that reduces localization of the charge in the trapping states. Nevertheless, aggressive efforts are needed to substantially increase the operational stability of OTFTs.

# REFERENCES

1. Raval, H. N.; Tiwari, S. P.; Navan, R. R.; Mhaisalkar, S. G.; Rao, V. R. "Solution processed bootstrapped organic inverters based on P3HT with a high-k gate dielectric material," *IEEE Electron Device Lett.* **2009**, 30(5), 484–486.

2. Spijkman, M. J.; Myny, K.; Smits, E. C. P.; Heremans, P.; Blom, P. W. M.; De Leeuw, D. M. "Dual gate thin film transistors, integrated circuits and sensors," *Adv. Mater.* **2011**, 23(29), 3231–3242.

3. Spijkman, M.; Smits, E. C.; Blom, P. W.; De Leeuw, D. M.; Come, Y. B. S.; Setayash, S.; Cantatore, E. J. "Increasing the noise margin in organic circuits using dual gate field effect transistors," *Appl. Phys. Lett.* **2008**, 92(14), 143304-1–143304-3.

4. Myny, K.; Beenhakkers, M. J.; Van Aerle, N. A. J. M.; Gelinck, G. H.; Genoe, J.; Dehaene W.; Heremans, P. "Unipolar organic transistor circuits made robust by dual-gate technology," *IEEE J. Solid-State Circuits* **2011**, 46(5), 1223–1230.

5. Klauk, H.; Zschieschang, U.; Pflaum, J.; Halik, M. "Ultralow-power organic complementary circuits," *Nature* **2007**, 445(7129), 745–748.

6. Yan, H.; Chen, Z.; Zheng, Y.; Newman, C.; Quinn, J. R.; Dötz, F. Facchetti, A. "A high-mobility electron-transporting polymer for printed transistors," *Nature* **2009**, 457(7230), 679–686.

7. Huang, T. C.; Fukuda, K.; Lo, C. M.; Yeh, Y. H.; Sekitani, T.; Someya, T.; Cheng, K. T. "Pseudo-CMOS: A design style for low-cost and robust flexible electronics," *IEEE Trans. Electron Devices* **2011**, 58(1), 141–150.

8. Brown, A. R.; Jarrett, C. P.; de Leeuw, D. M.; Matters, M. "Field effect transistors made from solution-processed organic semiconductors," *Sythetic Metals* **1997**, 88(1), 37–55.

9. Kim, H. S.; Won, S. M.; Won, M.; Ha, S. G.; Ahn, J. H.; Facchetti, A.; Marks, T. J.; Rogers, J. A. "Self-assembled nanodielectrics and silicon nanomembranes for low voltage, flexible transistors, and logic gates on plastic substrates," *Appl. Phys. Lett.* **2009**, 95(18), 183504-1–183504-3.

10. Guerin, M.; Bergeret, E.; Benevent, E.; Daami, A.; Pannier, P.; Coppard, R. "Organic complementary logic circuits and volatile memories integrated on plastic foils," *IEEE Trans. Electron Devices* **2013**, 60(6), 2045–2051.

11. Kane, M. G., Campi, J., Hammond, M. S., Cuomo, F. P., Greening, B., Sheraw, C. D., Jackson, T. N. "Analog and digital circuits using organic thin-film transistors on polyester substrates," *IEEE Electron Device Lett.* **2000**, 21(11), 534–536.

12. Wu, Q.; Zhang, J.; Qiu, Q. "Design considerations for digital circuits using organic thin film transistors on a flexible substrate," *Proc. 2006 IEEE Int. Symp. Circuits Syst.* **2006**, 1267–1270.

13. Klauk, H.; Gundlach, D. J.; Jackson, T. N. "Fast organic thin-film transistor circuits," *IEEE Electron Device Lett.* **1999**, 20(6), 289–291.

14. Ante, F.; Kälblein, D.; Zaki, T.; Zschieschang, U.; Takimiya, K.; Ikeda, M.; Klauk, H. "Contact resistance and megahertz operation of aggressively scaled organic transistors," *Small* **2012**, 8(1), 73–79.

15. Zschieschang, U.; Hofmockel, R.; Rödel, R.; Kraft, U.; Kang, M. J.; Takimiya, K.; Klauk, H. "Megahertz operation of flexible low-voltage organic thin-film transistors," *Organic Electron. Phys. Mater. Appl.*, **2013**, 14(6), 1516–1520.

16. Fortuna, L.; Frasca, M.; Gioffrè, M.; Rosa, M.; Malagnino, N.; Marcellino, A.; Vecchione, R. *et al.* "On the way to plastic computaion," *IEEE Circuits and Systems Magazine* 8(3), 6–18, 2008.

17. Schwartz, D. E.; Ng, T. N. "Comparison of static and dynamic printed organic shift registers," *IEEE Electron Device Lett.* **2013**, 34(2), 271–273.
18. Myny, K.; van Veenendaal, E.; Gelinck, G. H.; Genoe, J. "An 8b organic microprocessor on plastic foil," **2011**, 272–273.
19. Myny, K.; van Veenendaal, E.; Gelinck, G. H.; Genoe, J.; Dehaene, W.; Heremans, P. "An 8-bit, 40-instructions-per-second organic microprocessor on plastic foil," *IEEE J. Solid-State Circuits* **2012**, 47(1), 284–291.
20. Myny, K.; Smout, S.; Rockelé, M.; Bhoolokam, A.; Ke, T. H.; Steudel, S.; Heremans, P. "A thin-film microprocessor with inkjet print-programmable memory," *Sci. Rep.* **2014**, 4, 7398.
21. Gay, N.; Fischer, W. J.; Halik, M.; Klauk, H.; Zschieschang, U.; Schmid, G. "Analog signal processing with organic FETs," *Solid-State Circuits Conf.* **2006**, 6, 1070–1079.
22. Marien, H.; Steyaert, M.; van Veenendaal, E.; Heremans, P. "Analog techniques for reliable organic circuit design on foil applied to an 18dB single-stage differential amplifier," *Organic Electron.* **2010**, 11(8), 1357–1362.
23. Guerin, M.; Daami, A.; Jacob, S.; Bergeret, E.; Bènevent, E.; Pannier, P.; Coppard, R. "High-gain fully printed organic complementary circuits on flexible plastic foils," *IEEE Trans. Electron Devices* **2011**, 58(10), 3587–3593.
24. Chang, J.; Zhang, X.; Ge, T.; Zhou, J. "Fully printed electronics on flexible substrates: High gain amplifiers and DAC," *Organic Electron.* **2014**, 15(3), 701–710.
25. Sankhare, M. A.; Bergeret, E.; Pannier, P.; Coppard, R. "Full-printed OTFT modeling: Impacts of process variation," *Solid-State Integrated Circuit Tech.* **2014**, 14–16.
26. Taylor, D. M.; Patchett, E. R.; Williams, A.; Neto, N. J.; Ding, Z.; Assender, H. E.; Morrison, J. J.; Yeates, S. G. "Organic digital logic and analog circuits fabricated in a roll-to-roll compatible vacuum-evaporation process," *IEEE Trans. Electron Devices* **2014**, 61(8), 2950–2956.
27. Xiong, W.; Guo, Y.; Zschieschang, U.; Klauk, H.; Murmann, B. "A 3-V, 6-bit C-2C digital-to-analog converter using complementary organic thin-film transistors on glass," *IEEE J. Solid-State Circuits* **2010**, 45(7), 1380–1388.
28. Xiong, W.; Zschieschang, U.; Klauk, H.; Murmann, B. "A 3V 6b successive-approximation ADC using complementary organic thin-film transistors on glass," *ISSCC Dig. Tech. Pap.* **2010**, 134–135.
29. Marien, H.; Steyaert, M. S. J.; van Veenendaal, E.; Heremans, P. "A fully integrated ΔΣ ADC in organic thin-film transistor technology on flexible plastic foil," *IEEE J. Solid-State Circuits* **2011**, 46(1), 276-284.
30. Zaki, T.; Member, S.; Ante, F.; Zschieschang, U.; Butschke, J.; Letzkus, F. "Digital-to-analog converter using organic p-type thin-film transistors on glass," *IEEE J. Solid-State Circuits* **2012**, 47(1), 292–300.
31. Raiteri, D.; van Lieshout, P.; van Roermund, A.; Cantatore, E. "An organic VCO-based ADC for quasi-static signals achieving 1LSB INL at 6b resolution," *IEEE Int. Solid-State Circuits Conf. Dig. Tech. Pap.* **2013**, 108–109.
32. Prime, D.; Paul, S. "Overview of organic memory devices," *Phil. Trans. R. Soc. A* **2009**, 367(1905), 4141–4157.
33. Fukuda, K.; Sekitani, T.; Zschieschang, U.; Klauk, H.; Kuribara, K.; Yokota, T.; Sugino, T. *et al.* "A 4 V operation, flexible braille display using organic transistors, carbon nanotube actuators, and organic static random-access memory," *Adv. Funct. Mater.* **2011**, 21(21), 4019–4027..
34. Takamiya, M.; Sekitani, T.; Kato, Y.; Kawaguchi, H. "An organic FET SRAM with back gate to increase static noise margin and its application to braille sheet display," *IEEE J. Solid-State Circuits* **2007**, 42(1), 93–100.

35. Schroeder, R.; Majewski, L. A.; Grell, M. "All-organic permanent memory transistor using an amorphous, spin-cast ferroelectric-like gate insulator," *Adv. Mater.* **2004**, 16(7), 633–636.

36. Naber, R. C. G.; Tanase, C.; Blom, P. W. M.; Gelinck, G. H.; Marsman, A. W.; Touwslager, F. J.; Setayesh, S.; De Leeuw, D. M. "High-performance solution-processed polymer ferroelectric field-effect transistors," *Nat. Mater.* **2005**, 4(3), 243–248.

37. Katz, H. E.; Hong, X. M.; Dodabalapur, A.; Sarpeshkar, R. "Organic field-effect transistors with polarizable gate insulators," *J. Appl. Phys.* **2002**, 91(3), 1572–1576.

38. Singh, Th. B.; Marjanovic, N.; Matt, G. J.; Sariciftci, N. S.; Schwo-diauer, R.; Bauer, S. "Nonvolatile organic field-effect transistor memory element with a polymeric gate electret," *Appl. Phys. Lett.* **2004**, 85(22), 5409-1–5409-3.

39. Narayanan Unni, K. N.; De Bettignies, R.; Dabos-Seignon, S.; Nunzi, J. M. "A nonvolatile memory element based on an organic field-effect transistor," *Appl. Phys. Lett.* **2004**, 85(10), 1823-1–1823-3.

40. Wang, S.; Leung, C. W.; Chan, P. K. L. "Enhanced memory effect in organic transistor by embedded silver nanoparticles," *Org. Electron.* **2010**, 11(6), 990–995.

41. Sekitani, T.; Yokota, T.; Zschieschang, U.; Klauk, H.; Bauser, S.; Takeuchi, K.; Takamiya, M.; Sakurai, T.; Someya, T. "Organic nonvolatile memory transistors for flexible sensor arrays," *Science* **2009**, 326(5959), 1516–1519.

42. Ma, L.; Pyo, S.; Ouyang, J.; Xu, Q.; Yang, Y. "Nonvolatile electrical bistability of organic/metal-nanocluster/organic system," *Appl. Phys. Lett.* **2003**, 82(9), 1419-1–1419-3.

43. Chu, C. W.; Ouyang, J.; Tseng, J. H.; Yang, Y. "Organic donor–acceptor system exhibiting electrical bistability for use in memory devices," *Adv. Mater.* **2005**, 17(11), 1440–1443.

44. Mabrook, M. F.; Yun, Y. J; Pearson, C.; Zeze, D. A.; Petty, M. C. "A pentacene-based organic thin film memory transistor," *Appl. Phy. Lett.* **2009**, 94(17), 173302-1–173302-3.

45. Ma, L. P.; Liu, J.; Yang, Y. "Organic electrical bistable devices and rewritable memory cells," *Appl. Phys. Lett.* **2002**, 80(16), 2997–2999.

46. Ouyang, J. Y.; Chu, C. W.; Szmanda, C. R.; Ma, L. P.; Yang, Y. "Programmable polymer thin film and non-volatile memory device," *Nat. Mater.* **2004**, 3, 918–922.

47. Bozano, L. D.; Kean, B. W.; Deline, V. R.; Salem, J. R.; Scott, J. C. "Mechanism for bistability in organic memory elements," *Appl. Phys. Lett.* **2004**, 84(607), 607–609.

48. Jackson, T. N.; Lin, Y. Y., Gundlach, D. J.; Klauk, H. "Organic thin-film transistors for organic light-emitting flat-panel display backplanes," *IEEE J. Selected Topics Quantum Electron.*, **1998**, 4(1), 100–104.

49. Zhou, L.; Wanga, A.; Wu, S. C.; Sun, J.; Park, S.; Jackson, T. N. "All organic active matrix flexible display," *Appl. Phys. Lett.* **2006**, 88(3), 083502-1–083502-3.

50. Li, D.; Guo, L. J. "Organic thin film transistors and polymer light-emitting diodes patterned by polymer inking and stamping," *J. Phys. D: Appl. Phys.* **2008**, 41(10), 105115-1–105115-7.

51. Chu, C. W.; Chen, C. W.; Li, S. H.; Yanga, Y. "Integration of organic light emitting diode and organic transistor via a tandem structure," *Appl. Phys. Lett.*, **2005**, 86(25), 253503-1–253503-3.

52. Liu, P. T.; Chu, L. W. "Innovative voltage driving pixel circuit using organic thin-film transistor for AMOLEDs," *J. Display Techn.* **2009**, 5(6), 224–228.

53. Baude, P. F.; Enter, D. A.; Haase, M. A.; Kelley, T. W.; Muyres, D. V.; Thesis, S. D. "Pentacene based radio frequency identification circuitry," *Appl. Phys. Lett.* **2003**, 82(22), 3964-1–3964-3.

54. Cantatore, E.; Geuns, T. C. T.; Gelinck, G. H.; Veenendaal, E. V.; Gruijthuijsen, A. F. A.; Schrijnemakers, L.; Drews, S.; De Leeuw, D. M. "A 13.56-MHz RFID System based on organic transponders," *IEEE J. Solid-State Circuits* **2007**, 42(4), 84–92.

55. Myny, K.; Steudel, S.; Vicca, P.; Beenhakkers, M. J.; Van Aerle, N. A. J. M.; Gelinck, G. H.; Genoe, J.; Dehaene, W.; Heremans, P. "Plastic circuits and tags for HF radio-frequency communication," *Solid State Electron.* **2009**, 53(12), 1220–1226.

56. Myny, K.; Beenhakkers, M. J.; Van Aerle, N. A. J. M.; Gelinck, G. H.; Genoe, J.; Dehaene, W.; Heremans, P. "A 128b organic RFID transponder chip, including Manchester encoding and ALOHA anti-collision protocol, operating with a data rate of 1529b/s," *IEEE Int. Solid-State Circuits Conference-Digest of Tech. Pap.* **2009**, 206–207, doi: 10.1109/ISSCC.2009.4977380.

57. Cantatore, E.; Geuns, T. C. T.; Gelinck, G. H.; Veenendaal, E. V.; Gruijthuijsen, A. F. A.; Schrijnemakers, L.; Drews, S.; De Leeuw, D. M. "A 13.56-MHz RFID system based on organic transponders," *IEEE J. Solid-State Circuits* **2007**, 42(4), 84–92.

58. Mabeck, J. T.; Malliaras, G. G. "Chemical and biological sensors based on organic thin-film transistors," *Analy. Bioanal. Chem.* **2006**, 384(2), 343–353.

59. Andringa, A.-M.; Piliego, C.; Katsouras, I.; Blom, P. W. M.; de Leeuw, D. M. "NO2 detection and real-time sensing with field-effect transistors," *Chem. Mater.* **2014**, 26(1), 773–785.

60. Fraden, J., *Handbook of Modern Sensors: Physics, Designs and Applications*, 3rd ed., Springer, New York, 2004.

61. Mirza, M.; Wang, J.; Li, D.; Arabi, S. A.; Jiang, C. "Novel top-contact monolayer pentacene-based thin-film transistor for ammonia gas detection," *ACS Appl. Mater. Interfaces* **2014**, 6(8), 5679–5684.

62. Someya, T.; Dodabalapur, A.; Huang, J.; See, K. C.; Katz, H. E. "Chemical and physical sensing by organic field-effect transistors and related devices," *Adv. Mater.* **2010**, 22(34), 3799–3811.

63. Torsi, L.; Dodabalapur, A.; Sabbatini, L.; Zambonin, P. G. "Multi-parameter gas sensors based on organic thin-film-transistors," *Sensors Actuators B: Chem.* **2000**, 67(3), 312–316.

64. Torsi, L.; Marinelli, F.; Angione, M. D.; Dell'Aquila, A.; Cioffi, N.; Giglio, E. D.; Sabbatini, L. "Contact effects in organic thin-film transistor sensors," *Org. Electron.* **2009**, 10, 233–239.

65. Zan, H.-W.; Tsai, W.-W.; Lo, Y.-R.; Wu, Y.-M.; Yang, Y.-S. "Pentacene-based organic thin film transistors for ammonia sensing," *IEEE Sensors Journal* **2012**, 12(3).

66. Tanese, M. C.; Fine, D.; Dodabalapur, A.; Torsi, L. "Organic thin-film transistor sensors: Interface dependent and gate bias enhanced responses," *Microelectronics J.* **2006**, 37(8), 837–840.

67. Jagannathan, L.; Subramanian, V. "DNA detection using organic thin film transistors: Optimization of DNA immobilization and sensor sensitivity," *Biosensors and Bioelectronics* **2009**, 25(2), 288–293.

68. Locklin, J.; Bao, Z. "Effect of morphology on organic thin film transistor sensors," *Anal. Bioanal. Chem.* **2006**, 384(2), 336–342

69. Liao, F.; Chen, C.; Subramanian, V. "Organic TFTs as gas sensors for electronic nose applications," *Sensors and Actuators B: Chemical* **2005**, 107(2), 849–855.

70. Wang, L.; Yoon, M.; Lu, G.; Yang, Y.; Facchetti, A.; Marks, T. "High-performance transparent inorganic–organic hybrid thin-film n-type transistors," *Nat. Mater.* **2006**, 5, 893–900.

71. Xiong, Z. H.; Wu, D.; Vardeny, Z. V.; Shi, J. "Giant magnetoresistance in organic spin-valves," *Nature* **2004**, 427(6977), 821–824.

72. Dhandapani, D.; Morley, N. A.; Gibbs, M. R. J.; Kreouzis, T.; Shakya, P.; Desai, P.; Gillin, W. P. "The effect of injection layers on a room temperature organic spin valve," *IEEE Trans. Magnetics* **2010**, 46(6), 1307–1310.

73. Majumdar, S.; Majumdar, H. S.; Laiho, R.; Österbacka, R. "Comparing small molecules and polymer for future organic spin-valves," *J. Alloys Compd.* **2006**, 423(1–2), 169–171.

74. Santos, T. S.; Lee, J. S.; Migdal, P.; Lekshmi, I. C.; Satpati, B.; Moodera, J. S. "Room-temperature tunnel magnetoresistance and spin-polarized tunneling through an organic semiconductor barrier," *Phys. Rev. Lett.* **2007**, 98(1), 3–6.

75. Li, F.; Li, T.; Chen, F.; Zhang, F. "Excellent spin transport in spin valves based on the conjugated polymer with high carrier mobility," *Sci. Rep.* **2015**, 5, 9355.

76. Ikeda, S.; Hayakawa, J.; Ashizawa, Y.; Lee, Y. M.; Miura, K.; Hasegawa, H.; Tsunoda, M.; Matsukura, F., Ohno, H. "Tunnel magnetoresistance of 604% at 300 K by suppression of Ta diffusion in CoFeB/MgO/CoFeB pseudo-spin-valves annealed at high temperature," *Appl. Phys. Lett.* **2008**, 93(8), 082508.

77. Mermer, Ö.; Veeraraghavan, G.; Francis, T. L.; Sheng, Y.; Nguyen, D. T.; Wohlgenannt, M.; Khan, M. S. "Large magnetoresistance in nonmagnetic π-conjugated semiconductor thin film devices," *Phys. Rev. B* **2005**, 72(20), 205202.

78. Schraber, M. C.; Muhlbacher, D.; Koppe, M.; Denk, P.; Waldauf, C.; Heeger, A. J.; Brabec, C. J. "Design rules for donors in bulk-heterojunction solar cells-towards 10% energy-conversion efficiency," *Adv. Mater.* **2006**, 18(6), 789–794.

79. Chapin, D. M.; Fuller, C. S.; Pearson, G. L. "A new silicon *p-n* junction photo cell for converting solar radiations in to electrical power," *J. Appl. Phys.* **1954**, 25(5), 676.

80. Tang, C. W. "Two-layer organic photovoltaic cell," *Appl. Phys. Lett.* **1986**, 48(2), 183.

81. Ishihara, T. *Pervoskite Oxide for Solid Oxide Fuel Cells*, Springer, New York, **2009.**

82. Hoppe, H.; Sariciftci, N. S. "Organic solar cells: An overview," *J. Mater. Res.* **2004**, 19(7), 1924–1945.

83. Hauch, J. A.; Schilinsky, P.; Choulis, S. A.; Childersm, R.; Biele, M.; Brabec, C. J. "Flexible organic P3HT: PCBM bulk-heterojunction modules with more than 1 year outdoor lifetime," *Solar Energy Mater. Solar Cells* **2008**, 92(7), 727–731.

84. Yu, G.; Gao, J. C.; Hummelen, J. C.; Wudl, F.; Heeger, A. J. "Polymer photovoltaic cells: Enhanced efficiencies via a network of internal donor-acceptor heterojunctions," *Science* **1995**, 270(5243), 1789–1791.

# Appendix A: Simulation Examples

## EXAMPLE A.1

SINGLE GATE ORGANIC THIN FILM TRANSISTOR (SG_OTFT; SEE FIGURE A.1)

```
go atlas
#
Title: Organic Thin Film Transistor with Bottom Gate Top
Contact Device Analysis
# Define Structure File
      mesh    smooth=1 space.mult=1.0 width=500
#
      x.m                   l=0     spac=0.5
      x.m                   l=10    spac=0.5
      x.m                   l=100   spac=0.5
      x.m                   l=110   spac=0.5
#
      y.mesh l=-0.010       spacing=0.005
      y.mesh l=0.0          spacing=0.005
      y.mesh l=0.05         spacing=0.005
      y.mesh l=0.06         spacing=0.01
      y.mesh l=0.0625       spacing=0.01
      y.mesh l=0.0725       spacing=0.01
# Define Region Parameter
      region num=1    material=Pentacene  y.min=0.0     y.max=0.050
      name=Pentacene
      region num=2    material=Al2O3      y.min=0.050  y.max=0.06
      region num=3    material=SiO2       y.min=0.06   y.max=0.0625
      region num=4    material=air        y.min=-0.010 y.max=0.0
      x.min=10 x.max=100
# Define electrodes of the device
      elec    num=1    material=Gold    name=source  x.max=10.0
                                                     y.min=-0.010
                                                     y.max=0.0
      elec    num=2    material=alum    name=gate    y.min=0.0625
                                                     y.max=0.0725
      elec    num=3    material=Gold    name=drain   x.min=100.0
                                                     y.min=-0.010
                                                     y.max=0.0
```

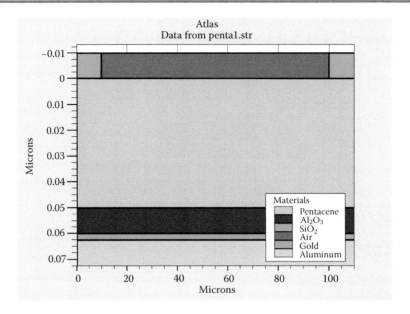

**FIGURE A.1**   Schematic structure for single gate organic thin film transistor (SG-OTFT).

```
# Defining Material Parameters and Application of the
Appropriate Physical Models
      doping     reg=1   uniform conc=4.1e17 p.type name=pentacene
#
      material name=Pentacene \
      eg300=2.8 nc300=1.0e21 nv300=1.0e21 permittivity=4.0 \
      mun=5e-5 mup=2.5
#
      material name=Al2O3 permittivity=4.5
      material name=SiO2  permittivity=3.9
# Defining Electrode Parameters
      contact name=source workfunc=5.1
      contact name=drain workfunc=5.1
      contact name=gate workfunc=4.28
# Defining the Defects in the Device
      defects   cont dfile=15don.dat afile=15acc.dat \
      nta=2.5e18 ntd=1.0e18 wta=0.129 wtd=0.5 \
      nga=0.0 ngd=0.0 ega=0.62 egd=0.78 wga=0.15 wgd=0.15 \
      sigtae=1.e-17 sigtah=1.e-15 sigtde=1.e-15 sigtdh=1
      .e-17 \
      siggae=2.e-16 siggah=2.e-15 siggde=2.e-15 siggdh=2
      .e-16
#  Defining the Model for the Device
      mobility deltaep.pfmob=1.792e-2  betap.pfmob=7.758e-5
      model    pfmob.p print
```

```
#
   output e.field j.electron j.hole j.conduc j.total
   e.velocity h.velocity \
   ey.field flowlines e.mobility h.mobility qss e.temp
   h.temp charge \
   recomb val.band con.band qfn qfp j.disp photogen impact
   tot.doping \
   u.srh u.rad
# Above Making Device and Applying Model and Defects
save outf=hagen.str
# Plot Device Structure using Solution File
tonyplot hagen.str
```

## OPERATIONAL BIAS CONDITIONS AND RUN THE SIMULATION (SEE FIGURE A.2)

```
# Transfer Characteristics (Ids versus Vgs) Biasing
solve init
#  Defining Method for Numerical Solution
method carriers=1 hole maxtrap=10
# Sweeping the Gate Voltage at Constant Drain Voltage
solve vdrain=-3
log outf=idvg.log
```

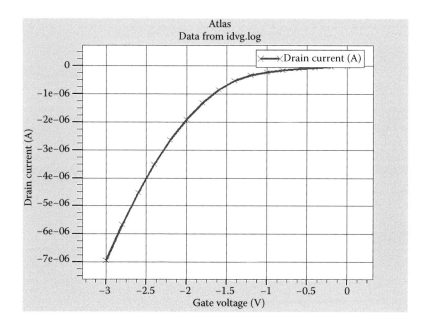

**FIGURE A.2** Transfer characteristics for single gate organic thin film transistor (SG-OTFT).

```
solve vgate=0.0   vstep=-0.2    vfinal=-3.0   name=gate
log off
tonyplot idvg.log
```

## OUTPUT CHARACTERISTICS ($I_{ds}$ VS. $V_{ds}$) BIASING (SEE FIGURE A.3)

```
Solve init
#
#solve prev  outfile=vg+00.bin
#solve vgate=0.0  vstep=5.0    vfinal=10   name=gate
outf=vg+10.bin onefile
#load  infile=vg+00.bin
          solve prev
# Sweeping the Drain Voltage at Fixed Gate Voltage
solve vgate=-0.0    name=gate  outf=vg-0.0.bin onefile
solve vgate=-1.0    name=gate  outf=vg-1.0.bin onefile
solve vgate=-2.0    name=gate  outf=vg-2.0.bin onefile
solve vgate=-3.0    name=gate  outf=vg-3.0.bin onefile
#solve vgate=-2.4    name=gate  outf=vg-2.4.bin onefile
#solve vgate=-2.7    name=gate  outf=vg-2.7.bin onefile
#solve vgate=-3.0    name=gate  outf=vg-3.0.bin onefile
#
load infile=vg-0.0.bin
solve prev
```

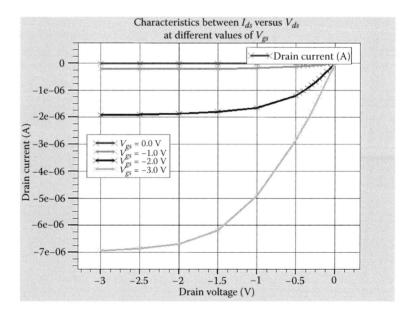

**FIGURE A.3** Output characteristics for single gate-organic thin film transistor (SG-OTFT).

```
# Saving the Solution File
log outf=vg-0.0mpfhm.log
solve vdrain=0.0  vstep=-0.05  vfinal=-0.5    name=drain
solve vdrain=-0.5 vstep=-0.5   vfinal=-3.0    name=drain
log off
#
load infile=vg-1.0.bin
solve prev
# Saving the Solution File
log outf=vg-1.0mpfhm.log
solve vdrain=0.0 vstep=-0.05   vfinal=-0.5    name=drain
solve vdrain=-0.5 vstep=-0.5   vfinal=-3.0    name=drain
log off
#
load infile=vg-2.0.bin
solve prev
# Saving the Solution File
log outf=vg-2.0mpfhm.log
solve vdrain=0.0 vstep=-0.05   vfinal=-0.5    name=drain
solve vdrain=-0.5 vstep=-0.5   vfinal=-3.0    name=drain
log off
#
load infile=vg-3.0.bin
solve prev
#Saving the Solution File
log outf=vg-3.0mpfhm.log
solve vdrain=0.0  vstep=-0.05  vfinal=-0.5    name=drain
solve vdrain=-0.5 vstep=-0.5   vfinal=-3.0    name=drain
log off
# Saving the Solution File
save outf=Staggered.str
log off
#   Plot Solution File
tonyplot vg-0.0mpfhm.log -overlay   vg-1.0mpfhm.log
                                    vg-2.0mpfhm.log
                                    vg-3.0mpfhm.log
quit
```

## EXAMPLE A.2

DUAL GATE ORGANIC THIN FILM TRANSISTOR (DG_OTFT; SEE FIGURE A.4)

```
go atlas
#
Title: DG OTFT Device TCAD Simulation
#
# Defining Structure File
mesh    smooth=1 space.mult=1.0 width=800
```

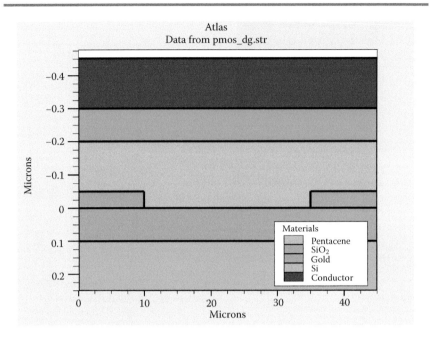

**FIGURE A.4**  Schematic structure for dual gate organic thin film transistor (DG-OTFT).

```
#
x.m     l=0     spac=0.2
x.m     l=10    spac=0.2
x.m     l=35    spac=0.2
x.m     l=45    spac=0.2
#
y.mesh l=-0.45    spacing=0.02
y.mesh l=-0.30    spacing=0.02
y.mesh l=-0.20    spacing=0.02
y.mesh l=-0.15    spacing=0.02
y.mesh l=-0.05    spacing=0.005
y.mesh l=0.0      spacing=0.005
y.mesh l=0.05     spacing=0.005
y.mesh l=0.10     spacing=0.02
y.mesh l=0.25     spacing=0.02
# Defining Region Parameter
region num=1    material=pentacene   y.min=-0.2 y.max=0.0
name=Pentacene
region num=2    material=SiO2 y.min=0.0 y.max=0.10
region num=3    material=SiO2 y.min=-0.3 y.max=-0.2
#
# Defining Electrodes of the Device
elec    num=1    material=Gold    name=source   x.max=10.0
y.min=-0.08 y.max=0.0
```

```
elec    num=2    material=Si        name=gate1    y.min=0.10
y.max=0.25
elec    num=3    material=Gold        name=drain    x.min=35.0
y.min=-0.08 y.max=0.0
elec    num=4    material=Aluminium name=gate      y.min=-0.45
y.max=-0.3
```

# Defining Material Parameters and Application of Appropriate Physical Model

```
doping    reg=3    uniform conc=1e16 p.type name=pentacene
#
material name=pentacene \
        eg300=2.2 nc300=2.0e21 nv300=2.0e21
        permittivity=6.0\mun=5e-5 mup=0.02


material    material=parylene   user.group=insulator user.
default=oxide permittivity=3.1
material    material=Al2O3        user.group=insulator user.
default=oxide permittivity=8
```

# Defining Electrode Parameters

```
contact num=1          workf=5.1
contact num=2          workf=4.4
contact num=3          workf=5.1
contact num=4          workf=4.4
contact name=gate      workf=4.33
contact name=gate1     workf=4.33    common=gate
```

# Defining the Defects in the Device

```
defects    cont dfile=l5don.dat afile=l5acc.dat \
        nta=2.5e18 ntd=1.0e18 wta=0.129 wtd=0.5 \
        nga=0.0 ngd=0.0 ega=0.62 egd=0.78 wga=0.15 wgd=0.15 \
        sigtae=1.e-17 sigtah=1.e-15 sigtde=1.e-15
        sigtdh=1.e-17 \
        siggae=2.e-16 siggah=2.e-15 siggde=2.e-15
        siggdh=2.e-16
```

# Defining the Model for the Device

```
mobility    deltaep.pfmob=1.792e-2    betap.pfmob=7.758e-5
  model          pfmob.p print


output    e.field j.electron j.hole j.conduc j.total
e.velocity h.velocity \
        ey.field flowlines e.mobility h.mobility qss e.temp
        h.temp charge \
        recomb val.band con.band qfn qfp j.disp photogen
        impact tot.doping\
        u.srh u.rad
```

# Above we are Making the Device and Applying Model and Defects

```
save outf=nikpmos1_dg.str
```

# Plot Solution File

```
tonyplot nikpmos1_dg.str
```

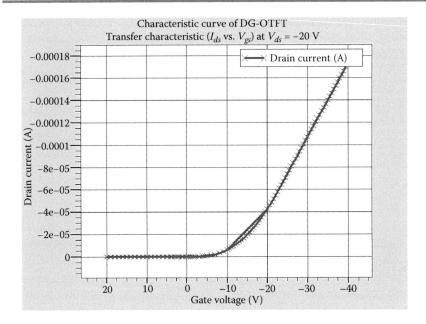

**FIGURE A.5** Transfer characteristics for dual gate-organic thin film transistor (DG-OTFT).

## OPERATIONAL BIAS CONDITIONS AND RUN THE SIMULATION (SEE FIGURE A.5)

```
# Transfer Characteristics (Ids vs Vgs) Biasing
solve init
#
# Defining Method For Numerical Solution

method carriers=1 hole maxtrap=2000
# Sweeping the Gate Voltage at Constant Drain Voltage
solve vdrain=-20
log outf=idvg1.log
solve    vgate=20     vstep=-1      vfinal=-20     name=gate
solve    vgate=-10    vstep=-0.5    vfinal=-40     name=gate

log off
tonyplot idvg1.log
```

## # OUTPUT CHARACTERISTICS ($I_{ds}$ VERSUS $V_{ds}$) BIASING (SEE FIGURE A.6)

```
Solve init
solve prev
```

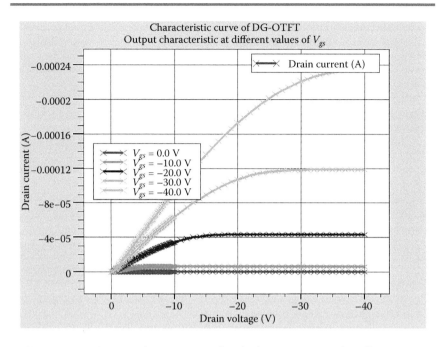

**FIGURE A.6**  Output characteristics for dual gate-organic thin film transistor (DG-OTFT).

```
# Sweeping the Drain Voltage at Fixed Gate Voltage
solve      vgate=-0.0      name=gate      outf=vg-0.bin onefile
solve      vgate=-10       name=gate      outf=vg-10.bin onefile
solve      vgate=-20       name=gate      outf=vg-20.bin onefile
solve      vgate=-30       name=gate      outf=vg-30.bin onefile
solve      vgate=-40       name=gate      outf=vg-40.bin onefile

load infile=vg-0.bin
solve prev
# Saving the Solution File
log outf=vg-0mpfhm.log
solve   vdrain=0.0     vstep=-0.1   vfinal=-10.0      name=drain
solve   vdrain=-10.0   vstep=-1     vfinal=-40.0      name=drain
log off
  load infile=vg-10.bin
solve prev
# Saving the Solution File
log outf=vg-10mpfhm.log
solve   vdrain=0.0     vstep=-0.1   vfinal=-10.0      name=drain
solve   vdrain=-10.0   vstep=-1     vfinal=-40.0      name=drain
log off

load infile=vg-20.bin
solve prev
```

```
# Saving the Solution File
log outf=vg-20mpfhm.log
solve   vdrain=0.0      vstep=-0.1    vfinal=-10.0    name=drain
solve   vdrain=-10.0  vstep=-1      vfinal=-40.0    name=drain
log off
      load infile=vg-30.bin
solve prev
# Saving the Solution File
log outf=vg-30mpfhm.log
solve   vdrain=0.0      vstep=-0.1    vfinal=-10.0    name=drain
solve   vdrain=-10.0  vstep=-1      vfinal=-40.0    name=drain
log off
load infile=vg-40.bin
solve prev
# Saving the Solution File
log outf=vg-40mpfhm.log
solve   vdrain=0.0      vstep=-0.1    vfinal=-10.0    name=drain
solve   vdrain=-10.0  vstep=-1      vfinal=-40.0    name=drain
save outf=dualgate1_dg.str
log off
# Plot Solution File
tonyplot  -overlay  vg-0mpfhm.log  vg-10mpfhm.log
vg-20mpfhm.log  vg-30mpfhm.log  vg-40mpfhm.log
quit
```

## EXAMPLE A.3

ORGANIC INVERTER CIRCUIT BASED ON SINGLE GATE
ORGANIC TRANSISTOR (SG-INVERTER; SEE FIGURE A.7)

```
# Defining the Device which is to be used in the Inverter
Circuit
go atlas
#
Title: Single Gate Organic Transistor Simulation
#
# Define Structure File
#mesh   smooth=1 space.mult=1.0 width=100
mesh    outf=penta_2.str master.out
#
x.m          l=0     spac=0.5
x.m          l=10    spac=0.5
x.m          l=100   spac=0.5
x.m          l=110   spac=0.5
#
y.mesh l=-0.010    spacing=0.005
y.mesh l=0.0       spacing=0.005
```

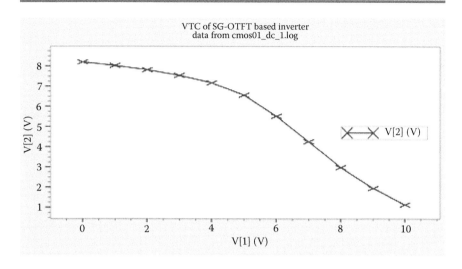

**FIGURE A.7**   Voltage Transfer Characteristics (VTC) of Organic Inverter Circuit with SG-OTFT.

```
y.mesh l=0.05       spacing=0.005
y.mesh l=0.06       spacing=0.01
y.mesh l=0.0625     spacing=0.01
y.mesh l=0.0725     spacing=0.01
#
# Define Region Parameter
region num=1    material=Pentacene   y.min=0.0 y.max=0.050
  name=Pentacene
region num=2    material=Al2O3       y.min=0.050 y.max=0.06
region num=3    material=SiO2        y.min=0.06 y.max=0.0625
region num=4    material=air         y.min=-0.010 y.max=0.0
                                     x.min=10 x.max=100
#
# Define Electrodes of the Device
elec    num=1   material=Gold   name=source   x.max=10.0
                                              y.min=-0.010
                                              y.max=0.0
elec    num=2   material=alum   name=gate     y.min=0.0625
                                              y.max=0.0725
elec    num=3   material=Gold   name=drain    x.min=100.0
                                              y.min=-0.010
                                              y.max=0.0
#
# Defining Material Parameters and Application of the
Appropriate Physical Models
doping      reg=1    uniform conc=4.1e17 p.type name=pentacene
#
```

```
material name=Pentacene \
        eg300=2.8 nc300=1.0e21 nv300=1.0e21 permittivity=4.0 \
        mun=5e-5 mup=2.5
#
material=Al2O3 permittivity=4.5
material=SiO2 permittivity=3.9
# Defining Electrode Parameters
contact name=gate workfunc= 4.28
#
# Defining the Defects in the Device
defects  cont dfile=l5don.dat afile=l5acc.dat \
   nta=2.5e18 ntd=1.0e18 wta=0.129 wtd=0.5 \
   nga=0.0 ngd=0.0 ega=0.62 egd=0.78 wga=0.15 wgd=0.15 \
   sigtae=1.e-17 sigtah=1.e-15 sigtde=1.e-15 sigtdh=1.e-17  \
   siggae=2.e-16 siggah=2.e-15 siggde=2.e-15 siggdh=2.e-16
#
#  Defining the Model for the Device
mobility deltaep.pfmob=1.792e-2  betap.pfmob=7.758e-5
model    pfmob.p print
#
output   e.field j.electron j.hole j.conduc j.total
e.velocity h.velocity \
   ey.field flowlines e.mobility h.mobility qss e.temp
   h.temp charge \
   recomb val.band con.band qfn qfp j.disp photogen impact
   tot.doping \
   u.srh u.rad
#
# Above we are Making Device and Applying Model and Defects
save outf=penta1.str
#tonyplot penta1.str

# Circuit Description for the Circuit
go atlas
.begin
#
#     OTFT inverters - DC point simulation
#     Circuit description
#
# Defining the Nodes and the Biasing Applied to V_{in}, V_{DD}
vin 1 0 0.
atft1 2=drain 1=gate 3=source infile=penta1.str width=5000.
atft2 0=drain 0=gate 2=source infile=penta1.str width=2500.
CL 2 0 6e-10f
vdd 3 0 10.
#
#  End of Circuit Description
#
.numeric vchange=5e-1.
.options print m2ln noshift
```

```
# Saving the solution file
.save outfile=cmos
.end
# Define Physical Models for ATLAS Device
#
#material material=P3HT user.group=semiconductor user.
default=pentacene

# Defining Material Parameters and Application of the
Appropriate Physical Models
material name=Pentacene \
   eg300=2.8 nc300=1.0e21 nv300=1.0e21 permittivity=4.0 \
   mun=5e-5 mup=2.5
 #
material=Al2O3 permittivity=4.5

mobility device=atft1 deltaep.pfmob=1.792e-2  betap.
pfmob=7.758e-5 mun=5e-5 mup=2.5
mobility device=atft2 deltaep.pfmob=1.792e-2  betap.
pfmob=7.758e-5 mun=5e-5 mup=2.5

contact device=atft1 name=gate workf=4.28
contact device=atft2 name=gate workf=4.28

model   device=atft1 reg=1 pfmob.p
model   device=atft2 reg=1 pfmob.p
#
#DC analysis for SG OTFT Inverter Circuit
go atlas
.begin
#
# OTFT inverters - DC Curve Simulation

#
#  Circuit description
#
# Defining the Nodes and the Biasing Applied to Vin, VDD
vin 1 0 0.
atft1 2=drain 1=gate 3=source infile=penta1.str width=5000.
atft2 0=drain 0=gate 2=source infile=penta1.str width=2500.
CL 2 0 6e-10f
vdd 3 0 10.
#
#    End of Circuit Description
#
.options print fulln noshift
#
.load infile=cmos
.log outfile=cmos01
# Saving the Solution File
.save master=cmos01
```

```
# DC Simulation Point
.dc vin 0. 10. 1.0
#
.end
#
#  Define Physical Models for ATLAS Device
#
#material material=P3HT user.group=semiconductor user.
default=pentacene
# Defining Material Parameters and Application of the
Appropriate Physical Models
material name=Pentacene \
   eg300=2.8 nc300=1.0e21 nv300=1.0e21 permittivity=4.0 \
   mun=5e-5 mup=2.5
#
material=Al2O3 permittivity=4.5
#
mobility device=atft1 deltaep.pfmob=1.792e-2  betap.
pfmob=7.758e-5 mun=5e-5 mup=2.5
mobility device=atft2 deltaep.pfmob=1.792e-2  betap.
pfmob=7.758e-5 mun=5e-5 mup=2.5
contact device=atft1 name=gate workf=4.28
contact device=atft2 name=gate workf=4.28
model   device=atft1 reg=1 pfmob.p
model   device=atft2 reg=1 pfmob.p
go atlas
# Plotting the Voltage Transfer Characteristics (VTC) Curve
tonyplot  cmos01_dc_1.log

#Transient Analysis for SG OTFT Inverter Circuit (see Figure
A.8)
 go atlas
.begin
#
#     Organic Inverters - Transient Simulation
#
#     Circuit description
#
#Defining the Nodes and the Biasing Applied to Vin, VDD
vin 1 0 0. PULSE 0 10 0s 150us 150us .8ms 1.8ms
atft1 2=drain 1=gate 3=source infile=penta1.str width=5000.
atft2 0=drain 0=gate 2=source infile=penta1.str width=2500.
CL 2 0 6e-10f
vdd 3 0 10.
#
#     End of Circuit Description
#
.numeric lte=0.05 dtmin=1e-15
.options print noshift
```

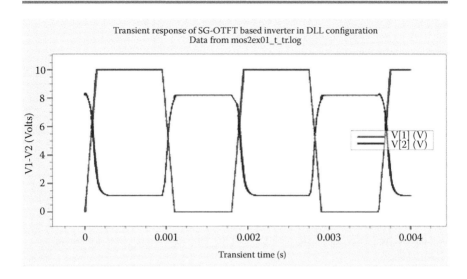

**FIGURE A.8**   Transient response of organic inverter based on SG-OTFT.

```
#
.load infile=cmos
.log outfile=mos2ex01_t
#
# Transient Simulation Point
.tran 200us 4ms
#
.end
#
#    ATLAS Device Models and Parameters
#
#  Define Physical Models for ATLAS Device
#material material=P3HT user.group=semiconductor user.default=
pentacene

# Defining Material Parameters and Application of the
Appropriate Physical Models
material name=Pentacene \
        eg300=2.8 nc300=1.0e21 nv300=1.0e21 permittivity=4.0 \
        mun=5e-5 mup=2.5
#
material=Al2O3 permittivity=4.5
mobility device=atft1 deltaep.pfmob=1.792e-2  betap.pfmob=
7.758e-5 mun=5e-5 mup=2.5
mobility device=atft2 deltaep.pfmob=1.792e-2  betap.pfmob=
7.758e-5 mun=5e-5 mup=2.5
contact device=atft1 name=gate workf=4.28
```

```
contact  device=atft2 name=gate workf=4.28
model    device=atft1 reg=1 pfmob.p
model    device=atft2 reg=1 pfmob.p
go atlas
# Plotting the Transient Response of the Organic Inverter
tonyplot mos2ex01_t_tr.log
quit
```

# EXAMPLE A.4

## ORGANIC INVERTER CIRCUIT BASED ON DUAL GATE
## ORGANIC TRANSISTOR (DG-OTFT; SEE FIGURE A.9)

```
# Defining the Device which is to be used in the Inverter
Circuit
go atlas
#
Title: DG-Organic Inverter Simulation
#
# Define Structure File
#mesh    smooth=1 space.mult=1.0    width=16000
mesh     outf=penta_2.str master.out
#
x.m         l=0         spac=1
x.m         l=10        spac=0.5
x.m         l=20        spac=0.5
x.m         l=30        spac=0.5
x.m         l=80        spac=0.5
x.m         l=90        spac=0.5
x.m         l=100       spac=0.5
x.m         l=110       spac=1
#
y.mesh l=0            spac=0.1
y.mesh l=0.05         spac=0.01
y.mesh l=0.20         spac=0.01
y.mesh l=0.297        spac=0.1
y.mesh l=0.347        spac=0.01
y.mesh l=0.35         spac=0.01
y.mesh l=0.45         spac=0.1
y.mesh l=0.50         spac=0.1
#
# Define Region Parameter
region num=1 user.material=parylene y.min=0.05   y.max=0.35
region num=2 material=Al2O3              y.min=0.35 y.max=0.50
region num=3 material=pentacene      x.min=20 x.max=90
y.min=0.20 y.max=0.35 name=pentacene
```

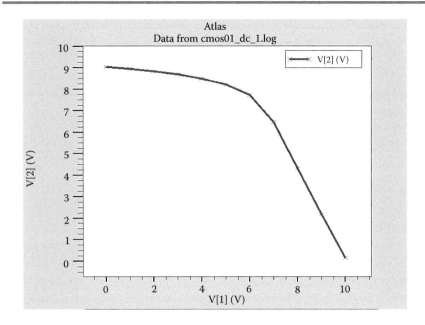

**FIGURE A.9**    VTC of organic inverter circuit based on dual gate organic transistor (DG-OTFT).

```
region num=4  material=air x.min=0    x.max=10 y.min=0
                                      y.max=0.05
region num=5  material=air x.min=100 x.max=110 y.min=0
                                      y.max=0.05
#
# Define Electrodes of the Device
elec   num=1   material=Gold      name=source   x.min=10 x.max=30
                                                y.min=0.297
                                                y.max=0.347
elec   num=2   material=Titanium     name=source1  x.min=10
                                                x.max=30
                                                y.min=0.347
                                                y.max=0.35
elec   num=3   material=Gold      name=drain    x.min=80 x.max=100
                                                y.min=0.297
                                                y.max=0.347
elec   num=4   material=Titanium     name=drain1   x.min=80
                                                x.max=100
                                                y.min=0.347
                                                y.max=0.35
elec   num=5   material=Titanium     name=gate    x.min=10
                                                x.max=100
                                                y.min=0
                                                y.max=0.05
```

```
elec   num=6   material=Titanium     name=gate1 x.min=10
                                                x.max=100
                                                y.min=0.45
                                                y.max=0.50
#
# Defining Material Parameters and Application of the
Appropriate Physical Models
doping reg=3 uniform conc=1e16 p.type name=pentacene
#
material name=pentacene  \
        eg300=2.2 nc300=2.0e21 nv300=2.0e21
        permittivity=6.0 \
        mun=5e-5 mup=0.02
#
material material=parylene user.group=insulator user.
default=oxide permittivity=3.1
material material=Al2O3 user.group=insulator user.
default=Si3N4 permittivity=8
# Defining Electrode Parameters
contact num=1        workf=5.1
contact num=2        workf=4.4
contact num=3        workf=5.1
contact num=4        workf=4.4
contact name=gate    workf=4.33
contact name=gate1   workf=4.33     common=gate
#
# Defining the Defects in the Device
defects   cont dfile=15don.dat afile=15acc.dat \
        nta=2.5e18 ntd=1.0e18 wta=0.129 wtd=0.5 \
        nga=0.0 ngd=0.0 ega=0.62 egd=0.78 wga=0.15 wgd=0.15 \
        sigtae=1.e-17 sigtah=1.e-15 sigtde=1.e-15
        sigtdh=1.e-17  \
        siggae=2.e-16 siggah=2.e-15 siggde=2.e-15
        siggdh=2.e-16

#  Defining the Model for the Device
mobility deltaep.pfmob=1.792e-2  betap.pfmob=7.758e-5
model    pfmob.p print
#
output   e.field j.electron j.hole j.conduc j.total
e.velocity h.velocity \
        ey.field flowlines e.mobility h.mobility qss e.temp
        h.temp charge \
        recomb val.band con.band qfn qfp j.disp photogen
        impact tot.doping \
        u.srh u.rad

## Above we are Making Device and Applying Model and Defects
save outf=penta1.str
tonyplot penta1.str
```

```
# Circuit Description for the Circuit
#
go atlas
.begin
#     OTFT inverters - DC Point Simulation
#     Circuit description
#
# Defining the Nodes and the Biasing Applied to Vin, VDD
vin 1 0 0.
atft1 2=drain 2=drain1 1=gate 1=gate1 3=source 3=source1
infile=penta1.str width=16000.
atft2 0=drain 0=drain1 0=gate 0=gate1 2=source 2=source1
infile=penta1.str width=3200.
CL 2 0 6e-10f
vdd 3 0 10.
#
#  End of Circuit Description
#
.numeric vchange=5e-1.
.options print m2ln noshift
#
# Saving the solution file
.save outfile=cmos
.end
#
# Define Physical Models for ATLAS Device
#
# Defining Material Parameters and Application of the
Appropriate Physical Model
material device=atft1 name=pentacene  \
        eg300=2.2 nc300=2.0e21 nv300=2.0e21
        permittivity=6.0 \
        mun=5e-5 mup=0.02
material device=atft2 name=pentacene  \
        eg300=2.2 nc300=2.0e21 nv300=2.0e21
        permittivity=6.0 \
        mun=5e-5 mup=0.02
#
material material=parylene user.group=insulator user.
default=oxide permittivity=3.1
material material=Al2O3    user.group=insulator user.
default=Si3N4 permittivity=8
mobility device=atft1 deltaep.pfmob=1.792e-2  betap.
pfmob=7.758e-5  mun=5e-5 mup=2.5
mobility device=atft2 deltaep.pfmob=1.792e-2  betap.
pfmob=7.758e-5  mun=5e-5 mup=2.5
#
contact device=atft1 name=gate  workf=4.33
contact device=atft1 name=gate1 workf=4.33
contact device=atft2 name=gate  workf=4.33
```

```
contact device=atft2 name=gate1 workf=4.33
model   device=atft1 reg=3 pfmob.p
model   device=atft2 reg=3 pfmob.p
#

#DC Analysis for DG OTFT Inverter Circuit
go atlas
.begin
# OTFT inverters - DC Curve Simulation
#     Circuit description
#Defining the Nodes and the Biasing Applied to Vin, VDD
vin 1 0 0.
atft1 2=drain 2=drain1 1=gate 1=gate1 3=source 3=source1
infile=penta1.str width=16000.
atft2 0=drain 0=drain1 0=gate 0=gate1 2=source 2=source1
infile=penta1.str width=3200.
CL 2 0 6e-10f
vdd 3 0 10.
#
# End of Circuit Description
#
.options print fulln noshift
.load infile=cmos
.log outfile=cmos01
# Saving the Solution File
.save master=cmos01
# DC Simulation Point
.dc vin 0. 10. 1.0
#
.end
#
#  Define Physical Models for ATLAS Device
#
# Defining Material Parameters and Application of the
Appropriate Physical Models
material device=atft1 name=pentacene   \
        eg300=2.2 nc300=2.0e21 nv300=2.0e21
        permittivity=6.0 \
        mun=5e-5 mup=0.02
material device=atft2 name=pentacene   \
        eg300=2.2 nc300=2.0e21 nv300=2.0e21
        permittivity=6.0 \
        mun=5e-5 mup=0.02
#
material material=parylene user.group=insulator user.default=
oxide permittivity=3.1
material material=Al2O3 user.group=insulator user.
default=Si3N4 permittivity=8
```

```
mobility device=atft1 deltaep.pfmob=1.792e-2  betap.
pfmob=7.758e-5  mun=5e-5 mup=2.5
mobility device=atft2 deltaep.pfmob=1.792e-2  betap.
pfmob=7.758e-5  mun=5e-5 mup=2.5
#
contact device=atft1 name=gate  workf=4.33
contact device=atft1 name=gate1 workf=4.33
contact device=atft2 name=gate  workf=4.33
contact device=atft2 name=gate1 workf=4.33
model   device=atft1 pfmob.p
model   device=atft2 pfmob.p
go atlas
# Plotting the Voltage Transfer Characteristics (VTC) Curve
tonyplot cmos01_dc_1.log

#Transient Analysis for DG OTFT Inverter Circuit
(see Figure A.10)
go atlas
.begin
#     Organic Inverters - Transient Simulation
#     Circuit description
#
```

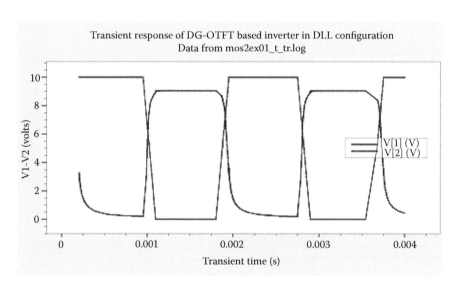

**FIGURE A.10**    Transient characteristics of organic inverter circuit based on dual gate organic transistor (DG-OTFT).

```
# Defining the Nodes and the Biasing Applied to Vin, VDD
vin 1 0 0. PULSE 0 10 0s 150us 150us 0.8ms 1.8ms
atft1 2=drain 2=drain1 1=gate 1=gate1 3=source 3=source1
infile=penta1.str width=16000.
atft2 0=drain 0=drain1 0=gate 0=gate1 2=source 2=source1
infile=penta1.str width=3200.
CL 2 0 6e-10f
vdd 3 0 10.
#
#     End of Circuit Description
#
.numeric lte=0.05 dtmin=1e-15
.options print noshift
#
.load infile=cmos
.log outfile=mos2ex01_t
# Transient Simulation Point
#
.tran 200us 4ms
.end
#


#     ATLAS Device Models and Parameters

# Defining Material Parameters and Application of the
Appropriate Physical Models
material device=atft1 name=pentacene  \
        eg300=2.2 nc300=2.0e21 nv300=2.0e21
        permittivity=6.0 \
        mun=5e-5 mup=0.02
material device=atft2 name=pentacene  \
        eg300=2.2 nc300=2.0e21 nv300=2.0e21
        permittivity=6.0 \
        mun=5e-5 mup=0.02

material material=parylene user.group=insulator user.
default=oxide permittivity=3.1
material material=Al2O3 user.group=insulator user.
default=Si3N4 permittivity=8
mobility device=atft1 deltaep.pfmob=1.792e-2  betap.
pfmob=7.758e-5  mun=5e-5 mup=2.5
mobility device=atft2 deltaep.pfmob=1.792e-2  betap.
pfmob=7.758e-5  mun=5e-5 mup=2.5

contact device=atft1 name=gate  workf=4.33
contact device=atft1 name=gate1 workf=4.33
contact device=atft2 name=gate  workf=4.33
contact device=atft2 name=gate1 workf=4.33
```

```
model     device=atft1 pfmob.p
model     device=atft2 pfmob.p
go atlas
# Plotting the Transient response of the Organic Inverter
tonyplot  mos2ex01_t_tr.log
quit
```

## EXAMPLE A.5

### SR FLIP-FLOP BASED ON DUAL GATE ORGANIC THIN FILM TRANSISTOR (DG_OTFT; SEE FIGURE A.11)

```
# Defining the Device which is to be used in the Circuit
go atlas
#
Title: DG OTFT Based SR FF in ZVLL Mode
#
# Define Structure File
mesh     smooth=1 space.mult=1.0 width=800  outf=SR_0.str
master.out
```

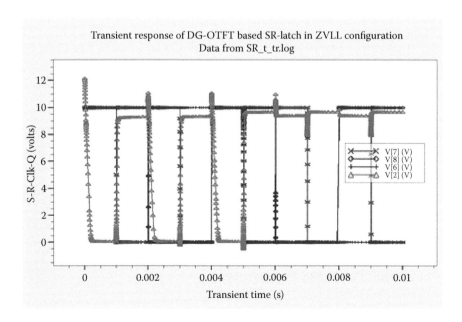

**FIGURE A.11**   Transient response of the SR flip flop circuit using DG-OTFT.

```
#
x.m            l=0      spac=0.5
x.m            l=10     spac=0.5
x.m            l=35     spac=0.5
x.m            l=45     spac=0.5
#
y.mesh l=-0.45    spacing=0.3
y.mesh l=-0.30    spacing=0.3
y.mesh l=-0.20    spacing=0.3
y.mesh l=-0.15    spacing=0.3
y.mesh l=-0.05    spacing=0.07
y.mesh l=0.0      spacing=0.07
y.mesh l=0.05     spacing=0.07
y.mesh l=0.10     spacing=0.3
y.mesh l=0.25     spacing=0.3
```

```
# Define Region Parameter
region num=1    material=pentacene  y.min=-0.2  y.max=0.0
name=Pentacene
region num=2    material=SiO2        y.min=0.0   y.max=0.10
region num=3    material=SiO2        y.min=-0.3  y.max=-0.2
#
# Define Electrodes of the Device
elec   num=1    material=Gold      name=source      x.max=10.0
y.min=-0.08  y.max=0.0
elec   num=2    material=Si        name=gate2       y.min=0.10
y.max=0.25
elec   num=3    material=Gold      name=drain       x.min=35.0
y.min=-0.08  y.max=0.0
elec   num=4    material=Aluminium  name=gate1
y.min=-0.45  y.max=-0.3
#
# Defining Material Parameters and Application of the
Appropriate Physical Models
doping reg=1    uniform conc=1e15 p.type name=pentacene
doping reg=2    uniform conc=1e21 n.type name= silicon
#
material name=Pentacene \
        eg300=1.2 nc300=1.0e21 nv300=1.0e21
        permittivity=4.0 \
        mun=5e-5 mup=0.5
 #

# Defining Electrode Parameters
contact name=gate1 workf=4.28
contact name=gate2 workf=4.6
#
# Defining the Defects in the Device
defects  cont dfile=15don.dat afile=15acc.dat \
        nta=2.5e18 ntd=1.0e18 wta=0.129 wtd=0.5 \
```

```
              nga=0.0 ngd=0.0 ega=0.62 egd=0.78 wga=0.15
              wgd=0.15 \
              sigtae=1.e-17 sigtah=1.e-15 sigtde=1.e-15
              sigtdh=1.e-17  \
              siggae=2.e-16 siggah=2.e-15 siggde=2.e-15
              siggdh=2.e-16
#
#  Defining the Model for the Device
mobility deltaep.pfmob=1.792e-2  betap.pfmob=7.758e-5
model    pfmob.p print
#
output   e.field j.electron j.hole j.conduc j.total
e.velocity h.velocity \
         ey.field flowlines e.mobility h.mobility qss e.temp
         h.temp charge \
         recomb val.band con.band qfn qfp j.disp photogen
         impact tot.doping \
         u.srh u.rad
#
# Above we are Making Device and Applying Model and Defects
#
#Plot Device Solution File to Call It in Net-list for Flip
Flop Circuit Analysis
save outf=pmos_dg.str
tonyplot pmos_dg.str
# Circuit Description for the Circuit
go atlas
.begin
#
#  DC Point Simulation
#
#    Circuit Description
#
# Defining the Nodes and the Biasing Applied to Vin, VDD
V1 7 0 0. PULSE 0 10 0s 5us 5us 1ms 2ms
V2 8 0 0. PULSE 0 10 0s 5us 5us 2ms 4ms
V3 6 0 0. PULSE 0 10 0s 5us 5us 5ms 10ms
#
C1 4 0 100pF
C2 1 0 100pF
#
atft1 4=drain 1=gate1 1=gate2 3=source infile=pmos_dg.str
width=400.
atft2 4=drain 2=gate1 2=gate2 3=source infile=pmos_dg.str
width=400.
atft3 0=drain 4=gate1 4=gate2 4=source infile=pmos_dg.str
width=2000.

atft4 1=drain 4=gate1 4=gate2 3=source infile=pmos_dg.str
width=400.
```

```
atft5 1=drain 5=gate1 5=gate2 3=source infile=pmos_dg.str
width=400.
atft6 0=drain 1=gate1 1=gate2 1=source infile=pmos_dg.str
width=2000.

atft7 2=drain 6=gate1 6=gate2 3=source infile=pmos_dg.str
width=400.
atft8 2=drain 7=gate1 7=gate2 3=source infile=pmos_dg.str
width=400.
atft9 0=drain 2=gate1 2=gate2 2=source infile=pmos_dg.str
width=2000.

atft10 5=drain 6=gate1 6=gate2 3=source infile=pmos_dg.str
width=400.
atft11 5=drain 8=gate1 8=gate2 3=source infile=pmos_dg.str
width=400.
atft12 0=drain 5=gate1 5=gate2 5=source infile=pmos_dg.str
width=2000.
#
vdd 3 0 10.
#
#      End of Circuit Description
#
.numeric vchange=5e-1.
.options print m2ln noshift
#
# Saving the Solution File
.save outfile=cmos
.end
#
#  Define Physical Models for ATLAS Device
#
# Defining Material Parameters and Application of the
Appropriate Physical Models
material device=atft1 name=Pentacene \
        eg300=1.2 nc300=1.0e21 nv300=1.0e21
        permittivity=4.0 \
        mun=5e-5 mup=0.5
material device=atft2 name=Pentacene \
        eg300=1.2 nc300=1.0e21 nv300=1.0e21
        permittivity=4.0 \
        mun=5e-5 mup=0.5
material device=atft3 name=Pentacene \
        eg300=1.2 nc300=1.0e21 nv300=1.0e21
        permittivity=4.0 \
        mun=5e-5 mup=0.5
material device=atft4 name=Pentacene \
        eg300=1.2 nc300=1.0e21 nv300=1.0e21
        permittivity=4.0 \
        mun=5e-5 mup=0.5
```

```
material device=atft5 name=Pentacene \
        eg300=1.2 nc300=1.0e21 nv300=1.0e21
        permittivity=4.0 \
        mun=5e-5 mup=0.5
material device=atft6 name=Pentacene \
        eg300=1.2 nc300=1.0e21 nv300=1.0e21
        permittivity=4.0 \
        mun=5e-5 mup=0.5
material device=atft7 name=Pentacene \
        eg300=1.2 nc300=1.0e21 nv300=1.0e21
        permittivity=4.0 \
        mun=5e-5 mup=0.5
material device=atft8 name=Pentacene \
        eg300=1.2 nc300=1.0e21 nv300=1.0e21
        permittivity=4.0 \
        mun=5e-5 mup=0.5
material device=atft9 name=Pentacene \
        eg300=1.2 nc300=1.0e21 nv300=1.0e21
        permittivity=4.0 \
        mun=5e-5 mup=0.5
material device=atft10 name=Pentacene \
        eg300=1.2 nc300=1.0e21 nv300=1.0e21
        permittivity=4.0 \
        mun=5e-5 mup=0.5
material device=atft11 name=Pentacene \
        eg300=1.2 nc300=1.0e21 nv300=1.0e21
        permittivity=4.0 \
        mun=5e-5 mup=0.5
material device=atft12 name=Pentacene \
        eg300=1.2 nc300=1.0e21 nv300=1.0e21
        permittivity=4.0 \
        mun=5e-5 mup=0.5
 #
mobility device=atft1 deltaep.pfmob=1.792e-2  betap.
pfmob=7.758e-5
mobility device=atft2 deltaep.pfmob=1.792e-2  betap.
pfmob=7.758e-5
mobility device=atft3 deltaep.pfmob=1.792e-2  betap.
pfmob=7.758e-5
mobility device=atft4 deltaep.pfmob=1.792e-2  betap.
pfmob=7.758e-5
mobility device=atft5 deltaep.pfmob=1.792e-2  betap.
pfmob=7.758e-5
mobility device=atft6 deltaep.pfmob=1.792e-2  betap.
pfmob=7.758e-5
mobility device=atft7 deltaep.pfmob=1.792e-2  betap.
pfmob=7.758e-5
mobility device=atft8 deltaep.pfmob=1.792e-2  betap.
pfmob=7.758e-5
```

```
mobility device=atft9 deltaep.pfmob=1.792e-2  betap.
pfmob=7.758e-5
mobility device=atft10 deltaep.pfmob=1.792e-2  betap.
pfmob=7.758e-5
mobility device=atft11 deltaep.pfmob=1.792e-2  betap.
pfmob=7.758e-5
mobility device=atft12 deltaep.pfmob=1.792e-2  betap.
pfmob=7.758e-5
 #
contact device=atft1 name=gate1 workf=4.28
contact device=atft1 name=gate2 workf=4.6
contact device=atft2 name=gate1 workf=4.28
contact device=atft2 name=gate2 workf=4.6
contact device=atft3 name=gate1 workf=4.28
contact device=atft3 name=gate2 workf=4.6
contact device=atft4 name=gate1 workf=4.28
contact device=atft4 name=gate2 workf=4.6
contact device=atft5 name=gate1 workf=4.28
contact device=atft5 name=gate2 workf=4.6
contact device=atft6 name=gate1 workf=4.28
contact device=atft6 name=gate2 workf=4.6
contact device=atft7 name=gate1 workf=4.28
contact device=atft7 name=gate2 workf=4.6
contact device=atft8 name=gate1 workf=4.28
contact device=atft8 name=gate2 workf=4.6
contact device=atft9 name=gate1 workf=4.28
contact device=atft9 name=gate2 workf=4.6
contact device=atft10 name=gate1 workf=4.28
contact device=atft10 name=gate2 workf=4.6
contact device=atft11 name=gate1 workf=4.28
contact device=atft11 name=gate2 workf=4.6
contact device=atft12 name=gate1 workf=4.28
contact device=atft12 name=gate2 workf=4.6
 #
model    device=atft1   pfmob.p
model    device=atft2   pfmob.p
model    device=atft3   pfmob.p
model    device=atft4   pfmob.p
model    device=atft5   pfmob.p
model    device=atft6   pfmob.p
model    device=atft7   pfmob.p
model    device=atft8   pfmob.p
model    device=atft9   pfmob.p
model    device=atft10  pfmob.p
model    device=atft11  pfmob.p
model    device=atft12  pfmob.p
 #
#Transient Analysis for Flip Flop Circuit
go atlas
.begin
```

```
#
# Circuit Description
#
# Defining the Nodes and the Biasing Applied to V_in, V_DD
V1 7 0 0. PULSE 0 10 0s 5us 5us 1ms 2ms
V2 8 0 0. PULSE 0 10 0s 5us 5us 2ms 4ms
V3 6 0 0. PULSE 0 10 0s 5us 5us 5ms 10ms
 #
C1 4 0 100pF
C2 1 0 100pF
 #
atft1 4=drain 1=gate1 1=gate2 3=source infile=pmos_dg.str
width=400.
atft2 4=drain 2=gate1 2=gate2 3=source infile=pmos_dg.str
width=400.
atft3 0=drain 4=gate1 4=gate2 4=source infile=pmos_dg.str
width=2000.

atft4 1=drain 4=gate1 4=gate2 3=source infile=pmos_dg.str
width=400.
atft5 1=drain 5=gate1 5=gate2 3=source infile=pmos_dg.str
width=400.
atft6 0=drain 1=gate1 1=gate2 1=source infile=pmos_dg.str
width=2000.

atft7 2=drain 6=gate1 6=gate2 3=source infile=pmos_dg.str
width=400.
atft8 2=drain 7=gate1 7=gate2 3=source infile=pmos_dg.str
width=400.
atft9 0=drain 2=gate1 2=gate2 2=source infile=pmos_dg.str
width=2000.

atft10 5=drain 6=gate1 6=gate2 3=source infile=pmos_dg.str
width=400.
atft11 5=drain 8=gate1 8=gate2 3=source infile=pmos_dg.str
width=400.
atft12 0=drain 5=gate1 5=gate2 5=source infile=pmos_dg.str
width=2000.
vdd 3 0 10.
#
# End of Circuit Description
#
.numeric lte=0.05 dtmin=1e-15
.options print fulln noshift
 #
.load infile=cmos
.log outfile=SR_t
# Saving the Solution File
.save master=SR_t
```

```
# Transient Simulation Point
.tran 10us 10ms
.end
#
# Define Physical Models for ATLAS Device
#
material device=atft1 name=Pentacene \
        eg300=1.2 nc300=1.0e21 nv300=1.0e21
        permittivity=4.0 \
        mun=5e-5 mup=0.5
material device=atft2 name=Pentacene \
        eg300=1.2 nc300=1.0e21 nv300=1.0e21
        permittivity=4.0 \
        mun=5e-5 mup=0.5
material device=atft3 name=Pentacene \
        eg300=1.2 nc300=1.0e21 nv300=1.0e21
        permittivity=4.0 \
        mun=5e-5 mup=0.5
material device=atft4 name=Pentacene \
        eg300=1.2 nc300=1.0e21 nv300=1.0e21
        permittivity=4.0 \
        mun=5e-5 mup=0.5
material device=atft5 name=Pentacene \
        eg300=1.2 nc300=1.0e21 nv300=1.0e21
        permittivity=4.0 \
        mun=5e-5 mup=0.5
material device=atft6 name=Pentacene \
        eg300=1.2 nc300=1.0e21 nv300=1.0e21
        permittivity=4.0 \
        mun=5e-5 mup=0.5
material device=atft7 name=Pentacene \
        eg300=1.2 nc300=1.0e21 nv300=1.0e21
        permittivity=4.0 \
        mun=5e-5 mup=0.5
material device=atft8 name=Pentacene \
        eg300=1.2 nc300=1.0e21 nv300=1.0e21
        permittivity=4.0 \
        mun=5e-5 mup=0.5
material device=atft9 name=Pentacene \
        eg300=1.2 nc300=1.0e21 nv300=1.0e21
        permittivity=4.0 \
        mun=5e-5 mup=0.5
material device=atft10 name=Pentacene \
        eg300=1.2 nc300=1.0e21 nv300=1.0e21
        permittivity=4.0 \
        mun=5e-5 mup=0.5
material device=atft11 name=Pentacene \
        eg300=1.2 nc300=1.0e21 nv300=1.0e21
        permittivity=4.0 \
        mun=5e-5 mup=0.5
```

```
material device=atft12 name=Pentacene \
        eg300=1.2 nc300=1.0e21 nv300=1.0e21
        permittivity=4.0 \
        mun=5e-5 mup=0.5
#
mobility device=atft1 deltaep.pfmob=1.792e-2  betap.
pfmob=7.758e-5
mobility device=atft2 deltaep.pfmob=1.792e-2  betap.
pfmob=7.758e-5
mobility device=atft3 deltaep.pfmob=1.792e-2  betap.
pfmob=7.758e-5
mobility device=atft4 deltaep.pfmob=1.792e-2  betap.
pfmob=7.758e-5
mobility device=atft5 deltaep.pfmob=1.792e-2  betap.
pfmob=7.758e-5
mobility device=atft6 deltaep.pfmob=1.792e-2  betap.
pfmob=7.758e-5
mobility device=atft7 deltaep.pfmob=1.792e-2  betap.
pfmob=7.758e-5
mobility device=atft8 deltaep.pfmob=1.792e-2  betap.
pfmob=7.758e-5
mobility device=atft9 deltaep.pfmob=1.792e-2  betap.
pfmob=7.758e-5
mobility device=atft10 deltaep.pfmob=1.792e-2  betap.
pfmob=7.758e-5
mobility device=atft11 deltaep.pfmob=1.792e-2  betap.
pfmob=7.758e-5
mobility device=atft12 deltaep.pfmob=1.792e-2  betap.
pfmob=7.758e-5
#
contact device=atft1 name=gate1 workf=4.28
contact device=atft1 name=gate2 workf=4.6
contact device=atft2 name=gate1 workf=4.28
contact device=atft2 name=gate2 workf=4.6
contact device=atft3 name=gate1 workf=4.28
contact device=atft3 name=gate2 workf=4.6
contact device=atft4 name=gate1 workf=4.28
contact device=atft4 name=gate2 workf=4.6
contact device=atft5 name=gate1 workf=4.28
contact device=atft5 name=gate2 workf=4.6
contact device=atft6 name=gate1 workf=4.28
contact device=atft6 name=gate2 workf=4.6
contact device=atft7 name=gate1 workf=4.28
contact device=atft7 name=gate2 workf=4.6
contact device=atft8 name=gate1 workf=4.28
contact device=atft8 name=gate2 workf=4.6
contact device=atft9 name=gate1 workf=4.28
contact device=atft9 name=gate2 workf=4.6
contact device=atft10 name=gate1 workf=4.28
contact device=atft10 name=gate2 workf=4.6
```

```
contact device=atft11 name=gate1 workf=4.28
contact device=atft11 name=gate2 workf=4.6
contact device=atft12 name=gate1 workf=4.28
contact device=atft12 name=gate2 workf=4.6
model   device=atft1  pfmob.p
model   device=atft2  pfmob.p
model   device=atft3  pfmob.p
model   device=atft4  pfmob.p
model   device=atft5  pfmob.p
model   device=atft6  pfmob.p
model   device=atft7  pfmob.p
model   device=atft8  pfmob.p
model   device=atft9  pfmob.p
model   device=atft10  pfmob.p
model   device=atft11  pfmob.p
model   device=atft12  pfmob.p
#
go atlas
tonyplot  SR_t_tr.log -set  SR_t_tr.set
quit
```

## EXAMPLE: A.6

ORGANIC MULTIPLEXER (2:1 MUX) CIRCUIT
PERFORMANCE ANALYSIS (SEE FIGURE A.12)

```
# Defining the Device That Is to Be Used in the Circuit
go atlas
#
Title: DG OTFT Based 2 to 1 MUX in ZVLL Mode
#
# Define Structure File
mesh smooth=1 space.mult=1.0 width=800 outf=MUX_0.str
master.out
#
x.m  l=0   spac=0.5
x.m  l=10   spac=0.5
x.m  l=35   spac=0.5
x.m  l=45   spac=0.5
#
y.mesh l=-0.45   spacing=0.3
y.mesh l=-0.30   spacing=0.3
y.mesh l=-0.20   spacing=0.3
y.mesh l=-0.15   spacing=0.3
y.mesh l=-0.05   spacing=0.07
y.mesh l=0.0     spacing=0.07
y.mesh l=0.05    spacing=0.07
```

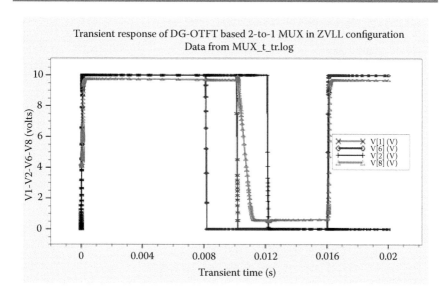

**FIGURE A.12**   Transient response of the 2-to-1 MUX in ZVLL configuration circuit using DG-OTFT.

```
y.mesh l=0.10    spacing=0.3
y.mesh l=0.25    spacing=0.3
# Define Region Parameter
region num=1  material=pentacene  y.min=-0.2 y.max=0.0
name=Pentacene
region num=2  material=SiO2  y.min=0.0 y.max=0.10
region num=3  material=SiO2  y.min=-0.3 y.max=-0.2
#
# Define Electrodes of the Device
elec num=1 material=Gold name=source x.max=10.0 y.min=-0.08
y.max=0.0
elec num=2 material=Si name=gate2 y.min=0.10 y.max=0.25
elec num=3 material=Gold name=drain x.min=35.0 y.min=-0.08
y.max=0.0
elec num=4 material=Aluminium name=gate1 y.min=-0.45
y.max=-0.3
#
# Defining Material Parameters and Application of the
Appropriate Physical Models
doping reg=1 uniform conc=1e15 p.type name=pentacene
doping reg=2 uniform conc=1e21 n.type name=silicon
#
material name=Pentacene \
eg300=1.2 nc300=1.0e21 nv300=1.0e21 permittivity=4.0 \
mun=5e-5 mup=0.5
#
```

```
# Defining Electrode Parameters
contact name=gate1 workf=4.28
contact name=gate2 workf=4.6
#
# Defining the Defects in the Device
defects cont dfile=15don.dat afile=15acc.dat \
        nta=2.5e18 ntd=1.0e18 wta=0.129 wtd=0.5 \
        nga=0.0 ngd=0.0 ega=0.62 egd=0.78 wga=0.15 wgd=0.15 \
        sigtae=1.e-17 sigtah=1.e-15 sigtde=1.e-15
        sigtdh=1.e-17 \
        siggae=2.e-16 siggah=2.e-15 siggde=2.e-15
        siggdh=2.e-16
#
# Defining the Model for the Device
mobility deltaep.pfmob=1.792e-2 betap.pfmob=7.758e-5
model pfmob.p print
#
output e.field j.electron j.hole j.conduc j.total e.velocity
h.velocity \
        ey.field flowlines e.mobility h.mobility qss e.temp
        h.temp charge \
        recomb val.band con.band qfn qfp j.disp photogen
        impact tot.doping \
        u.srh u.rad
#
# Above We Are Making Device and Applying Model and
Defects
#
#Plot Device Solution File to Call It in Net-list for Flip
Flop Circuit Analysis
save outf=pmos_dg.str
tonyplot pmos_dg.str
# Circuit Description for the Circuit
go atlas
.begin
#
# DC Point Simulation
#
# Circuit Description
#
# Defining the Nodes and the Biasing Applied to Vin, VDD
V1 1 0 0. PULSE 0 10 0s 100us 100us 10ms 20ms
V2 6 0 0. PULSE 0 10 0s 100us 100us 8ms 16ms
V3 2 0 0. PULSE 0 10 0s 100us 100us 12ms 24ms
#
C1 8 0 1pF

#C2 1 0 100pF
#
```

```
atft1 4=drain 1=gate1 1=gate2 3=source infile=pmos_dg.str
width=400.
atft2 4=drain 2=gate1 2=gate2 3=source infile=pmos_dg.str
width=400.
atft3 0=drain 4=gate1 4=gate2 4=source infile=pmos_dg.str
width=3000.
atft4 7=drain 5=gate1 5=gate2 3=source infile=pmos_dg.str
width=400.
atft5 7=drain 6=gate1 6=gate2 3=source infile=pmos_dg.str
width=400.
atft6 0=drain 7=gate1 7=gate2 7=source infile=pmos_dg.str
width=3000.
atft7 8=drain 7=gate1 7=gate2 3=source infile=pmos_dg.str
width=400.
atft8 8=drain 4=gate1 4=gate2 3=source infile=pmos_dg.str
width=400.
atft9 0=drain 8=gate1 8=gate2 8=source infile=pmos_dg.str
width=3000.
atft10 5=drain 2=gate1 2=gate2 3=source infile=pmos_dg.str
width=400.
atft11 0=drain 5=gate1 5=gate2 5=source infile=pmos_dg.str
width=3000.
#
vdd 3 0 10.
#
# End of Circuit Description
#
.numeric vchange=5e-1.
.options print m2ln noshift
#
# Saving the Solution File
.save outfile=cmos
.end
#
# Define Physical Models for ATLAS Device
#
# Defining Material Parameters and Application of the
Appropriate Physical Models
material device=atft1 name=Pentacene \
        eg300=1.2 nc300=1.0e21 nv300=1.0e21 permittivity=4.0 \
        mun=5e-5 mup=0.5
material device=atft2 name=Pentacene \
        eg300=1.2 nc300=1.0e21 nv300=1.0e21 permittivity=4.0 \
        mun=5e-5 mup=0.5
material device=atft3 name=Pentacene \
        eg300=1.2 nc300=1.0e21 nv300=1.0e21 permittivity=4.0 \
        mun=5e-5 mup=0.5
material device=atft4 name=Pentacene \
        eg300=1.2 nc300=1.0e21 nv300=1.0e21 permittivity=4.0 \
        mun=5e-5 mup=0.5
```

```
material device=atft5 name=Pentacene \
        eg300=1.2 nc300=1.0e21 nv300=1.0e21 permittivity=4.0 \
        mun=5e-5 mup=0.5
material device=atft6 name=Pentacene \
        eg300=1.2 nc300=1.0e21 nv300=1.0e21 permittivity=4.0 \
        mun=5e-5 mup=0.5
material device=atft7 name=Pentacene \
        eg300=1.2 nc300=1.0e21 nv300=1.0e21 permittivity=4.0 \
        mun=5e-5 mup=0.5
material device=atft8 name=Pentacene \
        eg300=1.2 nc300=1.0e21 nv300=1.0e21 permittivity=4.0 \
        mun=5e-5 mup=0.5
material device=atft9 name=Pentacene \
        eg300=1.2 nc300=1.0e21 nv300=1.0e21 permittivity=4.0 \
        mun=5e-5 mup=0.5
material device=atft10 name=Pentacene \
        eg300=1.2 nc300=1.0e21 nv300=1.0e21 permittivity=4.0 \
        mun=5e-5 mup=0.5
material device=atft11 name=Pentacene \
        eg300=1.2 nc300=1.0e21 nv300=1.0e21 permittivity=4.0 \
        mun=5e-5 mup=0.5
#
mobility device=atft1 deltaep.pfmob=1.792e-2 betap.
pfmob=7.758e-5
mobility device=atft2 deltaep.pfmob=1.792e-2 betap.
pfmob=7.758e-5
mobility device=atft3 deltaep.pfmob=1.792e-2 betap.
pfmob=7.758e-5
mobility device=atft4 deltaep.pfmob=1.792e-2 betap.
pfmob=7.758e-5
mobility device=atft5 deltaep.pfmob=1.792e-2 betap.
pfmob=7.758e-5
mobility device=atft6 deltaep.pfmob=1.792e-2 betap.
pfmob=7.758e-5
mobility device=atft7 deltaep.pfmob=1.792e-2 betap.
pfmob=7.758e-5
mobility device=atft8 deltaep.pfmob=1.792e-2 betap.
pfmob=7.758e-5
mobility device=atft9 deltaep.pfmob=1.792e-2 betap.
pfmob=7.758e-5
mobility device=atft10 deltaep.pfmob=1.792e-2 betap.
pfmob=7.758e-5
mobility device=atft11 deltaep.pfmob=1.792e-2 betap.
pfmob=7.758e-5
#
contact device=atft1 name=gate1 workf=4.28
contact device=atft1 name=gate2 workf=4.6
contact device=atft2 name=gate1 workf=4.28
contact device=atft2 name=gate2 workf=4.6
contact device=atft3 name=gate1 workf=4.28
```

```
contact device=atft3 name=gate2 workf=4.6
contact device=atft4 name=gate1 workf=4.28
contact device=atft4 name=gate2 workf=4.6
contact device=atft5 name=gate1 workf=4.28
contact device=atft5 name=gate2 workf=4.6
contact device=atft6 name=gate1 workf=4.28
contact device=atft6 name=gate2 workf=4.6
contact device=atft7 name=gate1 workf=4.28
contact device=atft7 name=gate2 workf=4.6
contact device=atft8 name=gate1 workf=4.28
contact device=atft8 name=gate2 workf=4.6
contact device=atft9 name=gate1 workf=4.28
contact device=atft9 name=gate2 workf=4.6
contact device=atft10 name=gate1 workf=4.28
contact device=atft10 name=gate2 workf=4.6
contact device=atft11 name=gate1 workf=4.28
contact device=atft11 name=gate2 workf=4.6
#
model  device=atft1  pfmob.p
model  device=atft2  pfmob.p
model  device=atft3  pfmob.p
model  device=atft4  pfmob.p
model  device=atft5  pfmob.p
model  device=atft6  pfmob.p
model  device=atft7  pfmob.p
model  device=atft8  pfmob.p
model  device=atft9  pfmob.p
model  device=atft10  pfmob.p
model  device=atft11  pfmob.p
#
# Transient Analysis for Flip Flop Circuit
go atlas
.begin
#
# Circuit Description
#
# Defining the Nodes and the Biasing Applied to Vin, VDD
V1 1 0 0. PULSE 0 10 0s 100us 100us 10ms 20ms
V2 6 0 0. PULSE 0 10 0s 100us 100us 8ms 16ms
V3 2 0 0. PULSE 0 10 0s 100us 100us 12ms 24ms
#
C1 8 0 100pF
#C2 1 0 100pF
#
atft1 4=drain 1=gate1 1=gate2 3=source infile=pmos_dg.str
width=400.
atft2 4=drain 2=gate1 2=gate2 3=source infile=pmos_dg.str
width=400.
atft3 0=drain 4=gate1 4=gate2 4=source infile=pmos_dg.str
width=3000.
```

```
atft4 7=drain 5=gate1 5=gate2 3=source infile=pmos_dg.str
width=400.
atft5 7=drain 6=gate1 6=gate2 3=source infile=pmos_dg.str
width=400.
atft6 0=drain 7=gate1 7=gate2 7=source infile=pmos_dg.str
width=3000.
atft7 8=drain 7=gate1 7=gate2 3=source infile=pmos_dg.str
width=400.
atft8 8=drain 4=gate1 4=gate2 3=source infile=pmos_dg.str
width=400.
atft9 0=drain 8=gate1 8=gate2 8=source infile=pmos_dg.str
width=3000.
atft10 5=drain 2=gate1 2=gate2 3=source infile=pmos_dg.str
width=400.
atft11 0=drain 5=gate1 5=gate2 5=source infile=pmos_dg.str
width=3000.
vdd 3 0 10.
#
# End of Circuit Description
#
.numeric lte=0.05 dtmin=1e-15
.options print fulln noshift
#
.load infile=cmos
.log outfile=MUX_t
# Saving the Solution File
.save master=MUX_t
# Transient Simulation Point
.tran 200us 20ms
.end
#
# Define Physical Models for ATLAS Device
#
material device=atft1 name=Pentacene \
        eg300=1.2 nc300=1.0e21 nv300=1.0e21 permittivity=4.0 \
        mun=5e-5 mup=0.5
material device=atft2 name=Pentacene \
        eg300=1.2 nc300=1.0e21 nv300=1.0e21 permittivity=4.0 \
        mun=5e-5 mup=0.5
material device=atft3 name=Pentacene \
        eg300=1.2 nc300=1.0e21 nv300=1.0e21 permittivity=4.0 \
        mun=5e-5 mup=0.5
material device=atft4 name=Pentacene \
        eg300=1.2 nc300=1.0e21 nv300=1.0e21 permittivity=4.0 \
        mun=5e-5 mup=0.5
material device=atft5 name=Pentacene \
        eg300=1.2 nc300=1.0e21 nv300=1.0e21 permittivity=4.0 \
        mun=5e-5 mup=0.5
```

```
material device=atft6 name=Pentacene \
        eg300=1.2 nc300=1.0e21 nv300=1.0e21 permittivity=4.0 \
        mun=5e-5 mup=0.5
material device=atft7 name=Pentacene \
        eg300=1.2 nc300=1.0e21 nv300=1.0e21 permittivity=4.0 \
        mun=5e-5 mup=0.5
material device=atft8 name=Pentacene \
        eg300=1.2 nc300=1.0e21 nv300=1.0e21 permittivity=4.0 \
        mun=5e-5 mup=0.5
material device=atft9 name=Pentacene \
        eg300=1.2 nc300=1.0e21 nv300=1.0e21 permittivity=4.0 \
        mun=5e-5 mup=0.5
material device=atft10 name=Pentacene \
        eg300=1.2 nc300=1.0e21 nv300=1.0e21 permittivity=4.0 \
        mun=5e-5 mup=0.5
material device=atft11 name=Pentacene \
        eg300=1.2 nc300=1.0e21 nv300=1.0e21 permittivity=4.0 \
        mun=5e-5 mup=0.5
#
mobility device=atft1 deltaep.pfmob=1.792e-2 betap.
pfmob=7.758e-5
mobility device=atft2 deltaep.pfmob=1.792e-2 betap.
pfmob=7.758e-5
mobility device=atft3 deltaep.pfmob=1.792e-2 betap.
pfmob=7.758e-5
mobility device=atft4 deltaep.pfmob=1.792e-2 betap.
pfmob=7.758e-5
mobility device=atft5 deltaep.pfmob=1.792e-2 betap.
pfmob=7.758e-5
mobility device=atft6 deltaep.pfmob=1.792e-2 betap.
pfmob=7.758e-5
mobility device=atft7 deltaep.pfmob=1.792e-2 betap.
pfmob=7.758e-5
mobility device=atft8 deltaep.pfmob=1.792e-2 betap.
pfmob=7.758e-5
mobility device=atft9 deltaep.pfmob=1.792e-2 betap.
pfmob=7.758e-5
mobility device=atft10 deltaep.pfmob=1.792e-2 betap.
pfmob=7.758e-5
mobility device=atft11 deltaep.pfmob=1.792e-2 betap.
pfmob=7.758e-5

#
contact device=atft1 name=gate1 workf=4.28
contact device=atft1 name=gate2 workf=4.6
contact device=atft2 name=gate1 workf=4.28
contact device=atft2 name=gate2 workf=4.6
contact device=atft3 name=gate1 workf=4.28
contact device=atft3 name=gate2 workf=4.6
contact device=atft4 name=gate1 workf=4.28
```

```
contact device=atft4 name=gate2 workf=4.6
contact device=atft5 name=gate1 workf=4.28
contact device=atft5 name=gate2 workf=4.6
contact device=atft6 name=gate1 workf=4.28
contact device=atft6 name=gate2 workf=4.6
contact device=atft7 name=gate1 workf=4.28
contact device=atft7 name=gate2 workf=4.6
contact device=atft8 name=gate1 workf=4.28
contact device=atft8 name=gate2 workf=4.6
contact device=atft9 name=gate1 workf=4.28
contact device=atft9 name=gate2 workf=4.6
contact device=atft10 name=gate1 workf=4.28
contact device=atft10 name=gate2 workf=4.6
contact device=atft11 name=gate1 workf=4.28
contact device=atft11 name=gate2 workf=4.6
model  device=atft1  pfmob.p
model  device=atft2  pfmob.p
model  device=atft3  pfmob.p
model  device=atft4  pfmob.p
model  device=atft5  pfmob.p
model  device=atft6  pfmob.p
model  device=atft7  pfmob.p
model  device=atft8  pfmob.p
model  device=atft9  pfmob.p
model  device=atft10  pfmob.p
model  device=atft11  pfmob.p
go atlas
tonyplot  MUX_t_tr.log  -set  MUX_t_tr.set
quit
```

## EXAMPLE A.7

REALIZATION OF STATIC RANDOM ACCESS MEMORY (SRAM) CIRCUIT USING P-TYPE PENTACENE AND N-TYPE A-SI:H SG-TFT (SEE FIGURE A.13)

```
# Defining the Device which is to be used in the SRAM
Circuit
go atlas
#
Title: Complementary SRAM Circuit Simulation

Title: P-type Device Analysis as Driver (see Figure A.14)
## Define Structure File
mesh   smooth=1 space.mult=1.0 width=800
#
x.m         l=0    spac=0.5
x.m         l=10   spac=0.5
```

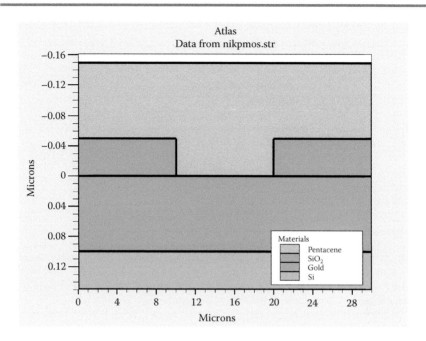

**FIGURE A.13**    Simulated schematic structure of n-type a-Si:H SG-TFT.

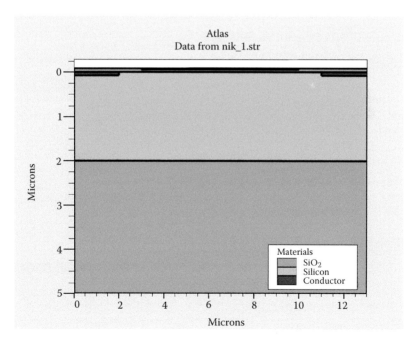

**FIGURE A.14**    Schematic structure of *p*-type pentacene based SG-OTFT.

```
x.m            l=20     spac=0.5
x.m            l=30     spac=0.5
#
y.mesh l=-0.15     spacing=0.05
y.mesh l=-0.05     spacing=0.005
y.mesh l=0.0        spacing=0.05
y.mesh l=0.10      spacing=0.05
y.mesh l=0.15      spacing=0.05
# Define Region Parameter
region num=1    material=pentacene   y.min=-0.15 y.max=0.0
name=Pentacene
region num=2    material=SiO2 y.min=0.0 y.max=0.10
# Define Electrodes of the Device
elec num=1 material=Gold name=source x.max=10.0 y.min=-0.05
y.max=0.0
elec num=2 material=Si name=gate y.min=0.10 y.max=0.15
elec num=3 material=Gold name=drain x.min=20.0 y.min=-0.05
y.max=0.0
## Defining Material Parameters and Application of the
Appropriate Physical Models
doping reg=1    uniform conc=1e14 p.type name=pentacene
doping reg=2    uniform conc=1e21 n.type   name= silicon
#
material name=Pentacene \
         eg300=2.2 nc300=1.0e21 nv300=1.0e21
         permittivity=4.0 \
         mun=5e-5 mup=0.14
# Defining Electrode Parameters
contact name=gate workf=4.6
# Defining the Defects in the Device
defects  cont dfile=15don.dat afile=15acc.dat \
         nta=2.5e18 ntd=1.0e18 wta=0.129 wtd=0.5 \
         nga=0.0 ngd=0.0 ega=0.62 egd=0.78 wga=0.15 wgd=0.15 \
         sigtae=1.e-17 sigtah=1.e-15 sigtde=1.e-15
         sigtdh=1.e-17  \
         siggae=2.e-16 siggah=2.e-15 siggde=2.e-15
         siggdh=2.e-16
# Defining the Model for the Device
mobility deltaep.pfmob=1.792e-2  betap.pfmob=7.758e-5
model    pfmob.p print
#
output   e.field j.electron j.hole j.conduc j.total
e.velocity h.velocity \
         ey.field flowlines e.mobility h.mobility qss e.temp
         h.temp charge \
         recomb val.band con.band qfn qfp j.disp photogen
         impact tot.doping \
         u.srh u.rad
# Above Making Device and Applying Model and Defects
save outf=nikpmos.str
```

```
# Plot Device Structure using Solution File
tonyplot nikpmos.str

# Defining the Device which is to be used in the SRAM
Circuit
go atlas
Title:    a-si TFT Simulation
 # Define Structure File
mesh
#
x.m           l=0     spac=0.5
x.m           l=2     spac=0.5
x.m           l=3     spac=0.25
x.m           l=4     spac=0.25
x.m           l=6.5   spac=1.5
x.m           l=9     spac=0.25
x.m           l=10    spac=0.25
x.m           l=11    spac=0.5
x.m           l=13    spac=0.5
#
y.mesh l=-0.10     spacing=0.05
y.mesh l=-0.06     spacing=0.05
y.mesh l=0.0       spacing=0.0075
y.mesh l=0.075     spacing=0.01
y.mesh l=2.0       spacing=0.0075
y.mesh l=3.0       spacing=0.25
y.mesh l=5.0       spacing=5
#
# Define Region Parameter
region        num=1  y.max=0.    oxide
region        num=2  y.min=0.    y.max=2.0    silicon
region        num=3  y.min=2.    oxide
#
#    1=gate   2=source   3=drain
# Define Electrodes of the Device
elec  num=1  x.min=3     x.max=10   y.min=-0.1 y.max=-0.06
name=gate
elec  num=2  x.min=0.    x.max=2.     y.min=0.     y.max=0.075
name=source
elec  num=3  x.min=11.0 x.max=13.0  y.min=0.     y.max=0.075
name=drain
#
# Defining Material Parameters and Application of the
Appropriate Physical Models
doping     reg=2  uniform conc=4.e12 n.type
doping     reg=2  gauss conc=3.e18 n.type x.right=4 char=0.3
doping     reg=2  gauss conc=3.e18 n.type x.left=9 char=0.3

# Defining Electrode Parameters
material region=2 mun=20 mup=1.5 nc300=2.5e20 \
```

```
 nv300=2.5e20 eg300=1.9
#
# Defining the Defects in the Device
defects nta=1.e21 ntd=1.e21 wta=0.033 wtd=0.049 \
  nga=1.5e15 ngd=1.5e15 ega=0.62 egd=0.78 wga=0.15
  wgd=0.15 \
  sigtae=1.e-17 sigtah=1.e-15 sigtde=1.e-15 sigtdh=1.e-17  \
  siggae=2.e-16 siggah=2.e-15 siggde=2.e-15 siggdh=2.e-16
#
# Defining Electrode Parameters
contact        num=1 alum
#  Defining the Model for the Device
models  temp=300
#
method itlimit=30
solve init
#
# Above we are Making Device and Applying Model and Defects
save outf=nik_1.str
tonyplot nik_1.str
# Circuit Description for the Circuit
go atlas
.begin
#
#     OTFT inverters - DC Point Simulation
#     Circuit description
#
# Defining the Nodes and the Biasing Applied to Vin, VDD
vdd 1 0 10
vs 3 0 0
#R2 4 0 1K

atftp1 3=drain 4=gate 1=source infile=nikpmos.str
width=80000.
atftn1 3=drain 4=gate 0=source infile=nik_1.str width=0.1.
atftp2 4=drain 3=gate 1=source infile=nikpmos.str
width=80000.
atftn2 4=drain 3=gate 0=source infile=nik_1.str width=0.1.
#
#  End of Circuit Description
#
.numeric vchange=5e-1.
.options print m2ln noshift
#
# Saving the solution file
.save outfile=cmos
.end
#
# Define Physical Models for ATLAS Device
#
```

```
# Defining Material Parameters and Application of the
Appropriate Physical Models
material device=atftp1 name=Pentacene \
        eg300=2.2 nc300=1.0e21 nv300=1.0e21
        permittivity=4.0 \
        mun=5e-5 mup=.14
material device=atftp2 name=Pentacene \
        eg300=2.2 nc300=1.0e21 nv300=1.0e21
        permittivity=4.0 \
        mun=5e-5 mup=.14
material device=atftn1 name=silicon mun=20 mup=1.5
nc300=2.5e20 \
 nv300=2.5e20 eg300=1.9
material device=atftn2 name=silicon mun=20 mup=1.5
nc300=2.5e20 \
 nv300=2.5e20 eg300=1.9

mobility device=atftp1 deltaep.pfmob=1.792e-2  betap.
pfmob=7.758e-5
mobility device=atftp2 deltaep.pfmob=1.792e-2  betap.
pfmob=7.758e-5
mobility device=atftn1 deltaep.pfmob=1.792e-2  betap.
pfmob=7.758e-5
mobility device=atftn2 deltaep.pfmob=1.792e-2  betap.
pfmob=7.758e-5
contact device=atftn1 name=gate workf=4.6
contact device=atftn2 name=gate workf=4.6
contact device=atftp1 name=gate n.poly
contact device=atftp2 name=gate n.poly
model   device=atftn1  temp=300
model   device=atftn2  temp=300
model   device=atftp1  pfmob.p
model   device=atftp2  pfmob.p
 #DC analysis for SG OTFT SRAM Circuit
go atlas
.begin
#
#     OTFT inverters - DC Curve Simulation
#
# OTFT SRAM - DC Curve Simulation
#
#  Circuit description
#Defining the Nodes and the Biasing Applied to Vin, VDD
vdd 1 0 10
vs 3 0 0
#R2 4 0 10K
atftp1 3=drain 4=gate 1=source infile=nikpmos.str width=80000.
atftn1 3=drain 4=gate 0=source infile=nik_1.str width=0.1.
atftp2 4=drain 3=gate 1=source infile=nikpmos.str width=80000.
atftn2 4=drain 3=gate 0=source infile=nik_1.str width=0.1.
```

```
#
#       End of Circuit Description
#
.options print fulln noshift
#
.load infile=cmos
.log outfile=cmos01
# Saving the Solution File
.save master=cmos01
 #R1 3 11 1k
# DC Simulation Point
.dc vs 0. 10. 0.5
#
.end
#
#  Define Physical Models for ATLAS Device
#
# Defining Material Parameters and Application of the
Appropriate Physical Models
material device=atftp1 name=Pentacene \
        eg300=2.2 nc300=1.0e21 nv300=1.0e21
        permittivity=4.0 \
        mun=5e-5 mup=.14
material device=atftp2 name=Pentacene \
        eg300=2.2 nc300=1.0e21 nv300=1.0e21
        permittivity=4.0 \
        mun=5e-5 mup=.14
material device=atftn1 name=silicon mun=20 mup=1.5
nc300=2.5e20 \
 nv300=2.5e20 eg300=1.9
material device=atftn2 name=silicon mun=20 mup=1.5
nc300=2.5e20 \
 nv300=2.5e20 eg300=1.9
mobility device=atftp1 deltaep.pfmob=1.792e-2  betap.
pfmob=7.758e-5
mobility device=atftp2 deltaep.pfmob=1.792e-2  betap.
pfmob=7.758e-5
mobility device=atftn1 deltaep.pfmob=1.792e-2  betap.
pfmob=7.758e-5
mobility device=atftn2 deltaep.pfmob=1.792e-2  betap.
pfmob=7.758e-5
contact device=atftn1 name=gate workf=4.6
contact device=atftn2 name=gate workf=4.6
contact device=atftp1 name=gate n.poly
contact device=atftp2 name=gate n.poly
model   device=atftn1  temp=300
model   device=atftn2  temp=300
model   device=atftp1  pfmob.p
model   device=atftp2  pfmob.p
go atlas
```

```
# Plotting the Voltage Transfer Characteristics (VTC) Curve
tonyplot  cmos01_dc_1.log
#Transient Analysis for SG OTFT SRAM Circuit
 go atlas
.begin
# OTFT inverters - DC Curve Simulation
#   Circuit description
# Defining the Nodes and the Biasing Applied to Vin, VDD
#
vdd 1 0 10
#vs 3 0 0
vs1 4 0 0
#R2 4 0 10K
#R1 3 0 10k
atftp1 3=drain 4=gate 1=source infile=nikpmos.str width=80000.
atftn1 3=drain 4=gate 0=source infile=nik_1.str width=0.1.
atftp2 4=drain 3=gate 1=source infile=nikpmos.str width=80000.
atftn2 4=drain 3=gate 0=source infile=nik_1.str width=0.1.
#
#     End of Circuit Description
.numeric vchange=5e-1.
.options print m2ln noshift
#
.save outfile=cmos1
.end
#Defining Material Parameters and Application of the
Appropriate Physical Models
#
material device=atftp1 name=Pentacene \
         eg300=2.2 nc300=1.0e21 nv300=1.0e21
         permittivity=4.0 \
         mun=5e-5 mup=.14
material device=atftp2 name=Pentacene \
         eg300=2.2 nc300=1.0e21 nv300=1.0e21
         permittivity=4.0 \
         mun=5e-5 mup=.14
material device=atftn1 name=silicon mun=20 mup=1.5
nc300=2.5e20 \
 nv300=2.5e20 eg300=1.9
material device=atftn2 name=silicon mun=20 mup=1.5
nc300=2.5e20 \
 nv300=2.5e20 eg300=1.9
#  Define Physical Models for ATLAS Device
# Defining Material Parameters and Application of the
Appropriate Physical Models
mobility device=atftp1 deltaep.pfmob=1.792e-2  betap.
pfmob=7.758e-5
mobility device=atftp2 deltaep.pfmob=1.792e-2  betap.
pfmob=7.758e-5
mobility device=atftn1 deltaep.pfmob=1.792e-2  betap.
```

```
pfmob=7.758e-5
mobility device=atftn2 deltaep.pfmob=1.792e-2  betap.
pfmob=7.758e-5
contact device=atftn1 name=gate workf=4.6
contact device=atftn2 name=gate workf=4.6
contact device=atftp1 name=gate n.poly
contact device=atftp2 name=gate n.poly
model   device=atftn1  temp=300
model   device=atftn2  temp=300
model   device=atftp1  pfmob.p
model   device=atftp2  pfmob.p
 #
 # OTFT SRAM - DC Curve Simulation
go atlas
.begin
#     OTFT inverters - DC Curve Simulation
 #     Circuit description
vdd 1 0 10
#vs 3 0 0
vs1 4 0 0
#R2 4 0 10K
#R1 3 0 10k
atftp1 3=drain 4=gate 1=source infile=nikpmos.str width=80000.
atftn1 3=drain 4=gate 0=source infile=nik_1.str width=0.1.
atftp2 4=drain 3=gate 1=source infile=nikpmos.str width=80000.
atftn2 4=drain 3=gate 0=source infile=nik_1.str width=0.1.
# End of Circuit Description
#
.options print fulln noshift
#
.load infile=cmos1
.log outfile=cmos02
# Saving the Solution File
.save master=cmos02
 #R1 3 11 1k
# DC Simulation Point
.dc vs1 0. 10. 0.5
#
.end
# Define Physical Models for ATLAS Device
# Defining Material Parameters and Application of the
Appropriate Physical Models
material device=atftp1 name=Pentacene \
        eg300=2.2 nc300=1.0e21 nv300=1.0e21
        permittivity=4.0 \
        mun=5e-5 mup=.14
material device=atftp2 name=Pentacene \
        eg300=2.2 nc300=1.0e21 nv300=1.0e21
        permittivity=4.0 \
        mun=5e-5 mup=.14
```

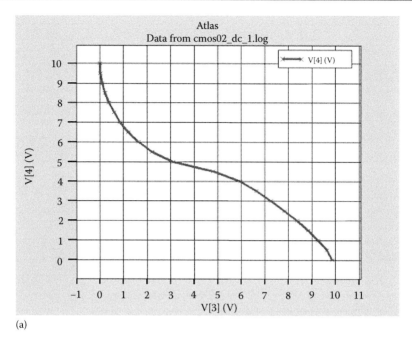

(a)

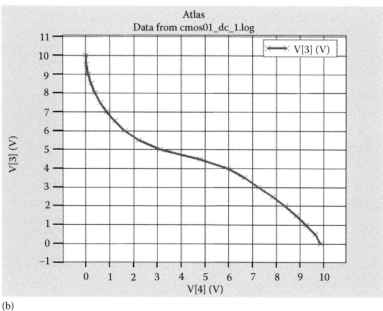

(b)

**FIGURE A.15**    VTC of organic inverter circuit based on SG-OTFT to extract value of static noise margin (SNM) of complementary SRAM.

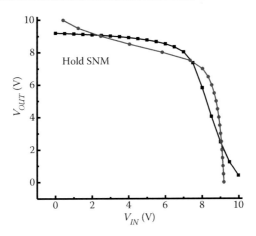

**FIGURE A.16**    Butterfly curves during standby and read mode for organic complementary SRAM cells based on SG-OTFT.

```
material device=atftn1 name=silicon mun=20 mup=1.5
nc300=2.5e20 \
 nv300=2.5e20 eg300=1.9
material device=atftn2 name=silicon mun=20 mup=1.5
nc300=2.5e20 \
 nv300=2.5e20 eg300=1.9
mobility device=atftp1 deltaep.pfmob=1.792e-2  betap.
pfmob=7.758e-5
mobility device=atftp2 deltaep.pfmob=1.792e-2  betap.
pfmob=7.758e-5
mobility device=atftn1 deltaep.pfmob=1.792e-2  betap.
pfmob=7.758e-5
mobility device=atftn2 deltaep.pfmob=1.792e-2  betap.
pfmob=7.758e-5
contact device=atftn1 name=gate workf=4.6
contact device=atftn2 name=gate workf=4.6
contact device=atftp1 name=gate n.poly
contact device=atftp2 name=gate n.poly
model    device=atftn1  temp=300
model    device=atftn2  temp=300
model    device=atftp1  pfmob.p
model    device=atftp2  pfmob.p
go atlas
# Plotting the Voltage Transfer Characteristics (VTC) Curve
(see Figures A.15 and A.16)
 tonyplot cmos02_dc_1.log
quit.
```

# Index

Page numbers ending in "f" refer to figures. Page numbers ending in "t" refer to tables.